AF787768

High Angular Resolution in Astrophysics

NATO ASI Series

Advanced Science Institutes Series

A Series presenting the results of activities sponsored by the NATO Science Committee, which aims at the dissemination of advanced scientific and technological knowledge, with a view to strengthening links between scientific communities.

The Series is published by an international board of publishers in conjunction with the NATO Scientific Affairs Division

A Life Sciences **B Physics**	Plenum Publishing Corporation London and New York
C Mathematical and Physical Sciences **D Behavioural and Social Sciences** **E Applied Sciences**	Kluwer Academic Publishers Dordrecht, Boston and London
F Computer and Systems Sciences **G Ecological Sciences** **H Cell Biology** **I Global Environmental Change**	Springer-Verlag Berlin, Heidelberg, New York, London, Paris and Tokyo

PARTNERSHIP SUB-SERIES

1. Disarmament Technologies	Kluwer Academic Publishers
2. Environment	Springer-Verlag / Kluwer Academic Publishers
3. High Technology	Kluwer Academic Publishers
4. Science and Technology Policy	Kluwer Academic Publishers
5. Computer Networking	Kluwer Academic Publishers

The Partnership Sub-Series incorporates activities undertaken in collaboration with NATO's Cooperation Partners, the countries of the CIS and Central and Eastern Europe, in Priority Areas of concern to those countries.

NATO-PCO-DATA BASE

The electronic index to the NATO ASI Series provides full bibliographical references (with keywords and/or abstracts) to more than 50000 contributions from international scientists published in all sections of the NATO ASI Series.
Access to the NATO-PCO-DATA BASE is possible in two ways:

– via online FILE 128 (NATO-PCO-DATA BASE) hosted by ESRIN, Via Galileo Galilei, I-00044 Frascati, Italy.

– via CD-ROM "NATO-PCO-DATA BASE" with user-friendly retrieval software in English, French and German (© WTV GmbH and DATAWARE Technologies Inc. 1989).

The CD-ROM can be ordered through any member of the Board of Publishers or through NATO-PCO, Overijse, Belgium.

Series C: Mathematical and Physical Sciences – Vol. 501

High Angular Resolution in Astrophysics

edited by

Anne-Marie Lagrange

Chargé de Recherche au CNRS,
Laboratoire d'Astrophysique de Grenoble,
Grenoble, France

Denis Mourard

Astronome-Adjoint,
Observatoire de la Côte d'Azur,
Département Fresnel,
Nice, France

and

Pierre Léna

Université Paris VII
& Observatoire de Paris,
Paris, France

Kluwer Academic Publishers

Dordrecht / Boston / London

Published in cooperation with NATO Scientific Affairs Division

Proceedings of the NATO Advanced Study Instiute on
High Angular Resolution in Astrophysics
Les Houches, France
9–19 April 1996

A C.I.P. Catalogue record for this book is available from the Library of Congress

ISBN-13: 978-94-010-6507-8

Published by Kluwer Academic Publishers,
P.O. Box 17, 3300 AA Dordrecht, The Netherlands.

Sold and distributed in the U.S.A. and Canada
by Kluwer Academic Publishers,
101 Philip Drive, Norwell, MA 02061, U.S.A.

In all other countries, sold and distributed
by Kluwer Academic Publishers,
P.O. Box 322, 3300 AH Dordrecht, The Netherlands.

Printed on acid-free paper

All Rights Reserved

ISBN-13: 978-94-010-6507-8 e-ISBN-13: 978-94-009-0041-7
DOI: 10.1007/978-94-009-0041-7

© 1997 Kluwer Academic Publishers
Softcover reprint of the hardcover 1st edition 1997

No part of the material protected by this copyright notice may be reproduced or utilized in any form or by any means, electronic or mechanical, including photocopying, recording or by any information storage and retrieval system, without written permission from the copyright owner.

TABLE OF CONTENTS

PREFACE

Since 1970, when Antoine Labeyrie first broke the seeing barrier and reached the diffraction-limited resolution of large telescopes observing from ground-based sites in the visible, the emergence of high angular resolution at optical (visible and infrared) has been progressive and is nowadays becoming spectacular. Speckle methods, first in the visible and soon in the infrared brought the first astrophysical results, mostly on double stars and circumstellar environments, using painful reconstruction methods to obtain an important but limited amount of information. But their systematic developement led to an unprecedented and detailed knowledge of the atmospheric turbulence and of its effects on astronomical images, including a better understanding of the parameters to be taken in account when a new astronomical site is selected, or when a large telescope is built . This was capital for the steps to follow.

The next step blossoms since the end of the eighties, with the remarkable development of adaptive optics, which capitalizes on the knowledge accumulated on the Earth's atmosphere and becomes to-day a mandatory sub-system in the conception of large or even moderate size telescopes. Scientific productivity of adaptive optics, in the near-infrared and now in the visible, is already demonstrated in almost every field of astrophysics, with imaging, spectro-imaging or polaro-imaging of planetary surfaces, solar system objects, multiple stars and star clusters, circumstellar environments, nuclei of galaxies and quasars. Comparison of such results with images obtained with the Hubble Space Telescope demonstrates the effective quality and great future of what is obtained from the ground. The next step for adaptive optics is to get rid of the intrinsic limitations in the magnitude of the objects it can actually correct. Artificial laser star programs, promising but expensive, may bring the ultimate solution within a few years.

Although multi-telescope interferometry has been first achieved in 1974, again by Antoine Labeyrie, progress in astrophysical uses of it has been much slower. The years 1980s have seen a number of prototype instruments implemented, which brought significant but limited results. This fact is a natural result of the increased complexity and the experimental challenges which have to be faced when one couples meter- size telescopes, tens or hundreds of meters apart, while keeping an optical path accuracy of a few tens of nanometers between the various light beams. But there is a fundamental point which has been achieved during these years of exploration : the physics of beam recombination at optical wavelengths, as well as the atmospheric effects on the phase of the incoming wave, are perfectly understood, as was demonstrated, among other results, by some remarkable achievements in astrometry obtained at Mt. Wilson by Michael Shao and coworkers. Therefore, difficulties are only of a technical nature, and can be solved when appropriate resources are allocated.

Facing this and perceiving the wealth of astronomical results -almost a revolution- which can result from a resolution gain of factors 10-30 compared to single telescope operation, the astronomical community embarked, at the end of the eighties, in the construction of powerful optical interferometers, remarkable either by the size of the telescopes to be combined -the giants one in the 10-m class currently under construction-, or by their number, or by various combinations of these two factors. If the European Very Large Telescope or the Keck-Telescopes represent the most impressive ambition in this respect, a number of other arrays demonstrate the confidence currently placed by astronomers in this new tool of astronomy, which will become available to them in the very first years of the next millenium. In addition, after years of uncertainties and adventurous concepts, optical interferometry in space appeared also to become a reality : an interferometric space mission is programmed in the United States, and at least one is seriously envisaged in the European space program.

It therefore appeared timely to hold at Les Houches a school on High Angular Resolution in Astrophysics, in order to prepare the exploitation of these powerful interferometers. It was then decided by the Scientific Organizing Committee to concentrate on multi-telescope interferometry at optical wavelengths : the justification of this choice immediatly results from the above described status of the field. The title of the school may be misleading, as it seems to cover the whole wavelengths range. Although it appeared more appropriate to restrict the topics to the optical, it was indeed recognized that many lessons remain to be learned from the long-lasting and elaborate practice of interferometry achieved at radio-wavelengths, which is therefore present in the lectures. The initial intent was to focus on science programs and they indeed represent a significant part of the present Volume. They would have been even more present if some of the authors had not finally declined to provide their manuscript, undoubtly overloaded by writing obligations of various kinds. This is regrettable, but a sufficient number of papers covering a great variety of astrophysical topics, written by authoritative scientists, is present for the main goal of the school to be properly represented here.

The lectures which present the existing or planned interferometers outline the performances an astronomer, not necessarily specialist in interferometry, can expect when he considers to use such a system for a definite goal of his own. They should help to bridge the gap between the quite special and yet esoteric language of distinguished interferometrists and the rest of the astronomical community.

The eagerness and youth of the attendance were certainly perceived by all, and give confidence in the creativity of the generation which will exploit the extraordinary instruments currently in construction. Les Houches in spring time carry their own special flavor, so different from the summer or the fall in this heart of the French Alps. So many great scientific ventures took off from this exceptional place that we all felt to be in good company !

It is a pleasure to thank here the colleagues who contributed by their lectures and active presence, the ones who gave short contributions, the students for their cheerful spirit during the school.

We are grateful to the NATO Scientific Affairs Division who generously supported this ASI. We also thank our Kluwer contact, Mrs W. Bruins for her help. We also thank all the institutions which participated to the funding of the school : the Ecole de Physique des Houches, the Centre National de la Recherche Scientifique, the European Space Agency and the Centre National d'Etudes Spatiales, and the french Programme National pour la Haute Résolution Angulaire en Astrophysique. The staff of the Laboratoire d'Astrophysique de l'Observatoire de Grenoble and that of the Ecole de Physique des Houches assisted us very efficiently. We thank them for their help.

We hope that students and researchers interested in High Angular Resolution in Astrophysics will find this book useful and stimulating.

Anne-Marie Lagrange
Denis Mourard
Pierre Léna

LIST OF PARTICIPANTS

ADCOCK Miles
Applied Optics
The Blackett Laboratory
LONDON SW7 2BZ
United Kingdom
miles@op.ph.ic.ac.uk

ALTARAC Sandrine
Observatoire de Lyon
9, Avenue Charles Andre
69561 Saint Genis Laval
France
altarac@obs.univ-lyon1.fr

BECKWITH Steven
MPI für Astronomie
Konigstuhl 17
69117 Heidelberg
Germany
svwb@mpia-hd.mpg.de

BELY Pierre
Space Telescope Science Inst.
3700 San Martin Drive
Baltimore Maryland 21218
United States of America
bely@stsci.edu

BERIO Philippe
OCA-GI2T
2130 Route de l'Observatoire
06460 Saint Vallier de Thiey
France
berio@gi2t.obs-azur.fr

BEUZIT Jean-Louis
Observatoire de Geneve
1290 Sauverny
Suisse
beuzit@gag.observ-gr.fr

BROCKIE Richard
University of Edimburgh
Physics and Astronomy
Edimburgh EH9 4HJ
United Kingdom
R.Brockie@roe.ac.uk

BUTLER David
Department of Physics
University College Galway
Galway
Ireland
dbutler@epona.physics.ucg.ie

BUTLER Ray
Department of Physics
University College Galway
Galway
Ireland
ray@epona.physics.ucg.ie

CHANG Mark
Dept of Physics, University of Durham
South Road, Durham DH1 3LE
United Kingdom
mark.chang@durham.ac.uk

COSTA Roberto
Instituto astronomico e geofisico USP
Av. Miguel Stefano 4200
04301-904, Sao-Paulo, SP
Bresil
costa@obs-nice.fr

COTTON Bill
NRAO
520 Edgemont Road
Charlottesville, VA 22903
United States of America
bcotton@nrao.edu

DAIGNE Gerard
Observatoire de Bordeaux
BP89, 33270 Floirac
France
daigne@observ.u-bordeaux.fr

DAMGOV Vladimir
Bulgaria
VDAMGOV@BGEARN.ACAD.BG

DAVIS John
Chatterton Astronomy Dept.
University of Sydney
NSW, 2006
Australia
davis@astron.physics.su.oz.au

DESSENNE Caroline
ONERA
BP 72
92322 CHATILLON CEDEX
France
dessenne@onera.fr

EISENHAUER Frank
MPI für extraterrestrische Physik
Giessenbachstrasse
85740 Garching
Germany
eisenhau@mpa-garching.mpg.de

EISLOEFFEL Jochen
Thueringer Landessternwarte
Karl-Schwarzschild-Observatorium
D-07778 Tautenburg
Germany
eisloeffel@tls-tautenburg.de

ENCRENAZ Therese
Observatoire de Meudon
92195 Meudon Principal Cedex
France
therese.encrenaz@obspm.fr

ESSLINGER Olivier
Dpt of Physics & Astronomy
University of Wales
PO BOX 913 Cardiff
United Kingdom
Olivier.Esslinger@astro.cf.ac.uk

FELDT Markus
Max-Planck-Research-Unit
Schillergaesschen 3
D-07745 Jena
Germany
mfeldt@astro.uni-jena.de

FOLHA Daniel
Physics Dpt
Queen Mary & Westfield College
London E14NS
United Kingdom
dfmfolha@qmw.ac.uk

FOY Renaud
Obs. de Lyon Groupe AIRI
Avenue Charles-Andre
St. Genis-Laval 69561 Cedex
France
foy@po.univ-lyon1.fr

GAMEIRO Jorge Filipe
Univ. Porto CAUP
R. Campo Alegre, 8234150 Porto
Portugal
jgameiro@astro.up.pt

GARCIA Paulo
Obs. de Lyon Groupe AIRI
Avenue Charles-Andre
St. Genis-Laval 69561 Cedex
France
pgarcia@image.univ-lyon1.fr

GUETH Frederic
IRAM
300, rue de la piscine
30406 St Martin d'Heres
France
gueth@iram.fr

GUILLOTEAU Stephane
IRAM
300, rue de la piscine
30406 St Martin d'Heres
France
guillote@iram.grenet.fr

HARDER Stephan
Observatoire de Grenoble
B.P. 53
F-38041 Grenoble Cedex 9 (France)
France
harder@gag.observ-gr.fr

HILDEBRAND John
CHARA, Georgia State University
Atlanta, GA
30303-3083
United States of America
jlh@chara.gsu.edu

HOLMBERG Johan
Lund Observatory
Box 43
S-221 00 LUND
Sweden
johan@astro.lu.se

HOLOHAN Liam
Imperial college
Prince consort road
London SW7 2Bz
United Kingdom
liam@op.ph.ic.ac.uk

HRYNEVYCH Michael
School of Physics
University of Sydney
NSW, 2006
Australia
mick@Physics.usyd.edu.au

IDIART Thais
Instituto astronomico e geofisico USP
Av. Miguel Stefano 4200
04301-904, Sao-Paulo, SP
Bresil
thais@obs-nice.fr

JUNOR Bill
Phillips Laboratory
Kirtland Air Force Base
New Mexico 87185
United States of America
bjunor@cezanne.plk.af.mil

KLUTTIG Annelie
Dpt. Physics & Astro., UCL
Crower St.
London WL1E 6BT
United Kingdom
awk@star.ucl.ac.uk

KOEHLER Bertrand
ESO VLTI System Engineer
Karl-Schwarzschild-Strasse 2,
D-W 8046 Garching bei Munchen
Germany
bkoehler@eso.org

KOEHLER Rainer
MPI für Astronomie
Koenigstuhl 17
D-69117 Heidelberg
Germany
koehler@mpia-hd.mpg.de

LAGARDE Stephane
Laboratoire d'Astrophysique
Univ. de Nice - Parc Valrose
06000 Nice
France
stephane@AYALGA.UNICE.FR

LAGRANGE Anne-Marie
Observatoire de Grenoble
BP 53X
38041 Grenoble Cedex
France
Lagrange@gag.observ-gr.fr

LANNES Andre
Observatoire Midi-Pyrenees
14 Avenue Edouard Belin
31400 Toulouse
France
lannes@obs-mip.fr

LAWSON Peter
MRAO, Cavendish Laboratory
Madingley Road
Cambridge CB3 0HE
United Kingdom
lawson@mrao.cam.ac.uk

LE COARER Etienne
Observatoire de Grenoble
BP 53X
38041 Grenoble Cedex
France
lecoarer@gag.observ-gr.fr

LEGER Alain
Institut d'Astronomie Spatiale
BP10
91371 Verrieres le Buisson Cedex
France
leger@iaslab.ias.fr

LE LOUARN Miska
ESO
Karl-Schwarzschild Strasse 2
D-85748 Garching bei Munchen
Germany
lelouarn@eso.org

LENA Pierre
Observatoire de Meudon
92195 Meudon Principal Cedex
France
lena@obspm.fr

LE MIGNANT David
European Southern Observatory
Casilla 19001
Santiago 19
CHILI
dlemigna@eso.org

LEVEQUE Samuel
ESO VLT Interferometry Group
Karl-Schwarzschild-Strasse 2
D-W 8046 Garching bei Munchen
Germany
sleveque@eso.org

LEWIS Roberts
CHARA, Georgia State University
Atlanta, GA
30303-3083
United States of America
lewis@chara.gsu.edu

LOISEAU Sacha
Observatoire de Meudon
5 Place Jules Janssen
92195 Meudon Principal Cedex
France
loiseau@bluenote.obspm.fr

LUKIN V.
Institute of Atmospheric Optics
1, Akademicheskii Ave.
Tomsk, 634055
Russia
zuev@iao.tomsk.su

MALBET Fabien
Obs. de Grenoble
B.P. 53X
F-38041 Grenoble Cedex 9
France
malbet@gag.observ-gr.fr

MARCO Olivier
Observatoire de Meudon
5 Place Jules Janssen
92195 Meudon Principal Cedex
France
marco@hplyot.obspm.fr

MAYOR Michel
Observatoire de Geneve
1290 Sauverny
Suisse
mayor@obs.unige.ch

MENARD Francois
Obs. de Grenoble
B.P. 53X
F-38041 Grenoble Cedex 9
France
menard@gag.observ-gr.fr

MOTTE Frederique
CEA/Service d'Astrophysique
Orme des Merisiers, Bat 709
F-91191 Gif-sur-Yvette
France
motte@ariane.saclay.cea.fr

MOUILLET David
Obs. de Grenoble
B.P. 53X
F-38041 Grenoble Cedex 9
France
mouillet@gag.observ-gr.fr

MOURARD Denis
Observatoire de Calern
2130 Route de l'Observatoire
06460 Saint Vallier de Thiey
France
mourard@obs-azur.fr

NURNBERGER Dieter
Astronomisches Institut
Am Hubland
D-97074 Wuerzburg
Germany
nurnberg@astro.uni-wuerzburg.de

OZEREN Ferhat Fikri
Ankara University
Astronomy Department
06100 Tandogan-ANKARA
TURKEY
gozlem-e@servis.net.tr

PARESCE Francesco
ESO VLTI
Karl-Schwarzschild-Strasse 2,
D-W 8046 Garching bei Munchen
Germany
fparesce@eso.org

PERAULT Michel
Ecole Normale Superieure
24 Rue Lhomond
75231 Paris Cedex 5
France
perault@obspm.fr

PETR Monika
MPI für Astronomie
Koenigstuhl 17
69117 Heidelberg
Germany
petr@mpia-hd.mpg.de

POLATIDIS Antonis
Alkiminis 47
54249 Thessaloniki
GREECE
apolatid@astro.auth.gr

RAGLAND Sam
large Physical Research Laboratory
Navrangpura
Ahmedabad 380009
India
sam@prl.ernet.in

RICCARDI Armando
Osservatorio di Arcetri
50125 Firenze, L.go Fermi 5
ITALY
riccardi@arcetri.astro.it

RIDGWAY Steve
NOAO
PO Box 26732
Tucson, Arizona 85726
United States of America
sridgway@noao.edu

RIGAUT Francois
CFHT
PO Box 1597, Kamuela, Hawaii 96743-1597
United States of America
rigaut@cfht.hawaii.edu

RODRIGUEZ-RAMOS Jose Manu
Instituto de Astrofisica de Canarias.
Via Lactea, s/n.
La Laguna. Tenerife. 38200
Spain
jmr@ll.iac.es

ROUAN Daniel
Observatoire de Meudon
92195 Meudon Principal Cedex
France
rouan@obspm.fr

ROUSSELET-PERRAUT Karine
LAS - CNRS
Traverse du Siphon - B.P. 8
13376 MARSEILLE Cedex 12
France
karine@astrsp-mrs.fr

RUYMAEKERS Goedele
Royal Observatory of Belgium
B-1180 Brussels
Belgium
Goedele.Ruymaekers@oma.be

SAMARKIN Vadim
NICTL
Dm. Ulyanov 4, bld. 2, apt. 13
Moscow, 117333
Russia
adopt@laser.nictl.msk.su

SHAO Mike
306-388 JPL
4800 OakGrove
Pasadena, CA 91109
United States of America
mshao@huey.jpl.nasa.gov

TRIFALENKOV Ilya
Sternberg Astronomical Institue
Universitetsky pr. 13, Moscow
Russia
galaxy@jet.msk.su

TSIOPA Olga
Pulkovo Observatory
196140 St.Petersburg
Russia
tsiopa@pulkovo.spb.su

TURNER Nils
CHARA, Georgia State University
Atlanta, GA
30303-3083
United States of America
nils@chara.gsu.edu

TYLER David
USAF Phillips Laboratory
2000 Randolph Rd. SE, St. 205
Albuquerque, New Mexico 87106
United States of America
dave@markov.plk.af.mil

VON DER LUHE Oskar
ESO VLTI
Karl-Schwarzschild-Strasse 2,
D-W 8046 Garching bei Munchen
Germany
ovdluhe@eso.org

HIGH ANGULAR RESOLUTION INSTRUMENTATION

AN INTRODUCTION TO ATMOSPHERIC TURBULENCE

PIERRE LENA
Université Paris VII-Denis Diderot & Observatoire de Paris
92195 Meudon, France
email: Pierre.Lena@obspm.fr

Abstract. Atmospheric turbulence seriously affects the process of image formation, through the corrugation of the phase in the wavefront coming from an astronomical source, or other sources, such as artificial laser stars, satellites, etc. observed through the terrestrial atmosphere.

Obvious examples of these effects are given by the appearance of a scene through the heated air of a barbecue, of remote cars on a hot road, the scintillation of stars, etc. Observing the edge of a cumulus cloud with binoculars shows constantly changing eddies of various sizes, as one also notices on the water flowing behind the piles of a bridge. It is the analysis and aerodynamic tunnel simulations of such wakes which helped to localize the four VLT telescopes on Cerro Paranal in Chili, in order to minimize the wake effect of one telescope on the others.

The same effects impact on the use of interferometric techniques, which are simply an other way to deal with the retrieval of spatial information from the observed source. Although an extremely complicated phenomenon (Lesieur 1987a), turbulence may in some circumstances be described in simple terms, which at least give some intuitive understanding, provide an insight on the physics, allow some analytical formalism to be introduced and can establish, at least to first order, a quantitative and statistical description of images formed through the atmosphere (Lesieur 1987b).

1. What is turbulence ?

Flows of classical fluids may be classified as either irrotational (also called laminar), or rotational. The velocity field v(r,t) at position r and instant t is then such that:

A-M. Lagrange et al. (eds.), High Angular Resolution in Astrophysics, 3–12.
© 1997 *Kluwer Academic Publishers.*

$$laminar : curl[v(r,t)] = 0 \tag{1}$$

$$rotational : curl[v(r,t)] \neq 0 \tag{2}$$

In simpler terms, rotational flows show eddies of various sizes, which are either stable or constantly change their size and position with time. Taking over a certain volume V the average <> of the curl, we may find values of V for which the average is non-zero. Such values define the size or sizes of the eddies :

$$< curl[v(r,t)] >_V \neq 0 \tag{3}$$

A turbulent flow is characterized by a large and continuous span of values for the scales of rotational motions : eddies of all sizes exist in the fluid, their distribution being random. This distinguishes turbulence from chaos, where only a few degrees of freedom are excited, but turbulence, like chaos, is not predictable beyond a certain interval of time. Turbulent eddies must have a maximum size, fixed either by the process generates the turbulence, or by the volume of the fluid itself. They will also have a minimum size, since the viscous dissipation for a given kinetic energy in the eddy increases as its size decreases (for air under normal conditions, this scale is in the millimeter range). Outer scale L_O and inner scale l_0 designate these boundaries. Turbulence is homogeneous when, statistically, the velocity field is invariant by translations in space, isotropic when invariant by rotations, and stationary when invariant by translations in time. We shall restrict the discussion to cases where the fluid may be considered as incompressible : this is satisfied, even in compressible fluids like the air, when velocities remain well below the velocity of sound.

When does a flow become turbulent? Kelvin theorem states that if vorticity in fluid is zero at a given instant, it stays so at any time. But this is only true for a fluid without any viscosity, and tells then that turbulence can not develop in a free flow. The matter changes when obstacles are present : a simple rule is then given by the value of the Reynolds number $Re = LV/\nu$, a dimensionless quantity formed from the velocity of the flow V, the kinematic viscosity ν (ratio of the dynamic viscosity μ over the specific mass) of the medium and a length L, which characterizes the size of the obstacle encountered by the flow, or the dimension of the pipe carrying it, or any similar spatial quantity affecting the motion of the flow. When $Re \leq 1$, viscous dissipation occurs but when $Re \geq 2000$, the laminar flow becomes turbulent due to the development of instabilities. The value of n for air is 1.3810^{-5} in S.I. units. Physically, the Reynolds number may be understood as a measure of the ratio between the kinetic energy in the fluid at scale L, and the dissipated energy as heat by viscosity at this same scale.

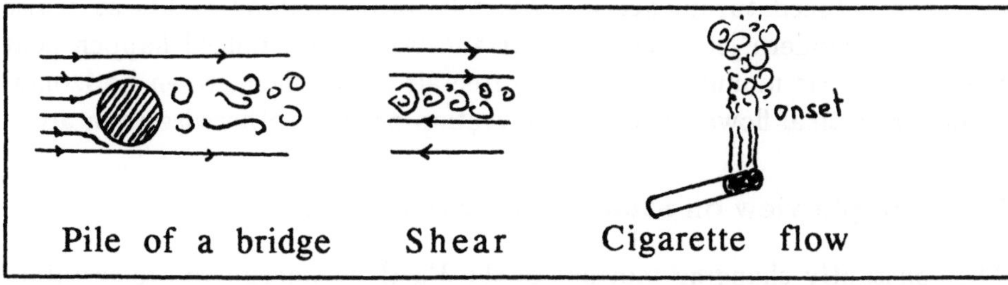

Figure 1. Turbulence and transition regime

As we consider here tri-dimensional turbulence, the number of degrees of freedom in the turbulent motion is then approximately equal to $Re^{9/4}$.

Near the critical value of Re, one observes a transition between the laminar and the turbulent regimes, often called the onset of turbulence and characterized by a sporadic regime. The quiet, then suddenly agitated, upgward flow of cigarette smoke (caution ! smoking may be dangerous for health) demonstrates immediatly this transition : the hot air raises in a column, and accelerates upwards due to the buoyancy force. Scattering the ambient light, the smoke particles finely materialize the velocity field. The initial flow is laminar, but at some point breaks into very unstable patterns, which come and go. Higher up, these patterns tend to stabilize : the velocities appear random, but with some conservation of the global aspect over time.

In contrast to the purely random velocity field of established turbulence, this transition regime may show stable rotational structures (the well-known Bénard-von Karman eddies) in the wake of a bridge pile or at the wing tip of an aircraft (easily seen from inside the plane when condensation near landing materializes the wake). In daytime convective plumes develop above the heated ground and similar structures may be observed by atmospheric radars .

The dimension L is an order of magnitude : for a given value of L, aerodynamical studies seek to reduce the turbulence induced by the wing of an aircraft or the shape of a bridge. Proper choice of the shape can move the onset of turbulence to higher values of Re.

Obstacles in the flow produce turbulence. But it may also result from shearing layers of the fluid in motion. Shear occurs at the boundary between two layers having very different relative velocities, as may be the case in river flows (a famous example is the mixing of the Rio Negro and Amazon rivers). Shear also happens near the jet streams at the tropopause, where a tube of air, driven by the general circulation of the atmosphere (equator to poles and back), moves at high velocities (ca. 200 km/h) through calmer air : the induced turbulence is often felt during bumpy flights at cruise altitude in modern jet aircraft. Shear and wakes may indeed happen near telescopes, near mountains or in the free air above them, when horizontal winds or vertical flows driven by convection occur in otherwise calm air.

2. A simple view on atmospheric turbulence

The constantly changing motions of the Earth's atmosphere are certainly a challenge to any attempt to describe its turbulent state by a universal model. There nevertheless exist some constant patterns of flows created by the general circulation of the air, driven by zonal temperature and pressure gradients. Hence, unless very specific and local circumstances exist (as, for example, an inversion layer characterized by a positive vertical temperature gradient dT/dz), one can describe some general properties.

During daytime, convection often occurs, producing a near-ground ($\leq 1km$) turbulence which affects for instance solar observations at visible wavelengths, but also observations with millimeter radiotelescopes : this will progressively die away as night proceeds. Convection constantly changes with insolation or the nature of the ground cover, as glider pilots know well. At night, convection rapidly disappears and the dominant effects are due to the horizontal circulation of air. Negative effects may nevertheless subsist locally : if the inside of a telescope dome is warmer than the outside, buoyant air may raise through the slit and generate the onset of turbulence (leading to dome seeing). The same may happen on the primary mirror surface (mirror seeing). Stabilizing against convection is therefore recommended : one may select a tree environment for the telescopes rather than dark ground, cool the inside of the dome and control the primary mirror temperature. As the low atmosphere (troposphere) is easily convective because of its negative temperature gradient ($dT/dz < 0$), one will search for inversion layer conditions, where dT/dz is forced to low or even positive values, for instance by a cold lake or sea (Mauna Kea in the Pacific Ocean, Paranal near the El Nino Pacific current, etc.) surrounding the telescope site.

The boundary layer near the ground extends up to an altitude of roughly one kilometer, depending on the local orography and the strength of the wind; the jet stream zonal winds often produce turbulence at the tropopause level (10 to 17 km, depending on latitude); at intermediate altitudes (1 to 10 km), remote obstacles (mountains) can produce a wake persisting over hundreds of kilometers downstream.

The so-called Hufnagel or standard profile of the atmospheric turbulence describes these general effects in a quantitative manner (Fig. 2) : it is

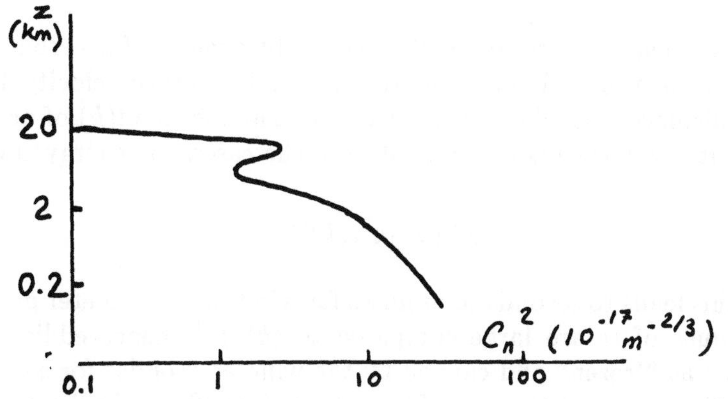

Figure 2. **The intensity (abcissa) of atmospheric turbulence versus height z (ordinates). Intensity is expressed by the quantity C_{n^2}, which is introduced in Sec.4. This average intensity model is called the Hufnagel model. The dominant layers are indicated.**

tabulated and plots some average quantity (deduced from a physical model to be discussed in Sec. 4) which measures the intensity of the turbulence, or equivalently the kinetic energy stored in it at any given time, versus the altitude z above the ground. This profile results from accumulated measurements and is convenient as a first-order quantitative reference for an average atmosphere, to which significant departure may occur.

Atmospheric turbulence profiles are measured either by balloon sounding (the balloons usually carry temperature sensors), or by similar sensors placed on masts, or by optical methods observing double stars scintillation from the ground (Scidar method).

Fig.2 shows the benefit of a high elevation site, properly isolated (such as a mountain peak) to escape the ground boundary layer. The South Pole plateau is not affected by high altitude winds and is therefore a low turbulence site for observations made a few tens of meters above the ground, as recent surveys indicate. In general, the importance of turbulence near the ground shows that the horizontal propagation of optical beams (e.g. on the long baseline of an optical interferometer) can be very badly perturbed if they are not suitably stabilized.

The velocity field $v(r,t)$ in a turbulent flow is doubly random : in space and in time. In the particular circumstances of atmospheric motions submitted to the circadian rythm, the phenomenon may be stationnary in time but not in the absolute and formal statistical sense : over short periods of time (from seconds to possibly hours, depending of the conditions), quasi-stationarity may be achieved. If so, ergodicity generally applies, meaning that averages taken over time or on independent realizations in space are identical.

It is interesting to compare the typical lifetime t_e of a turbulent eddy (size l) to the time l/V it takes for the wind (average velocity V of the flow) to displace this eddy by its own size. The lifetime $t(k)$ of an eddy at wave number k (size $1/k$) is derived from the cascading energy $\epsilon(k)$

$$t(k) = [k^3 \epsilon(k)]^{-1/2} \tag{4}$$

and this leads to seconds to minutes for winds of a few meter per second. These values of te are large compared to $(kV)^{-1}$, hence eddies may be considered as "frozen" and carried by the wind as "solids" for some length of time. This apparent freezing of turbulence for sufficiently short times has applications in adaptive optics.

3. Turbulence and light propagation

The purpose of this short introduction is neither to describe in detail the effects of a turbulent medium on the propagation of light, nor to give the formalism required to evaluate the quantitative effects. They may be found in specialized references (Roddier 1991, Roddier 1997). A few remarks may nevertheless be useful.

The propagation of light is affected by fluctuations on the index of refraction of "air", which may be an inhomogeneous medium. An index variation Δn , which is almost achromatic, creates a phase delay (phase corrugation) $\Delta\Phi = 2\pi L \Delta n/\lambda$, at wavelength λ and for a path of length L. The achromaticity of these optical path perturbations enables for instance to measure them at a given wavelength and to correct for them at another wavelength, as is done in adaptive optics, or for fringe tracking in interferometry.

The residual, very small chromaticity of the index of refraction may have some applications in adaptive optics. For instance, a bi-chromatic artificial laser star may be produced by a laser exciting two lines of the sodium mesospheric layer, and the return light paths become slightly different because of this chromaticity : this may allow a measurement of the wavefront tilt induced by the atmosphere to be made, leading to a proper correction on the natural observed star.

When almost perfectly transparent, at wavelengths located far from absorption bands, the air has an index of refraction which depends on pressure and temperature. Index variations due to local fluctuations of the pressure in a turbulent atmospheric flow are usually negligible. Similarly, when the turbulence is mixing air masses of uniform temperatures, temperature fluctuations are negligible : this isothermal turbulence has no impact on light propagation.

But most often, because of the horizontal transport, the vertical temperature gradients in the atmosphere and the presence of the ground, large temperature differences exist in the air masses which are mixed by the turbulence. These differences do not become quickly homogenized because of the poor thermal conductivity of the air : such turbulence has an important effect on propagation. For air at pressure $P(Pa)$ and temperature $T(K)$, index of refraction fluctuations Dn are simply proportional to temperature fluctuations ΔT as given by the simple formula relating the variances

$$\sigma_{n^2} = 7.510^{-7} \frac{P}{T^2} \sigma_{T^2} \tag{5}$$

The theory of turbulence tells us that the temperature exactly follows the fluctuations of the velocity field. It is also the case for any passive (not reactive) constituent : this is known as the Obukhov law. Atmospheric water, in the somewhat transparent wings of strong absorption bands, is responsible of the phase perturbations $\Delta\Phi$ in the mid- infrared (ca. 20 mm) and even more in the millimeter domain. Millimeter radio telescopes or interferometers must then allow for similar phase corrugation effects. At longer (centimeter and beyond) wavelengths, the propagation is unaffected by the troposphere, but is sensitive to turbulent index fluctuations in the ionosphere, where the electron density varies. It is also at these longer wavelengths that interstellar turbulence creates fluctuations of the local electronic density and produces the phenomenon of pulsar scintillation, exploited to measure the distances of these objects.

So far, the propagation is only affected by phase changes, as the medium is considered fully or at least partly transparent. How then can intensity fluctuations be observed, as is seen in scintillation of stars ? As a detailed model of propagation would demonstrate, scintillation results from global, integrated effects of the propagation across the whole atmosphere, where inhomogeneities act essentially as converging or diverging lenses, enhancing or reducing the amplitude of the local field.

4. A quantitative model

The random velocity field $v(r,t)$ is described as a stochastic quantity, which is assumed to be stationary (certainly an idealization compared with real situations, which extend seldom over periods of time exceeding minutes) and ergodic : averages $<>$ taken over space r or time t are equal. It is clear that the eddy structure produces $< v >= 0$, while $< |v^2| >\neq 0$: average turbulent kinetic energy is present at any scale.

The Kolmogorov description of turbulence involves this kinetic energy constantly cascading from the scale where it is generated (the outer scale L_0) to smaller scales, until it reaches the inner scale l_0 where the value of Re becomes unity and where all the kinetic energy is transformed into heat by viscous friction. In between the scales L_0 and l_0, there is no dissipation of energy but simply transfer by elastic collisions in the fluid. This is known as a scale invariant process.

Simple reasoning tells that the velocity at any given scale x can only depend from the power input ϵ per unit mass of the fluid (measured in $Wattkg^{-1}$) at scale L, and of this scale x. The single quantity having the dimensions of a velocity and which can be made with these two quantities is $(\epsilon x)^{1/3}$: this tells us that the velocity at scale x varies as $x^{1/3}$, or, if one prefers to think in terms of wave numbers k, as $k^{-1/3}$. This expression forms the basis of all the analytical derivations which follow naturally from the Kolmogorov model.

In particular, it is possible to express the power spectral density $E(k)$ of the kinetic energy at wave number k as

$$E(k) \propto k^{-11/3} \tag{6}$$

or taking isotropy into account (satisfied when turbulence is established),

$$E(k) \propto k^{-5/3} \tag{7}$$

Fig. 3 shows the Kolmogorov power spectrum $E(k)$. Obviously its analytical expression as a power law has to be bounded at small and large scales. Analytical, physically sound extrapolations of the Kolmogorov spectrum for $k > {l_0}^{-1}$ or $k < {L_0}^{-1}$ have been proposed, as sketched in the figure.

A change in injected power ϵ will simply translate vertically the whole power spectrum. One can demonstrate that temperature fluctuations $\Theta(r,t) = T(r,t) - \langle T \rangle$ have a similar power spectrum, as do the index of refraction fluctuations which are proportional to Θ. Since the covariance of temperature fluctuations $\langle \Theta(r+\rho,t)\Theta(r,t) \rangle$ diverges if one does not take into account finite scales, it is usually preferable to use the structure function defined by:

$$D_\Theta(r) = \langle |\Theta(r+\rho,t) - \Theta(r,t)|^2 \rangle \tag{8}$$

Computed using the Kolmogorov spectrum, this leads to

$$D_\Theta(\rho) = C_{\Theta^2} \rho^{2/3} \tag{9}$$

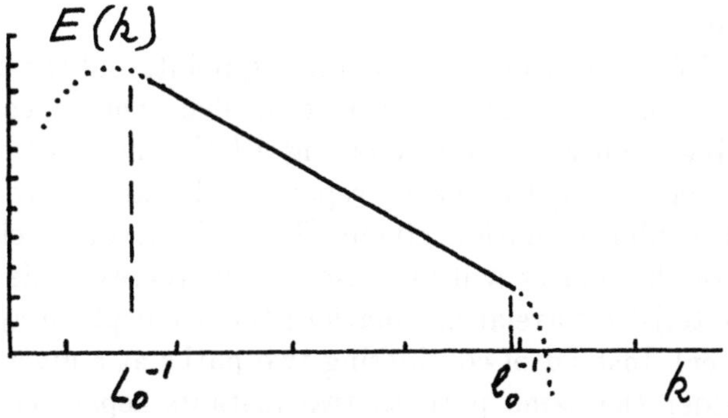

Figure 3. Power spectral density of the kinetic energy $E(k)$ for isotropic turbulence. The dotted lines show the low and high frequency ends of the spectrum, where the inner and outer scales bound the turbulence.

or similarly for the index of refraction fluctuations

$$D_n(r) = C_{n^2} \rho^{2/3} \tag{10}$$

The proportionality factors C_Θ^2 or C_{n^2} are constants which simply define the intensity of the turbulence. In Fig.2, $C_{n^2}(z)$ is plotted versus the altitude for the Hufnagel model. The physical meaning of the last equation is clear : the mean quadratic amplitude of temperature differences between two points increases with their separation, as they belong to increasingly decorrelated air masses.

When a large volume of fluid, such as the overall local atmosphere above an observatory, is considered, it can be treated as made up of independent (or almost independent as far as turbulence is considered) layers at various altitudes, each characterized by a value of the quantity $C_{n^2}(z)$, and hence parametrized by the altitude z. The knowledge of $C_{\Theta^2}(z)$ is therefore a complete measure of turbulence at a given time.

Is this simple Kolmogorov model a good representation of actual atmospheric turbulence? Turbulence is such a complicated, somewhat erratic and unstable phenomenon in the Earth's atmosphere that there is no universally valid answer to this question. Quantitative criteria can be established to study this issue. For example, it is reasonably easy to predict, on the basis of the Kolmogorov model and of an atmospheric profile $C_{n^2}(z)$, what will be the variance of successive spatial modes of a deformed optical wavefront (e.g. Zernike polynomials). This variance may also be measured by a system analyzing the spatial behaviour of the wavefront coming from a star, versus time : such a system, common in adaptive optics, is called a wavefront sensor. A comparison of measured and Kolmogorov predicted variances sometimes shows excellent agreement, but at other times significant departures.

Using the Kolmogorov model as a starting point, and the above defined quantity $C_{n^2}(z)$, one may compute the total phase shift encountered by a wave originating from a star and crossing all the layers of the turbulent atmosphere. Considering two paths separated by an horizontal distance r and using the above relation giving $D_n(r)$, it becomes clear that the decorrelation of the phases will increase as r increases : this leads one to define a characteristic correlation length of the atmosphere r_0 (the famous Fried parameter). Instead of considering two paths at a given instant, one can also consider the same path at two instants separated by time t : again there will be a characteristic correlation time t_0. Both r_0 and t_0 are wavelength dependent (as $\lambda^{6/5}$) : this wavelength dependence, which is very important in all astronomical applications (adaptive optics, interferometry) can be derived from a calculation of the phase shift, induced by Kolmogorov turbulence, integrated over the whole optical path across the atmosphere.

5. Conclusion

Turbulence is a fascinating phenomenon, but a one which seriously limits astronomical observations, especially when the preservation of the wavefront is critical, as is the case in imaging and interferometry. Knowledge of atmospheric turbulence has progressed, and is still progressing from the work astronomers expend to fight its detrimental effects. The advent of artificial laser stars, requested for the most efficient use of adaptive optics, will undoubtly stimulate new progress.

References

Lesieur, M. (1987a), Turbulence in fluids, Nighoff.

Lesieur, M. (1987b), “Hydrodynamic turbulence“, in Astrophysical fluid dynamics (Zahn, J.-P., Zinn-Justin, J. Eds.), Les Houches, 1987 Series XLVII, 33-92, North Holland.

Roddier, F. (1981) The effects of atmospheric turbulence in optical astronomy, Progress in Optics, XIX, 281.

Roddier, F. (ed.), (1997) Adaptive optics in astronomy, Cambridge University Press, in press.

DESIGN OF ARRAYS FOR GROUND BASED OPTICAL INTERFEROMETRY

STEPHEN T. RIDGWAY
National Optical Astronomy Observatories
P.O. Box 26732
Tucson, Arizona 85726
U.S.A.

1. Introduction

The objective of this article is to summarize most of the topics you need to consider in order to design an optical interferometric array. Since a detailed treatment would obviously require many more pages, this discussion is in the nature of an outline of options. Also, the truth is that in many areas of array design astronomers really don't understand very well yet what the optimum choices are.

Naturally an array design should ultimately flow down from clear ideas about the science objectives. But since interferometry remains technology rather than science driven at its current stage of development, we will recognize that fact by deferring science discussion to a brief resumé at the end.

In the following, optical will be understood to include both visible and infrared through about 12–20 μm, and interferometry will be understood to refer primarily to direct rather than heterodyne beam combination.

2. *U–V* Plane Issues

The topic of *U–V* coverage and image synthesis has been extensively studied and exploited in radio astronomy (*e.g.* Perley *et al.*, 1985). In the radio, the speed of an array increases approximately with the square of the number of telescopes, since the amplified signal can be combined on $n(n-1)/2$ baselines with no noise penalty. Also in the radio, supersynthesis has been used with great success.

In the optical regime, additional telescopes added to an array bring additional photon noise as well, leaving a gain more nearly proportional to the

A-M. Lagrange et al. (eds.), High Angular Resolution in Astrophysics, 13–38.
© 1997 *Kluwer Academic Publishers.*

number of telescopes than to the square. Furthermore, in the optical, the seeing deteriorates rapidly with increasing zenith distance, so supersynthesis at a ground based array is intrinsically less effective at optical than at radio wavelengths. Also, many interesting optical phenomena are relatively short-lived, especially at high angular resolution. For these reasons, optical interferometry may develop in the direction of either very large numbers of telescopes (snapshot capability), or of a smaller number of telescopes with a capability for array reconfiguration (perhaps even very rapid reconfiguration — hypersynthesis).

A systematic determination of the optimum layout of an array for snapshot *U–V* coverage leads to a circular configuration (Cornwell, 1988), at least for a number of telescopes greater than about 8–10. The circle also has another advantage for optical interferometry. In order to obtain image detail a source must naturally be resolved. For typical extended sources (*e.g.* a star) the visibility beyond the first zero in the *U–V* plane will drop rapidly. So for a telescope baseline required to resolve detail on the stellar disk, the visibility will be of order 1% or less. This can be extremely difficult to detect. A possible technique for maintaining cophasing of such telescopes is "chaining" of intermediate baselines, on which the source is unresolved and high fringe visibility is obtained. An ideal arrangement to accomplish this is a generally circular array geometry, in which every telescope has nearest neighbors which have a separation small enough that the visibility for a compact source will be high. This will allow the array to be cophased, while the telescope separations across the circle provide resolution several to many times higher.

Image reconstruction techniques may favor a partially redundant array for observation of complex sources, though for simple collections of point sources non-redundancy may be preferred (Panel, 1989).

For an array, it is not very convenient to position telescopes within a few telescope aperture diameters of each other. This results in insufficient *U–V* coverage at low spatial frequencies. Since for many sources most of the power is in the low frequencies, this is a significant problem. The most attractive solution may be to combine complex visibilities from an array with those from a large monolithic or multi-mirror telescope, perhaps utilizing aperture masking and adaptive optics (Nakajima and Haniff, 1993). The advantage of having the large aperture as part of the array facility is obvious.

For fewer than 6–7 fixed telescopes, with the limited supersynthesis capability available in the optical range, it is virtually impossible to get uniform *U–V* coverage, and choices must be made.

3. Site Selection

The performance of optical interferometry is critically dependent on seeing, and good seeing sites often have the character of steep terrain projecting into laminar wind flow. Such sites tend to have interesting topography, and this becomes an interesting factor in site selection, especially for baselines much greater than 100 meters. Site seeing surveys tend to emphasize the optimum telescope locations, while a seeing study directed toward array interferometry might better treat average locations. While a major telescope might reside on a pier which raises it above ground turbulence, a smaller, perhaps transportable array telescope might be closer to the ground and in the boundary flow.

Ground stability cannot be taken for granted. High mountains tend to be seismically active, with a background of minor tremors. Waves breaking on distant shores are detected far inland. High mountains also have high median wind speeds. Wind is a significant but not fatal problem for conventional telescopes up to wind speeds of order 10 m/sec. Wind speeds this high may be a serious problem for array telescopes.

Developed sites are far less expensive than new sites. However, they may have vehicular and foot traffic, rotating telescope domes or enclosures, and mechanical equipment of many types, as well as buildings which couple wind buffeting into the ground.

No comprehensive optical array site search has been attempted, though limited characterization of several sites has been reported (Panel, 1989; Walters *et al.*, 1990).

4. Fields of View

There are several relevant fields of view. Some interferometric techniques work essentially inside the Airy disk of the unit telescopes. Interferometric operation is possible over a much larger FOV in which the atmospheric piston error is nearly constant. However, operation with useful efficiency also requires a good image quality. If adaptive optics are employed, the FOV will most likely not exceed the isoplanatic patch, which is on the order of a few arcsec in the visible. The implementation of adaptive optics will require access to a wavefront reference star, and dual beam interferometry will require access to two sources within the field. These fields are all relatively small by conventional standards. However, in interferometry the beams may be projected over hundreds of meters, or injected into single mode fibers. In either case, the FOV must be considered, and multiple pickoffs may be required in order to obtain interferometric operation with an AO equipped, dual interferometric beam system.

5. Optics

The efficiency of an optical interferometer is closely related to the Strehl ratio observed on a point source, and in many detection schemes signal-to-noise depends on the square of this quantity. Thus any factor which causes loss in the Strehl ratio is extremely deleterious.

An interferometer may incorporate as many as 50 optical surfaces in each interferometric beam, and optical quality is a critical issue. The surfaces must be accurate, and equally important they must be stable, if the transmitted beam is to have acceptable characteristics. Low CTE materials are absolutely required where any temperature variation is allowed. Even small optics require careful mount design, and optics should be tested in the mounts and in the operating orientation.

Even with diffraction limited tolerances on each optical element, the wavefront will be significantly aberrated after a large number of surfaces. To the extent that this aberration is constant (ie excluding moving elements) it may be possible to reduce the aberration by means of corrective elements in each beam. Although adaptive optics may offer the possibility to implement such a correction, in practice it will be difficult. The adaptive optics wavefront sensor will best be located at the telescope (to avoid problems of pupil conjugation, loss of light, and FOV limitations). The correction of optical surfaces after the AO system would then be implemented open-loop, unless a second wave-front-sensor is employed to monitor the optics in the beam path.

6. Modulation Transfer Factor

Many optical and mechanical effects will cause a reduction in the contrast of observed interferometric modulation. A convenient way of describing this is the modulation transfer factor, T. Physically, it can be thought of as the fringe modulation which would be observed for a point source. In most cases, a T factor can be assigned to each source of fringe modulation loss. For example, the optical train will have a modulation transfer factor equal to the Strehl ratio it can deliver. The efficiency of most subsystems can be combined multiplicatively to compute the net efficiency of the array, *i.e.*,

$$T_{array} = T_{atmosphere}\, T_{optics}\, T_{opd} \;..... \qquad (1)$$

It is important to appreciate how severe this rule is. There are many potential causes of reduced efficiency — it can be extremely difficult to ensure that any of them have an efficiency near 1.00, and many will have a much smaller value. As a most obvious example, a telescope operating with a r_o size aperture and perfect tilt correction is expected to yield a Strehl ratio of only 0.87. The optimum aperture diameter chosen to maximize signal-to-noise ratio in the photon-limited-case (Buscher, 1988), is approximately $3r_o$ (expected tilt-corrected Strehl approximately 0.43). This explains why many prototype interferometers operate with observed point source visibilities smaller than 0.8 and as small as 0.1–0.2. It is also worth noting that adaptive optics systems commonly provide Strehl ratios only in the range 0.1–0.9 (depending strongly on wavelength), and the higher values are only achieved for very favorable situations.

7. Angular Dispersion

The prismatic dispersion of the atmosphere deviates blue light more than red light. At a zenith distance of 50 degrees (for a particular model) the deviation of 500 nm light is 68.5 arcsec, while the deviation of 900 nm light is 67.5 arcsec. Since this dispersion will be approximately the same for all telescopes in an array, it does not impede detection of interference fringes. However, for a beam propagated over several hundred meters, the relative physical displacement of red and blue light is a significant consideration.

Atmospheric dispersion can be partially corrected with a compensating prism. Since the dispersion varies with zenith distance, the compensating prism must have a variable dispersion, and since the dispersion of glasses differs from the dispersion of the atmosphere, it may be necessary to use a multi-element prism in order to partially achromatize the correction. A widely utilized design employing pairs of wedged elements is called Risley (or Herschel) prisms (*e.g.* Simon, 1966). The achromatic implementation requires 4 glass elements. Optical contacting of some elements and use of an index matching fluid for others may be possible in order to reduce the light losses. The Risley prisms must naturally be located at or near the telescope in order to obtain the reduction in beam spread. In order to track the varying magnitude (and in some cases relative direction) of atmospheric dispersion, the prisms are rotated. The accuracy of this rotation and the smoothness are both important concerns, since rotation also changes the optical path difference.

8. Polarization

Light which is initially non-polarized will be partially polarized by reflections in an array. Differences in polarization between different beams will result in a reduction in the modulation transfer factor and reduced array efficiency. The simplest approach for maintaining equal polarization properties in all beams is to use an optical design in which all beams experience the same sequence of reflections (as described by the direction cosines) — most readily achieved by homology of optical paths (Traub, 1988). Polarization differences could be readily corrected monochromatically, but it would be challenging to devise a wide-band achromatic correction. It is also notable that the silver coatings with dielectric overcoats which are widely used for their high reflectivity have relatively severe polarizing properties.

If the effort to control polarization is imperfect, it is possible to split the beam by polarization, detecting orthogonal polarizations independently. In fact, many prototype instruments simply use a polarizer to increase the observed modulation (by rejecting the out of phase polarization).

9. Longitudinal (Phase) Dispersion

Both transmitting and reflecting optical elements produce a wavelength dependent phase delay. As long as these are identical in all beams the zero path difference condition will still be achromatic. Otherwise, the zero path position will be wavelength dependent. Chromatism of ZPD position generates a relation between the bandwidth and the visibility transfer factor. In the case of detection in a dispersive (*e.g.* grating) spectrometer, there will be a visibility transfer factor associated with the bandwidth per pixel.

Phase dispersion also arises due to differences in the total air path for different beams. Of course the air path through the atmosphere to each telescope is nearly the same, but in general each telescope will require a different path length to the beam combiner in order to equalize optical delay. There are several possible strategies for minimizing this dispersion. The simplest conceptually is to evacuate the light paths. A second possibility is to construct an array in which the telescopes are continuously movable and may be displaced in order to equalize both the optical delay and the air paths (Vivekanand *et al.*, 1988).

If the design allows a residual phase dispersion, it may be negligible (over a narrow bandwidth or in the infrared). It can also be compensated with blocks of glass (3 mm of BK–7 glass approximately compensates the visible dispersion of 10 m of air). The compensation must be continuously variable, and in the visible an achromatic design may be required, analogous to the requirements for Risley prisms.

10. Wavelength Range of Operation

It is very desirable to allow concurrent operation of the array in the visible and the infrared for operational reasons, irrespective of the science emphasis. The visible photons can be detected with low noise silicon detectors for cophasing the telescopes when observing in the infrared, and the higher visibility infrared interference pattern can be used to track the optical path difference while observing in the visible. Multiple wavelength operation imposes additional complications, however. Anti-reflection coatings have limited bandwidth, and Risley prism design is more strongly constrained.

11. Support for Dual Beams

Simultaneous detection of interference for two sources in the field of view is a powerful technique for extending interferometry to fainter sources. A bright source can be used to track the ZPD, while information is acquired about a nearby faint source.

It will usually be necessary to isolate the two sources at the telescope — transmission over hundreds of meters would otherwise excessively limit the separation of the two sources. This requires a relatively complex optical and mechanical system — an image plane, X–Y stage for pickoff, probably recollimation, and the whole must rotate, probably at a variable rate.

12. Complexity and Throughput

It should be clear by now that an interferometric array, particularly general purpose, is relatively complex. A straight-forward implementation of all the required subsystems leads to designs with 30–55 optical surfaces per beam. Such a system will not have a large throughput. Is this complexity really necessary? It is difficult to make a large reduction in the complexity without a corresponding loss in functionality, and some approaches to reducing the optics count add difficulty or expense in other areas.

The alt-alt telescope mount can save several reflections compared to the standard coudé arrangements, at the price of a driven tertiary mirror and a sometimes oblique tertiary angle. Delay lines could be dispensed with if the telescopes could be moved continuously, but at considerable cost. Adaptive optics could be implemented with few additional surfaces if the secondary of the telescope were the adaptive element, but this technology has not yet been developed. Longitudinal dispersion of the air can be neglected if enough of the optical paths are in vacuum, but it is probably not possible to anti-reflection coat the windows to optimize throughput in both visible and IR. In practice, the number of optical elements will also tend to grow for reasons of "convenience", such as placing the optical delay lines along an azimuth dictated by terrain rather than by efficiency, or shifting the beam heights to a level convenient for working on an optical table.

Partial compensation for optical complexity is in the use of efficient coatings. Fresh silver coatings have a reflectivity of 98–99% at wavelengths of primary interest for optical arrays. Since they are attacked readily by air-borne sulphur (which may enter the observatory by devious routes such as in rubber or lubricant), the commercial version of these coatings is usually multilayer with a protective overcoat. The characteristics of reflection and phase delay for these items are complex and difficult to obtain, and the variation between pieces is not known. The use of bare silver coatings appears more favorable, and this is another good argument in favor of a vacuum beam path to increase longevity of coatings.

13. Atmosphere

The atmosphere contributes a visibility transfer factor which is approximately equal to the Strehl ratio of a point source viewed through the atmosphere with a perfect telescope. By using small apertures, the atmospheric Strehl and transfer factor may be kept relatively high. However, the restriction in the visible to apertures of a few cm diameter is a distinct limitation. A limited but important gain is possible by subdividing an oversize aperture into subapertures which are each small enough to maintain a high Strehl. The S/N is increased somewhat, and the probability of losing the fringe position is reduced (ten Brummelaar, 1996). The atmospheric wavefront errors are more benign at longer wavelengths, and r_o size apertures are considerably larger. The more systematic approach to working with large apertures is to utilize adaptive optics to improve the wavefront quality.

14. Telescopes for Interferometry

Acquisition of reliable telescopes is a primary concern in building a good interferometric array. In several prototype array projects of the last decade, weak telescope systems have crippled the interferometric performance for years. It is easy to underestimate the challenge of designing and building telescopes which have the optical quality, pointing and tracking performance desired — research quality telescopes are not trivial systems.

Over the last 20 years, finite element analysis has permitted a revolution in telescope engineering, resulting in a continuing trend toward light structures. Interferometric optical testing techniques, numerically controlled machines and other advances have allowed more economical fabrication of fast telescope mirrors. The use of computer controlled active structures is also becoming wide spread. Thus modern telescopes, especially large ones, are smaller and lighter than ever before. However, this trend should not be endorsed blindly for interferometric arrays. These telescopes are much more susceptible to wind shake due to the reduction in length and the reduction in mass (the ratio of wind torque to moment of inertia is increased). Active control systems have a conflict between accuracy and stability — actively removing low frequency errors enhances higher frequency errors. Also, the larger the errors to be corrected actively, the higher the required signal-to-noise in the error detection — a potential limitation when a natural star is the ultimate reference source. Sufficient modeling of these modern methodologies to ensure suitability for interferometric applications may be very challenging.

Selection of telescope aperture size for an array can be driven by the requirement to reach a faint source limit (requiring large apertures). Otherwise, the objective to obtain good *U–V* coverage rapidly will push the array design towards a larger number of smaller apertures and/or reconfiguration capability.

An afocal telescope will produce a collimated beam which can be directed to the central laboratory with a minimum number of optical surfaces. In that case, the magnification of the telescope (ratio of the telescope aperture to the afocal beam diameter) is a critical parameter. If the afocal beam is small, diffraction effects will occur. These are of two types. Light will be lost due to diffraction out of the beam. Also, phase structure in the wavefront due to diffraction over different distances in different beams will reduce the interference visibility. These effects can be minimized by having a large afocal beam, but at the cost of larger and more expensive optics.

With the afocal telescope the options for separating beams in a dual beam system are limited, and some adaptive optics concepts are more difficult to implement. For these reasons, a focal telescope may be preferred.

It is desirable to minimize the pupil obstruction in an array telescope in order to reduce diffraction effects and light loss. In addition, the coupling efficiency of a telescope into an optical fiber depends not only on f/# and fiber parameters (Shaklan and Roddier, 1988) but on pupil central obstruction (Coudé du Foresto, 1996). A large central obstruction may also weaken the wavefront reconstruction algorithms used with AO.

Telescopes for interferometry differ from conventional telescopes in part because stability of the optical path is critical. The telescope rotations required to point and track should be decoupled from optical path changes, ruling out off-axis mounts. More subtly, the most important structural vibration modes should be decoupled from optical path changes to the extent possible.

Telescope vibrations will inevitably cause OPD changes. The goal will be to minimize the disturbances which cause vibrations, and to damp the resulting vibrations effectively. After eliminating equipment vibrations, and after waiting for any vibrations caused by dome rotation to damp out, the major sources of telescope vibration should be wind.

Protection from wind is a difficult problem area, and the first message should be to think carefully before selecting a windy site for interferometry. In controlling wind shake, a challenge is to protect the telescope from wind (telescope enclosure or wind baffles) while not causing excessive dome seeing. Some telescope designs are inherently more stable against wind. For example, a design with an axis at the midpoint of the telescope tube will experience lower angular acceleration from wind than a tube with the axis near one end.

An accurate and drift-free focus is critical for efficient interferometric operation. The most satisfactory solution is an opto-mechanical structure with a CTE near zero. Use of expansion-free materials (optics and primary-secondary spacers) is the most sure design solution, though the use of optics spacers may introduce ancillary complications. Passive thermal compensated designs will depend on an assumption of small temperature gradients in the structure, which may require careful design and insulation of the mount. Active focus control based on temperature sensing will depend on thermal models or empirical calibration, and may be risky due to the OPD change associated with any focus motion. Adaptive optics will provide a good focus error signal (from the wavefront sensor) but does not in any way substitute for a thermally stable design.

Tilt correction is of course required, even in the absence of higher order AO. Ideally this correction should be implemented near the pupil or a pupil image. It is also wise to locate the tilt correcting element in or near the telescope to avoid wandering of the beam across subsequent optical elements (for the thermal IR, 8–12 μm, this is especially important). The primary mirror will probably be too large to control with adequate bandwidth (tens of Hz at least). The secondary mirror may be a good choice. For a large secondary and/or a lightweight telescope structure a momentum compensated mirror may be required, while for a small secondary on a massive telescope structure, momentum compensation may possibly be neglected. As an alternative to carrying out an elaborate structure analysis, tilt-correction can be implemented at the secondary at low cost without momentum compensation on a "best-effort" basis. A bandwidth of just a few Hz will reduce the residual error to a low level so that a small, high bandwidth incremental correction may be introduced downstream in a small optical element. Ideally the rotation axes of the tilting elements should intersect in the center of the optical surface, though the tolerances are not severe. Tilt detection can be located at the telescope, which has the advantage of greater optical throughput to the tilt sensor, or at the central laboratory which has the advantage of removing both drifts and vibrations in the intervening optical train.

For an array of any size, fully remote operation of the telescopes and enclosures will be mandatory if operation by a small staff is intended. For the operators to have sufficient feedback about the status of each telescope, one or more video camera views of each may be a useful adjunct to computer displays.

15. Damping of Telescope Vibrations

A classical telescope mount and tube of bolted or welded metal members will have many poorly damped resonances, as may be verified by tapping on the structure. Numerous strategies may be employed to damp these resonances mechanically. Passive techniques start with the use of composite structural materials (graphite-epoxy types) which may be specified with some damping, as well as favorable CTE, though at risk of considerable expense. The commonly used hollow structural steel tubes may be filled with damping materials (*e.g.* foamed in place). Structural members called damping struts (*e.g.* Honeywell Corporation in Glendale Arizona) can be installed in place of normal struts, and the mechanical design can be adjusted to profit from these. (Design aside, such members can cost less than $1K to fabricate.) Viscous mass tuned dampers (Honeywell) can be effective in damping specific resonances, *e.g.* tuning fork mode vibration of a mount. After fabrication, sticky backed damping materials can be added to exterior surfaces of the structure (*e.g.* Scotchdamp from 3M Corporation). In desperation, members can be retrofitted to suppress particularly troublesome modes. Although it may not be intuitively obvious, most of these mechanisms do damp effectively at the nanometer level.

Active mechanical damping is available with proof-mass-actuators, in which accelerometers detect undesired motions and counter them by accelerating a floating mass. (This technique is used to reduce wind sway in tall buildings, employing masses of many tons.) There is a limited commercial trade in devices suitable for telescopes, with likely costs in the few $10K's per unit.

Vibrations of the dynamic structure (that is vibrations in the rotations utilized in pointing and tracking) are more difficult to suppress. Mechanical damping (*e.g.* vanes in viscous oil) is not normally used in telescopes, perhaps because a release mechanism would be necessary in order to slew.

The telescope drive will possess a mechanical inertia which will resist accelerations. The effective inertia will be the inertia of the drive motor multiplied by the square of the drive reduction factor. Modern telescope designs tend to have low drive reduction factors, in order to achieve high drive bandwidth, but this is at the cost of drive inertia, increasing the risk of high frequency drive servo jitter.

16. Telescope Auxiliary Instrumentation

Although interferometry appears to require "bare telescopes", in fact a considerable amount of instrumentation is required at, or most logically located at, the telescope. This presents a conundrum. Should the equipment be located in the ambient environment, designed to tolerate the relatively hostile observing conditions, or should the telescope enclosure incorporate a controlled laboratory environment, raising other costs and risk of dome seeing?

The auxiliary instrumentation may include acquisition camera and detector, the tilt detection sensor and tilting optic, possibly a higher order wavefront sensor and adaptive element, an acquisition camera, dual beam pickoff, and Risley prisms. Following this equipment, the beam must be collimated for transmission to the central lab, unless some kind of optical relay is imagined. Implementing all of this equipment offers opportunity for a wide range of schemes.

Adaptive optics has the potential to provide significantly improved wavefronts, with correspondingly improved Strehl ratios. Unfortunately, adaptive optics with a natural guide star are limited to sources not much (if at all) fainter than can be reached with optimized interferometric systems with tilt correction alone. A substantial further gain may require laser beacon reference systems, if these prove as effective as is currently hoped.

17. Beam Transfer Issues

Vacuum is useful to reduce turbulence, to reduce longitudinal dispersion, to extend the life of mirror coatings and to maintain cleanliness. It may be worth noting that a long, straight run of evacuated beam pipe offers an unusual hazard — in the event of a massive vacuum rupture, a column of air will rush down the tube with considerable kinetic energy — see the literature on shock tube design. Flow baffles or burst disks may be advisable.

In limited applications (infrared, modest baselines) it is possible to dispense with vacuum, but pipes will be wanted in any event to reduce turbulence in the transmission of beams to the lab. Windows (perhaps optional for non-vacuum pipes) must be tilted enough to deviate secondary reflections out of the beam.

In addition to diffraction losses and visibility transfer efficiency of the beam transfer, the increased thermal emissivity should be recognized. In the 8–12 μm range the thermal background is so enormous that even a very small fractional modulation due to AO and tilt correction will produce a potentially crippling spurious signal - for this reason it may be neccessary to oversize the optics significantly to reduce the sensitivity (through diffraction) to thermal or emissivity gradients outside the beam.

The use of homologous paths to maintain equality of polarization dictates that the beam transport system will follow paths satisfying additional constraints, which may rule out direct paths "as the crow flies".

The use of single mode fibers for beam transport has been demonstrated at 2 μm for small telescope separations (Coudé du Foresto *et al.*, 1993).

Fibers are discussed further below, but in beam transport, requiring tens to hundreds of meter lengths, a major issue must be quality control and the equality of longitudinal dispersion achieved in different fibers.

18. Optical Delay Design

Adjustable optical delay in a propagating afocal beam can be implemented with a long continuously variable scan mechanism, with a short continuous scan plus a long slew and clamp (that is, slew to position and then hold position), or with a short continuous scan plus switchable segments. Continuously variable delay is by far the most expensive of these. Slew and clamp allows use of a simple, non-precision mechanism for translating the retroreflector, but still requires that the optic be located with appropriate alignment tolerances when halted at the desired location. Selectable segments requires a mechanism (*e.g.* tilting mirror) which selects the segments with suitable tolerances.

Most continuously variable delay mechanisms are either commercial scanning tables, or have their heritage in the scanning mechanisms developed for Fourier transform spectrometers — typically a wheeled carriage on rails, with a multi-level servo system controlling the carriage at low servo bandwidth, and one or more successively higher bandwidth servo loops on smaller components. Production of these units at Jet Propulsion Laboratory (Pasadena, California) for several arrays has resulted in a thoroughly refined design.

There are several options for retroreflectors in an optical trombone. The reflectors can be flats, two-mirror dihedrals (corner mirrors), three mirror catseyes, or three mirror corner cubes. Each has its share of advantages and disadvantages, with respect to efficiency, polarization, sensitivity of the wavefront to translation and tilt. Suffice it to say here that none are ideal but any may be made to work.

Interestingly, SM fibers have been proposed for optical delay also. The suggestion is to use switchable segments for a discretely variable delay. This can be very efficient in principle, since the junction between two fiber segments, with index matching fluid between fiber faces, can be almost lossless. A mechanized switching capability is assumed. For the continuously variable delay, a fiber can be wrapped in many winds around a piezoelectric cylinder and stretched by of order 1%. This technology has been demonstrated in the laboratory for short delays (Zhao *et al.*, 1992). The complications in extending this technique to wide use include mastering polarization and dispersion properties in wrapped and stretched fiber.

19. Interferometry Lab Design

Interferometers employ large numbers of optical elements for beam transport, delay, combination and detection, and these are commonly distributed over many tens to hundreds of meters of terrain and/or laboratory. It is unlikely that the site will be completely free of vibrational disturbances. Protection from ambient vibrations is achieved in large part with inertia and isolation, but note that slow drift is a common companion of isolation. Reasonable stability for interferometric components might be for the net effect of vibrations to remain smaller than atmospheric piston, and for alignment to remain stable at least for 24 hours, and preferably for a season or a year. There are many issues to consider in assuring this performance.

An interesting choice is construction on bedrock *vs* construction on soil. Bedrock offers greater stability, but bedrock offers little damping. Soils offer internal damping of $D \simeq 0.03 - 0.07$ (normalized to critical damping). This may not seem like much, but dynamic magnification is $1/2D$, and a smaller value of D is much worse.

A standard technique for partial isolation of sensitive equipment is to place it on isolated, massive foundation blocks which are carefully protected from building vibrations. These Inertial Blocks function by simulating (poorly) a free floating mass suspended on springs. Both soils and bedrock are deformable and have effective spring constants, and a construction directly on them will have a natural resonant frequency. This frequency is weakly dependent on the weight/unit area and weakly dependent on the total area (Tschebotarioff's law — Richart et al, 1970). For typical inertia blocks (1m $\times$ 1m $\times$ 1m) the resonant frequency will be in the range 1400 Hz (sand) to 2000 Hz (basalt), hence are not likely to be readily excited. This also shows that an inertia block sitting on soil or rock is coupled rather stiffly to ground vibrations. The isolation can be improved by placing the inertia blocks on engineered materials with a smaller spring constant. The standard installation requires placing a concrete slab on rock or sand, then above it an inertia block supported on special materials. There are several considerations. These specially engineered materials (*e.g.* Unisorb Technologies in Jackson, Michigan) are relatively expensive, in addition to entailing considerable installation cost. They also introduce a new, low frequency resonance (which may be damped, however). They are only effective at rejecting vibrational frequencies above a few Hz. Finally, they are designed for isolation and damping, and the long-term drift characteristics may not be known.

Several types of drift can be expected. But first, it should be realized that while foundation settling of a few millimeters is small by civil engineering standards, it can be large for an array if it was not foreseen.

Interestingly, if differential settlement of foundations is a concern, standard construction practice is to excavate a meter below the surface, mix and recompact the material. This is to promote homogeneity. But since the usual standard of compaction is 90–98% of maximum density, the potential for settling is obvious.

It might be hoped that use of a monolithic cement block would allow only bulk settling without bending, but cement is susceptible to slow relaxation in response to stress — large aspect ratios should be avoided. Also, concrete will experience an extended period of initial shrinkage and an even longer period of initial creep. Foundations may also undergo small environment related shifts. If there is any clay fraction in the soil, infiltration of moisture will cause it to expand due to capillary action and hydrostatic pressure. Any such soil should be removed to a generous depth and replaced with coarser materials.

Wave propagation in half-space consists of P-waves and S-waves, both of which drop in amplitude as the inverse square of the distance, plus R-waves (surface or Rayleigh-waves) which drop in amplitude only as the inverse square root of distance (Richart *et al.*, 1970). R-waves have a strong depth dependence and isolation by barrier is effective. The R-wave amplitude is 5× lower at a depth of 1.2 wavelengths. The speed of sound in the ground ranges from ≈ 200 meter/sec for sand to ≈ 600 meter/sec for basalt. So a trench depth of 2 meters effectively isolates for frequencies above about 100 Hz (sand) or 300 Hz (basalt). At a minimum, the inertia blocks must be free of rigid contact with surrounding building structure, and ideally there should be a gap (air or non-rigid filler) separating the inertia blocks on all sides from surrounding soil as well.

Any heavy rotating or reciprocating equipment poses the inverse problem. It must be mounted in a way that minimizes transmission of vibrations into the building and ground. Such equipment includes fans for air circulation, vacuum pumps, and large transformers. Motors and other rotating equipment can often be obtained with improved balancing of rotating components for of order 20% additional cost. The problem of isolating vibrating equipment is a very common one in industry, and extensive resources are available, including suspended pads and damping mounts, and it is reasonable to plan for continuous operation of this kind of equipment, provided precautions are taken in selection, placement, isolation and damping. For example, air circulation should be provided with blowers mounted externally and isolated, air circulation should be at low velocity, and ducts should have a low rigidity or high impedance section to suppress transmission of vibrations.

Of course a lot of optics will finally be installed on optical tables, and vibration isolation supports are available for these. However, they should be used with caution. Although good isolation may be available at low cost, simple mechanisms also tend to drift in absolute tilt and height.

Vacuum systems introduce a special problem. At each end or turning point, any vacuum pipe chain will experience uncompensated atmospheric pressure forces. It may be unwise to apply these to foundation blocks which also support sensitive equipment. First, resisting the pressure forces may require rigid coupling of the footing to the soil or larger building foundation. Second, varying pressure will introduce large, variable forces which may couple into displacement or creep of precisely aligned optics.

Thermal effects can easily cause diurnal and seasonal movement of equipment. It is easy to verify that with steel instrument mounts, temperature drifts of less than 1 K can cause significant misalignments. It is difficult to achieve temperature stability to a fraction of 1 K over an extended time, and it is probably too costly to fabricate all optical mounts from low expansion materials. However, it may be reasonable to aim for temperature drifts of less than 1 K per day. A construction technique employed at several prototype interferometers has been the double enclosure. Within a conventional building, a second structure (building-like, but without the requirement to withstand the external elements) is installed. The primary characteristics of the interior building are some degree of insulation, thermal inertia, and a reasonable seal against air circulation. This inner space will trend asymptotically to the temperature of the inter-wall space, and will be protected to a large degree against exterior changes. There is an interesting issue of whether or not to insulate beneath the inner building. Use of insulation will improve the homogeneity of the laboratory space. However, insulation under inertia blocks will result in higher construction costs and possibly in long term settling due to compression of the insulating material. Since the earth is an effectively infinite sink, the sub-surface temperature swings are considerably moderated compared to the outside ambient. For example, at a depth of 2 meters, the annual temperature range will be in the vicinity of 4–6 K, rather than a more typical 30K on the surface. A good design strategy is to provide a vertical insulating barrier at the outer building wall, extending 1–2 meters into the soil. This thermal buffer could be conveniently combined with the suggested gap for vibration isolation.

In most likely observatory locations, the mean subsurface temperature will be below comfortable laboratory temperatures. Therefore, even in the absence of a thermal model, we can recognize that the laboratory floor will be cooler than walls and air, and the air will tend to stratify when undisturbed. This stratification is a mixed blessing — it reduces the likelihood of convection, but any activity will stir up mixing of inhomogeneous air packets which will take some time to subside. Several implementations of double enclosure construction have accepted this condition, rather than try to eliminate stratification.

With the relatively benign temperature drifts described above, some attention to selecting temperature stabilized optical mounts, and the use of low expansion materials and additional insulation to isolate particularly sensitive sections of the instrumentation, should provide a cost effective solution for stability of optics.

20. Single Mode Fiber Optics Issues

Single mode fibers have some very attractive qualities, offering the possibility of "plug and play" multi-beam interferometers, in which complex interferometric equipment can be readily configured with reduced attention to interferometric tolerances.

Light from the telescope Airy disk can be coupled into the fiber's guided mode — light outside the Airy disk will be rejected or ejected from the fiber within a short distance. Thus the array will operate interferometrically only within the Airy disk of the unit telescopes. This limitation certainly impacts the science capability. Also, the impact of guiding errors on interferometry of sources within the Airy disk has not been adequately evaluated. The injection efficiency for optical coupling schemes which have been studied is only around 70%. Possible methods for improved phase matching of the Airy disk to guided mode have not been reported but may be possible.

As glasses are dispersive, and dispersion is a problem for interferometry, transmission through many meters of glass may appear bizarre. Fibers for the communication industry (quartz primarily, limited to wavelengths shorter than about 1.6 μm) are regularly fabricated with a nominally zero dispersion at a single wavelength. Concepts exist for flattening the dispersion to near zero over a relatively wide band (Coudé du Foresto *et al.*, 1995). The single mode performance of a fiber in any event is limited to a modest bandwidth.

Even if circularly symmetric fibers were perfectly homogeneous, any bends would introduce crosstalk between polarizations. There are developmental techniques for controlling this polarization, through intentional added twists and wrappings, which may be useful. A more certain solution is the use of non-circular core fibers which are polarization preserving (at the cost of ejecting the other polarization). Couplers have complex properties, are difficult to improvise, and the available options may not be well suited for astronomy.

The inherent spatial filtering nature of single mode fibers is very interesting. The difference between pupil and image plane loses meaning when transmitting through a fiber (or equivalent pin hole). All spatial evidence for wavefront irregularites before the fiber is lost, though the time dependence of wavefront aberations at the input will result in time dependent variations of the fiber throughput. SM fibers have been a critical component in beam combiners which have achieved very high accuracy in visibility measurements (Perrin *et al.*, 1996).

In summary, optical fibers have great potential utility in interferometric arrays, though fibers and fiber techniques are in need of further development for application in this area.

21. Engineering Science and Interferometry

Interferometry is a strange blend of the large (telescopes, earthworks, buildings, plumbing) and the small (fractional wavelength tolerances). Engineering, which is theory tempered by experience, has a very limited tradition of examples and solutions for interferometry, and little of the experience has been reported and reviewed.

The requirement to achieve both isolation from vibration and stability entails a careful selection of strategies. In pursuing a purely deterministic design approach, costs can potentially get out of control. For example, modeling and optimizing the dynamic response of structures can absorb limitless resources, without ever producing a guaranteed result. Since many array requirements are soft, in the sense that array efficiency will degrade gently with less than optimum performance, there will be frequent need for judgment that a design is acceptable, even in the absence of definitive analysis. In many cases, the astronomer may be better prepared to guide the engineering tradeoffs. Furthermore, the astronomer will most likely be faced with the prospect of optimally allocating a fixed budget. In no field is it more true that Good is the enemy of Good Enough. A close and comfortable collaboration is critical.

22. Beam Combination Design for Optical Interferometry

Unlike radio interferometry, in which each telescope signal can be recorded for mixing at leisure, in direct optical interferometry all beam combination must be done by combining the incident wavefronts prior to detection. For an array, the beam combination strategy will determine the signal-to-noise of the detection, the limiting sensitivity, and the speed with which U–V coverage is acquired. The U–V plane will be covered most rapidly if all beam pairs are measured simultaneously. This may be achieved by combining all beams on a single detector (or array), by subdividing each beam into $n(n-1)/2$ parts and forming all pairs simultaneously, or by an intermediate strategy. In the source photon noise limited case, it is slightly advantageous to combine all beams on the same detector (Buscher, 1988).

However, at very low photon rates it may be advantageous to exploit an array as a set of subarrays (Kulkarni *et al.*, 1991). A general quantitative analysis may depend on both signal and noise sources, array cophasing, and on the analysis strategy (visibility amplitudes, phase closure, etc.), and even the relative size of telescope apertures (if they are not all equal). In practical terms, pair-wise combination may be considerably easier (and less costly) than some schemes for multi-beam combination, but in pair-wise combination the options for phase determination are limited.

Calibration of visibilities requires observing a point source or source of known visibility within the time constant of instrumental or atmospheric change. The atmospheric time constant can be short, but the impact of atmospheric change can be alleviated by monitoring the wavefronts. Spatial filtering (*e.g.* SM fibers) converts the wavefront variation to an intensity variation, which is easy to track for calibration purposes (Perrin *et al.*, 1996).

Beam combination techniques which are intended to investigate a field of view larger than the unit telescope Airy disk must preserve the pupil configuration. Since the pupil geometry changes with sky position (for fixed telescopes) the beam combination geometry must change with pointing of the array. This can lead to a requirement for sophisticated mechanisms for continuously reconfiguring the instrumentation (Panel, 1989).

23. Phase Measurement Strategy

The description of a source in the *U–V* plane is complex, while the classical two-telescope determination of visibility amplitude measures only a real number. Direct measurement of the phase requires phasing of the array, in practice by reference to another source — the dual beam system, with appropriate metrology, can achieve this (Colavita et al, 1994). However, there are many indirect techniques of phase determination.

Phase closure (the sum of phases around a closed path in the *U–V* plane) can be obtained with three or more telescopes (Baldwin *et al.*, 1996). The number of paths increases as the cube of the number of telescopes, so this technique appears to favor simulaneous combination of a large number of telescopes. Phase information is also available indirectly by the technique of self-calibration (Cornwell, 1985) which works best with a richly populated array. Further, most interferometric detection schemes will involve more than one spectral channel, and phase relations between these spectral channels can be useful in image restoration. These and other techniques are elaborately developed in radio interferometry.

Ideally, a closure phase measurement is self-calibrating, in the sense that it is independent of uncertain phase on each array arm. But some phase measurement techniques will depend on an assumed instrument stability at the fraction of a wavelength level. These will require careful design and implementation for stability, perhaps internal monitoring of OPD, and provisions for suitable calibration on known sources.

24. Fringe Detection and Optical Path Length Control Strategy

If the array metrology is uncertain by more than the coherence length it may be necessary to scan the OPD over a range, finding the ZPD condition by searching for fringe modulation. If a single discrete detector is used for a relatively broad spectral band, the coherence length will be small. In this case modulation will be observed over a small path difference range (perhaps only a few wavelengths), although in the case of differential phase dispersion the position will be wavelength dependent.

The OPD can be scanned over an extended fringe packet to provide both spatial and spectral information by the technique of Double Fourier Interferometry (Mariotti and Ridgway, 1988). Alternatively it can be scanned over a short OPD of order one wavelength, to determine only the modulation and relative phase (Shao *et al.*, 1988). An extremely simple technique is to allow the OPD to vary randomly (within the coherence envelope) and to determine only the modulation amplitude (Tango and Twiss, 1980). In all of these techniques, the OPD varies with the atmospheric time constant (tens to hundreds of milliseconds in the visible or infrared), so the integration time per exposure is severely limited.

If the flux is dispersed in a spectrograph (Traub and Lacasse, 1988), the coherence length (corresponding to the bandwidth on a single pixel) may be large. When the OPD is within the coherence length, a fringed spectrum will be observed (channel spectrum). The OPD can be determined by analysis of this fringe pattern. Unless controlled actively, the OPD will vary due to the atmosphere, and the exposure time is again limited.

For a bright source (or with a bright reference source) it is possible to measure the OPD rapidly enough to feedback the OPD error signal and control the OPD (to zero or a selected value). In this case, the channel fringe pattern, for example, could be integrated in a long exposure. By cophasing the interferometer in this way on a bright reference source, interferometric information can be obtained on very faint sources. In the Double Fourier technique, accurate knowledge of the OPD can be used to achieve high spectral resolution.

25. Detector Options

The principle detector issues are detective quantum efficiency, detector noise (or amplifier or read noise), dark current, linearity, and dynamic range. Practical issues, in addition to cost, include operating temperature and availability (not always obvious). With the proliferation of array type detectors which have been developed in recent years, cost and availability become issues primarily when attempting to reach the last magnitude in sensitivity.

Photon "amplification" devices — which actually multiply the number of electrons produced by photoelectric effect emission — are available only for wavelengths in the red to near infrared and shorter. They have the advantage of increasing the signal well above the noise of following detectors, so to a good approximation detector noise is negligible. In some cases dark current may be significant, but with the short integration times commonly required in interferometry this may be inconsequential. Quantum efficiency and dynamic range may be more important limitations. Photomultipliers are a classic and well developed photocathode implementation, with a wide range of devices available. The implementation in image tubes is more specialized and there is a smaller selection of products. For decades photocathode devices have offered a strongly wavelength dependent quantum efficiency of only a few percent. Higher peak efficiencies (of order 40%) are regularly rumored to be "around the corner".

A recently developed and increasingly popular photon "multiplier" is the avalanche photo diode (APD). These have a peak detective QE of about 40%, a relatively broad visible-red response, and convenient packaging. Solid state photomultipliers (SSPMs) are experimental devices which have potentially interesting characteristics but are not commercially available. Recently, niobium superconducting tunnel junctions have been reported to provide photon counting with 50energy discrimination over the 200-1000 nm range, though these features have not yet been incorporated into a useful detector.

Integrating solid state arrays have been extensively developed in recent years. Charge coupled devices (CCDs) have a wavelength response characteristic of silicon detectors, extending from the blue to near 1 μm, and peaking in the red. Peak QEs over 80% are available, and read noise of a few electrons, and even one electron or less, have been reported, although it may not be possible to find all the desired qualities combined in a commercial detector. Fortunately the market for detectors for adaptive optics wavefront sensing may be sufficiently large to spur development of appropriate devices.

Several types of array detector technologies are available in the infrared. High QEs are available over large wave bands. In recent years, read noise levels of detector arrays available for the near-IR have been slowly dropping from several hundred electrons to a few tens of electrons. Costs still are rather high.

26. Alignment

There are good reasons to plan conservatively for an alignment capability to, effectively, correct small instrument drifts easily and automatically. This is because, in spite of the best efforts, drifts may occur on an inconveniently rapid time scale. In addition to the slow compaction and settling that may arise in foundations, residual temperature and other drifts may arise on a diurnal time scale. Concrete may be expected to creep, typically by an amount similar to the inital static deflection, but this can be minimized by aging for one year or more, or by sonic or other destressing techniques. Optical mounts often drift due to flow of lubricant in the threads of adjusting screws. (Some combinations of flex mounts and inching mechanisms may avoid or greatly reduce this at additional cost with other drawbacks.) Machined and welded structures may undergo slow relaxation if internal stresses were not sufficiently alleviated — selection of materials and proper thermal cycling can reduce these effects. However, fearing the worst, in general it is very desirable to have sufficient remote actuation of test sources, detectors, and a critical subset of optical adjustments to permit some level of realignment without physically accessing the instrumentation. Interferometric alignment by combination of two autocollimated array arms is a particularly strong and convenient alignment technique. This is achieved naturally with siderostat systems, and can be more or less adequately designed into telescope systems which do not employ siderostats.

27. Metrology

In order to detect interferometric modulation, the OPD must be within the coherence length. Equalizing optical paths to within on the order of one part in 10^6 is a demanding requirement, but need not be difficult with modern measuring instruments. The internal optical paths can be readily measured to within a few millimeters with a laser ranging device (as used by surveyors) which can be rented for less than \$100/day. Due to the hazard of spurious reflections (HST effect), this technique will be most satisfactory if confirmed to first order with a measuring tape. The more difficult challenge is determining the relative locations of the telescopes. The same laser ranging meter may be used for triangulation from known benchmarks. However, differential GPS may now be adequate and much easier.

Once stellar fringes have been detected, the vector separations of the telescopes can be fully determined by the rate at which the OPD changes as a function of position in the sky, or in practice simply be measuring the optical delay changes required to reach the ZPD condition for a selection of stars around the sky. Accurately tracking the relative length of the internal optical delay is thus the essential.

Relative internal delay length can be determined to an accuracy of a few hundred wavelengths with a mechanical scale or encoder — sufficient to detect fringes by setting the delay a little ahead and fixing it while the terrestrial rotation changes the OPD through the ZPD condition (Dyck *et al.*, 1993). For more accurate OPD control, and especially for smooth continuous OPD control, nothing can replace a laser metrology system. A system based on Zeeman splitting, developed about 20 years ago for control of machine tools, which is sensitive only to differential motion, is very satisfactory. In a dual beam interferometer metrology will be required for the optical delays in both beams — and this metrology will be sufficient for differential astrometry between the two beams.

For wide angle astrometry additional internal metrology will be required to monitor directly changes in the vector separation of the telescopes (Shao *et al.*, 1990; Elias, 1994). Thus far, wide angle astrometry by interferometry has been accomplished only with siderostat optical feeds, in which the separation can be monitored rather directly via retroreflectors at the intersection of rotation axes. It is questionable whether or not more complex telescopes can be utilized for wide angle astrometry.

In some situations differential metrology may not be satisfactory — for example for an array of frequently or continuously moving telescopes where triangulation may be required to characterize the array geometry. Absolute laser metrology can be achieved with multi-wavelength laser systems which have been demonstrated in a laboratory setting but apparently are not available commercially (Gursel, 1995).

28. Current and Planned Arrays

Eleven optical telescope arrays (most with only two telescopes) have become operational through 1996. All but 3 are still in operation. All but 2 can reasonably be described as prototypes.

Three potentially very significant array projects are currently under way.

- The expansion of the Berkeley Infrared Spatial Interferometer from 2 telescopes to 3 (each of 1.65–meter aperture), equipped with heterodyne detection capability for operation near 11 μm, optimized for study of circumstellar shells around bright, cool supergiant stars.
- The extension of the Naval Prototype Optical Interferometer at Anderson Mesa to include the Big Optical Array, consisting nominally of six telescopes of 0.66–meter aperture, optimized for imaging of stellar surfaces.

- The Center for High Angular Resolution Astronomy Array at Mt. Wilson, consisting nominally of 5 telescopes of 1–meter aperture, particularly designed for binary star studies.

The longer term is less clear. Two major programs are currently in the plans of major agencies. At both the ESO Very Large Telescope and at the California Association for Research in Astronomy's Keck telescope facility, the civil infrastructure for interferometry (beam combining labs, light tunnels and delay line rooms) have already been built into the facilities.

- ESO may restore a limited part of the more ambitious VLTI plan. The initial limited interferometric operation may consist of one or two 8–meter telescopes (possibly without adaptive optics) plus one or two new auxiliary telescopes of 1.8–meter aperture. The VLTI science plan emphasized stellar astronomy.
- NASA (a 1/6 partner in the Keck facility) is developing a plan for JPL to implement an interferometric capability between the two Keck 10–meter equivalent aperture telescopes with adaptive optics, and possibly including two or more auxiliary telescopes of nominal aperture in the range 0.8 to 2–meters. The initial science plan emphasizes measurement of zodiacal dust around stars and detection of exo-planets.

Although optical interferometry is in its infancy, the parallels with the development of radio interferometry (and the differences) are already fairly clear. It is reasonable to begin planning for large, dedicated optical arrays for general astronomy. Detailed planning for such a project could begin almost immediately. Conceptual studies at a fairly high level (Labeyrie, 1988; Roddier, 1995) have already charted a number of possible courses. There is a need for significant engineering effort — what are the opportunities and challenges of constructing a substantial number (say, 10–30) medium to large telescopes? The applicability of modern capabilities for fast optics and light structures should be explored. Newly developed AO systems should be evaluated for interferometric utility. Further investigation of site selection for interferometry is badly needed. Nevertheless, technical readiness is fairly high.

29. Interferometric Science Programs and Array Design

Although it is not widely realized by the general astronomy public, or even by many interferometrists, there has been a steadily increasing flow of excellent science results from prototype arrays. These results (more than 50 science papers) are overlooked by many because the topics (stellar diameters, binary orbits, circumstellar dust shells) tend to be in "niche" science areas outside the mainstream of astronomy, which increasingly emphasizes study of very faint sources in very large numbers.

Planned science objectives for arrays currently operating or in planning overwhelmingly emphasizes stellar astronomy and physics (eg Panel, 1989). This is appropriate for three reasons. First, the performance of interferometry with bright, compact sources can be predicted with reasonable confidence. Second, specific science goals for stellar astronomy can be described in considerable detail. Third, interferometry promises to truly revolutionize stellar physics with an abundance of observational detail in parameter spaces previously the exclusive domain of theory, and this at a time when stellar physics is fairly bursting with outstanding problems, including solar neutrino production and the ages of cluster stars, to mention two particularly grave issues.

Extension of optical interferometry to a variety of extragalactic programs (beyond characterization of a few nuclear regions) will require a combination of aperture diameter, site seeing, and adaptive optics which will extend AO correction to the faintness limit and wavelength range required. This probably requires apertures larger than 2–meters, and possibly larger than 4–meters. Further experience with AO and perhaps even additional site testing will be needed to pin down the aperture requirement.

There are so many choices and parameters in array design that quite possibly no two really similar arrays will ever be built. An array designed for a specific science objective, such as measuring angular diameters (SUSI), absolute astrometry (NPOI) or study of binary stars (CHARA) can be readily optimized. These and other small array projects currently underway have the potential to carry out limited imaging of stellar surfaces and circumstellar shells. The VLT and Keck projects have the potential of supporting at least limited characterization of considerably fainter sources. The deciding factor in the progress which is achieved with all of these projects is likely to be, not the basic facilities of telescopes and delay lines, but the level of support available for on-going development of beam combination and detetection systems and observing programs.

A truly general purpose interferometric facility — if possible at all — will require large telescopes in considerable number and possibly reconfigurable, adaptive optics, dual beam operation, will have to operate at visible, near-infrared and thermal infrared wavelengths.

A major facility, comparable in cost to the VLA or VLT, will be justified partially by the science we know it can do, partially by the science we hope it can do, and partially by the expectation that such a dramatic gain in measurement technology will reveal a host of new questions which we haven't yet been able to imagine.

30. Acknowledgements

Much of the array planning experience summarized here was carried out while in residence at the Center for High Angular Resolution Astronomy at Georgia State University, on sabatical and as a Visiting Scholar. I am grateful to NOAO and GSU for making these visits possible.

References

Baldwin, J.E., 1996, *Astron. Astrophys.* **306**, 13.
Buscher, D., 1988, *Mon. Not. R. Astr. Soc.* **235**, 1203–1226.
Colavita, M.M., Shao, M., at al, *Proc. SPIE* **2200**, 89.
Cornwell, T., 1985, in *Synthesis Imaging*, ed. Perley, R.A., Schwab, F.R., and Bridle, A.H., NRAO Workshop No. 13, 109–120.
Cornwell, T.J., 1988, *IEEE Transactions on Antennas and Propagation* **36**, 1165–1167.
Coudé du Foresto, V., Maze, G., and Ridgway, S.T. 1993, in *Fiber Optics in Astronomy II*, P.M. Gray, ed., *PASP Conference Series* **37**, 285–294.
Coudé du Foresto, V., Perrin, G, and Boccas, M., 1995, *Proc. SPIE* **2476**, 108.
Coudé du Foresto, V., 1996, Thesis (U. Paris, Observ. de Meudon).
Dyck, H.M., Benson, J.A., Mason, W.L., Howell, R.R., and Ridgway, S.T., 1993, *P.A.S.P.* **105**, 610.
Elias, N.M., 1994, *Proc. SPIE* **2200**, 71.
Gursel, Y, 1995, *Proc. SPIE* **2477**, 240.
Kulkarni, S., Prasad, S., and Nakajima, T., 1991, *J. Opt. Soc. Am. A* **8**, 499–510.
Labeyrie, A., 1988, in *High Resolution Imaging by Interferometry*, ed. F. Merkle (ESO Conference Proceedings, No.29), 669–694
Mariotti, J.-M. 1992, in *Targets for Space Based Interferometry* (ESA SP–354).
Mariotti, J.-M., and Ridgway, S.T., 1988, *Astron. Astrophys.* **195**, 350.
Nakajima, T. and Haniff, C.A., 1993, *P.A.S.P.* **105**, 509.
Panel, I., 1989, *The VLT Interferometer Implementation Plan*, VLT Report No. 59b.
Perley, R.A., Schwab, F.R., and Bridle, A.H. 1985, *Synthesis Imaging*, NRAO Workshop No. 13.
Perrin, G., Coudé du Foresto, V., Mariotti, J.-M., Ridgway, S.T., Carleton, N.P., and Traub, W.A., 1996, *Astron Astrophys*, in press.
Richart, F.E., Hall, J.R., and Woods, R.D., 1970, *Vibrations of Soils and foundations.*
Roddier, F., 1995, *Astrophysics and Space Sciences* **223**, 109–118.
Shaklan, S., and Roddier, F., 1988, *Appl. Opt.* 27, 2334–2338.
Shao, M., Colavita, M.M., Hines, B.E., Staelin, D.H., Hutter, D.J., 1988, *Astron. Astrophys.* **193**, 357.
Shao, M., Colavita, M.M., Hines, B.E., Hershey, J.L., and Hughes, J.A., 1990, *A.J.* **100**, 1701.
Simon, G.W. 1966, *Astron. J.* **71**, 190–194.
Tango, W.J., and Twiss, R.Q., 1980, *Prog. Opt.* **17**, 239–277.
ten Brummelaar, T.A., 1996, in prepartion.
Traub, W.A. 1988, in *High Resolution Imaging by Interferometry*, ed. F. Merkle (ESO Conference Proceedings, No.29), 1029–1038.
Traub, W.A. and Lacasse, M.G.,1988, in *High Resolution Imaging by Interferometry*, ed. F. Merkle (ESO Conference Proceedings, No.29), 947–958.
Vivekanand, M., Morris, D., and Downes, D., 1988, in *High Resolution Imaging by Interferometry*, ed. F. Merkle (ESO Conference Proceedings, No. 29), 1071–1074.
Walters, D.L., Vaucher, G.T., and Vaucher, C.A., 1990, *Proc. SPIE* **XXX**, 2–9.
Zhao, P.Mariotti, J. M., Léna, P.,; Coudé du Foresto, V., 1992, *Proc. SPIE* **2476**, 108–119

KILOMETRIC BASELINE SPACE INTERFEROMETRY

PIERRE Y. BELY
European Space Agency
Space Telescope Science Institute
3700 San Martin Drive
Baltimore, Md 21218, USA

Abstract. Two versions of a space-based kilometric baseline interferometer with equivalent science capabilities have been studied, one located on the Moon and the other operating as a free-flyer. It has been found that the Moon is not the ideal site for interferometry because of the need for long delay lines and the large temperature swings from day to night. Automatic deployment of the Moon-based interferometer would be difficult and site preparation and assistance by man appear to be essential. The free-flyer would be implemented as a very accurately controlled cluster of independent satellites placed in a halo orbit around the 2nd Lagrange point of the Sun-Earth system.

Both versions could attain the required scientific performances and each one needs the same type of metrology control. The free-flyer is intrinsically advantageous because of its reconfiguration flexibility, quasi-unlimited baseline length and observation efficiency (the Moon-based interferometer cannot be operated during the lunar day because of stray light.

1. Introduction

With an angular resolution 1000 times that of the Hubble Space Telescope, kilometric baseline interferometry from space at visible and near infrared wavelengths has the potential to revolutionize astrophysics. In our own galaxy, interacting binaries of all types could be examined, the pulsation of Cepheids could be observed, and surface features on stars could be mapped directly. In external galaxies, beamed synchrotron jets, gas motions in broad emission line regions of quasars and active galactic nuclei in

A-M. Lagrange et al. (eds.), High Angular Resolution in Astrophysics, 39–48.
© 1997 *Kluwer Academic Publishers.*

nearby galaxies could be imaged. The impact of the resulting discoveries on our understanding of physical processes in the universe could truly be extraordinary.

Going into space provides the same advantages for interferometry as for conventional telescopes, i.e. the elimination of the turbulent and partially opaque atmosphere and provides a significant increase in resolution and sensitivity and an unlimited isoplanatic patch size. A number of connected-element space interferometers have been proposed (Traub and Gursky 1980, Noordam et al. 1987, Bély et al. 1989, Nein and Morgan 1989) but their practical baseline limit is probably around 100 meters. Free-flying (separated) spacecraft have also been proposed (Labeyrie et al. 1984, Stachnik et al. 1984, Kulkarni et al. 1994), but are widely expected to present daunting difficulties in station keeping at the wavelength scale. The moon, on the other hand, has long been considered a suitable platform for a space interferometer because it combines the advantages of space with the baseline stability conferred by firm ground (Burns 1989, Labeyrie 1992).

In order to assess the relative merits of these two approaches, the European Space Agency (ESA) sponsored a comparative engineering of a free-flying and a lunar-based interferometer, which was performed by Matra-Marconi Space under the guidance of a team of scientists and interferometry specialists. We present here the principal results of this study, the details of which can be found in an ESA report.

Based on a detailed analysis of potential science programs, the instrumental characteristics of a nominal long baseline interferometer were established as listed in table 1. Most of the astronomical objects of interest at very high angular resolution have relatively complex morphologies which require imaging capability, a large dynamic range, and adequate sensitivity. We thus determined that the unit telescope size should be at least 1 meter for sufficient sensitivity and the number of telescopes should be at least 6 to allow for imaging with adequate phase closure and rapid coverage of the uv-plane.

A critical element of any interferometer is that optical pathlengths between each subaperture and the beam recombiner have to be maintained to a fraction of the wavelength of light (typically $\lambda/44$). We found that passive stabilization of the interferometer elements to that accuracy was extremely difficult both in space or on the moon, and that active control of the optical pathlengths using laser metrology and a reference star was required. By monitoring the fringes from a reference star, the positioning tolerances on the interferometer elements are relaxed by several orders of magnitude. However, stars used as reference need to display a high apparent fringe visibility which means that their diameter must not be resolved. For a 5 km interferometer operating at visible wavelengths (20 μarcseconds

TABLE 1. Desired scientific requirements

Resolution	0.02 milliarcsecond (in V band)
Imaging	up to 30 × 30 maps†
Limiting magnitude	up to m_v=30 per res. elemt.†
Dynamic range	up to 100†
Spectral range	0.4 to 12 μm
Spectral resolution	up to 30,000
Temporal resolution	< 1 day*
Field	0.1 arcsecond
Instrumental thermal background	< zodiacal light at 5 μm
Straylight	< zodiacal light
Instantaneous sky coverage	≥10 %
Cumulated sky coverage	≥60%

†interdependant parameters - maximum values shown.
* not applicable to the lunar interferometer which relies on moon rotation for aperture synthesis.

resolution), this implies that stars should be at about 400 parsecs on average, i.e. with a magnitude of about 18.5. Based on traditional star count statistics, a field of 2.5 arcminutes in radius is enough to find at least one such suitable reference star at the galactic poles with 95% probability.

2. Moon-based version

The general arrangement of the six-telescope moon based interferometer is shown in Fig. 1 The observatory is arranged in a 'Y' configuration tilted a few degrees away from the North/South direction with two telescopes in each arm. The observatory would be located near the moon limb to avoid earth straylight and at about 45 degree latitude to provide a good uv plane coverage as the moon rotates.

If the moon offers a relatively stable platform for interferometry and variable baseline thanks to its rotation, a major drawback is that delay lines have to be used to compensate for the changing optical pathlengths. These delay lines are of a length comparable to the baseline in order to permit the observation of targets close to the horizon. We selected to use mobile carts with independent traction systems because of their much higher optical efficiency compared to banks of fixed mirrors and lighter mass compared to railed systems. These carts do not move during single observations, but must be repositioned from one observation to the next. A delay line inside the hub allows for optical pathlength changes during a given observation.

Another drawback of the moon as an observatory platform is that heat

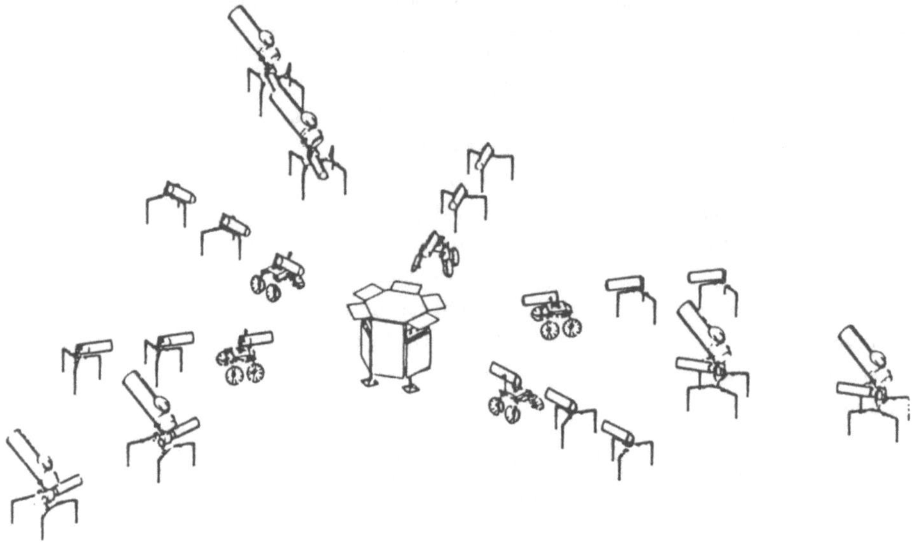

Figure 1. General arrangement of the 6 telescope array (not to scale). The telescopes are the larger elements arranged in a Y configuration and shown with their 2 meter long baffles deployed. The smaller mirrors, 5 of which are on mobile carts constitute the delay lines systems.

and straylight during the lunar day prevent observations, and that a non-solar power system has to be used for supplying electrical power during the lunar night. For power, we baselined regenerative fuel cells which make use of the propulsion hydrogen and oxygen tanks of the landers and which are regenerated during lunar day with solar arrays.

The entire observatory can be delivered to the moon surface in two upgraded Ariane 5 launches, about 3500 kg landed payload mass for each.

3. Free-flyer version

An extremely low disturbance orbit is a key ingredient of the free-flyer solution. Fuel economical array reconfiguration and station keeping during observations require that the cluster be sited on an orbit where the gravity gradient is very small. In addition, the observed sky field must be continuously visible during the full observing time which typically lasts several hours. These two conditions exclude low earth orbits. Among higher orbits such as a high geocentric orbit or the libration points of the earth-moon or sun-earth systems, a halo orbit close to the second Lagrange point of the sun-earth system offers the most advantage.

A powerful feature of the free-flying interferometer is its ability to be reconfigured to best match the mapping requirements of the observed source.

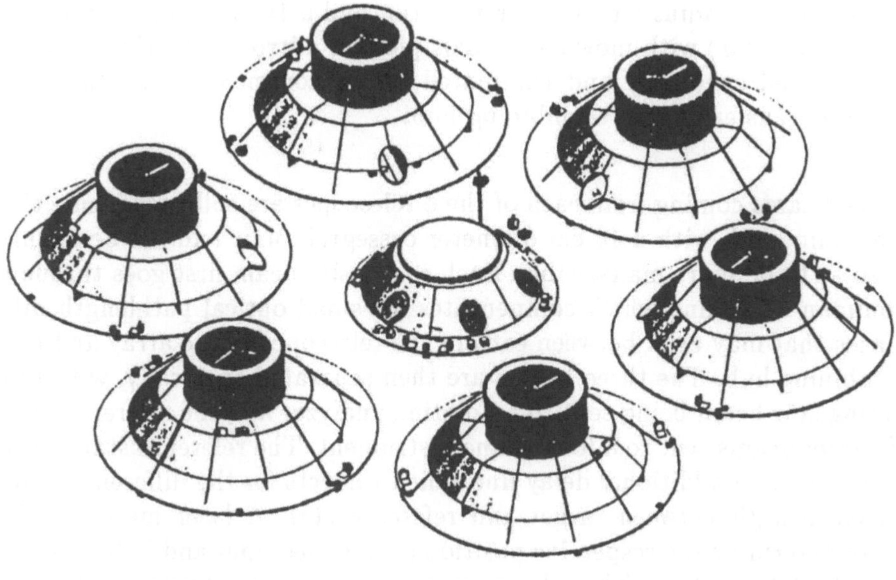

Figure 2. Free-flying interferometer shown in close formation for clarity. Under normal conditions the distance between the recombining hub and the individual telescopes varies from 50 to 2500 meters.

It is indeed possible to adjust both the maximum baseline and the number of intermediary measuring points to optimize resolution and distribution of the visibility measurement points in the Fourier plane. This optimization should be done on a case by case basis but, as a rule, the majority of targets requires uniform mapping which can be obtained with one of the configurations worked out by Cornwell (Cornwell 1988). In the case of a 6 element array, the angles are ±39.7, ±8.2, and ±159.7 degrees. Fuel economy dictates that the telescopes be displaced continuously rather than in a stop and go fashion. To cover the uv-plane more or less uniformly, the radial expansion or contraction should be accomplished with a velocity varying as the inverse distance from the hub. A view of the free flying cluster is shown in Fig. 2.

The six 1-meter telescopes in the array simultaneously observe the science target, a reference star and a pointing star. The reference star is an unresolved field star, typically of magnitude 18.5, which is used to cophase the array. The integration time required for the reference star is on the order of 10 seconds, which is too slow for pointing control. The pointing star is a brighter star, typically of magnitude 12, which will be used at a much higher refreshing rate to maintain the pointing of the telescopes and to coalign the six individual beams in the recombining hub. The reference star and pointing star are collected in the telescope's field of view (to within a 2.5 arcminute radius for the reference star, and a 15 arcminute radius for the pointing star) with movable off-axis pick-off mirrors. The three beams are separated spectrally and collimated into a common 40 cm diameter beam for transfer to the recombining hub.

The beams coming from each of the 6 telescopes are collected inside the recombining hub with a 40 cm diameter cassegrain unit and are demagnified so as to be more manageable. Each composite beam first goes through a common delay line which compensates for small optical pathlength differences that may exist between each of the telescopes in the array and the recombining hub. The three beams are then separated spectrally, with the pointing star beam being sent to a pointing analyzer and the reference star and target beams sent to a focal plane instrument. The reference star beam goes through an additional delay line which corrects for the differential optical pathlength between target and reference star. A laser metrology is used to determine the respective position of the telescopes and hub so as to ensure that they stay within the common reference-star/target cophasing range.

Due to continuous presence of the sun in the selected orbit, it is indispensable that the telescopes and recombining hub be well protected. This is accomplished by keeping in shadow the portions of the spacecraft in the view of all other spacecraft, and by extensive baffling of the output and input transfer telescopes. Because of this self shadowing, the portion of the sky that can be observed at any one time is about 14%. Over a one year period, however, the band between ± 45 deg of ecliptic latitude is observable, which represents about 72% of the entire sky. Station keeping and attitude control is performed by a Field Emission Electric Propulsion system (FEEP) (Gonzales et al. 1993) which is very compact and provides extremely smooth actuation over a wide range of thrust, from a few μN to a few mN. FEEPs' specific impulse is very high (6000 s) and maximum thrust is limited only by their large power consumption (600 Watts peak). The estimated total fuel requirement for the strawman 5 year mission is 8 kg per telescope.

The six telescope units, hub and transfer stage can be accommodated in a single Ariane 5 launch. Total mass of the cluster and transfer stage is estimated at about 5 metric tons (4260 kg for the interferometer and 800 kg for the wet mass of the transfer stage). This is beyond the capability of the current Ariane 5, but well within the expected capability of Ariane 5-E (5200 kg on the liberation orbit for $V_\infty = 0$).

4. Comparison of the two versions

The study of the two versions of the interferometer indicate that they are both feasible, do not exhibit major show stoppers and can essentially meet the scientific requirements we had established. Clearly, nonetheless, there are differences as indicated in table 2.

TABLE 2. Free-flyer vs Moon-based comparison

Characteristic	Free-flyer	Moon-based
Science related		
Maximum practical baseline	>5 km	5 km
uv coverage	Good	adequate
Minimum time required for *uv* coverage	1 day	14 days
Wavelength coverage	0.4 to 12 μm	0.4 to 12 μm
Optical throughput (in V)	0.75	0.50
Need for reference star	Yes	Yes
Optics temperature	60 K	100 K
Yearly sky coverage	72%	86%
Instantaneous sky coverage	14%	45%
Observing efficiency	80 %	45%
Engineering aspects		
Complexity of optical path control	-	long throw delay line
Thermal stability	excellent	adequate (night)
Radiation environment	-	slightly better
Micrometeorids flux	-	slightly less
Maintenance & upgrading	unit level	module level
Lifetime	10 years	indefinite
Power generation	solar	solar + fuel cells
Communications	adequate	same
Payload mass	4200	4400
Wet launch mass	5000 kg	15200
Launch requirement	1 Ariane 5E launch	2 Ariane 5E+H10

Perhaps the main point is that locating an interferometer on the moon does not bring any obvious simplification in construction or operation. One could argue, on the contrary, that the moon-based system has to cope with a much more severe environment than its space-based counterpart: large temperature swings (day/night), ground disturbances, the impossibility of observing during the lunar day, dust, and lack of solar-generated power during the lunar night.

From the observational point of view, too, the space-based system has the advantage. There is no real limit to its baseline length. Its efficiency is about twice as much because of the impossibility of the moon-based counterpart to observe during the lunar day. Its array can always be configured optimally, while the lunar based has to contend with whatever source declination and geographic latitude provide. And for some scientific programs, the moon simply rotates too slowly: super synthesis mapping requires half a lunar day or about 14 earth days, which is too long compared to the rotation or time scale of physical processes of many sources.

On the logistical side, our analysis has also shown how difficult the automatic deployment of a major facility on the moon would be. Although not strictly necessary, astronaut supervision and site preparation would be highly desirable. The availability of man at a lunar base means that maintenance and upgrading would be possible. The higher investment in a moon-based interferometer would thus result in a longer lifetime facility with continuous performance enhancement possible.

Finally, with regard to cost, although only rough orders of magnitude were estimated it is clear that the moon-based scheme would be more expensive. The free-flyer version is estimated at about 700 MAU including transfer stage and launch. The moon-based system, which requires two launches and a sophisticated landing module, could cost as much as 1200 MAU, with some savings possible if the lunar lander can be simplified thanks to astronaut assisted deployment.

We conclude that the free-flyer is better suited for implementation in the near or mid-term future, but that the moon-based version may be considered in the long term in conjunction with a manned lunar base.

References

Traub W.A. and H. Gursky, Coherent optical arrays for space astronomy, SPIE Vol 228, 1980.

Noordam J.E., A.H Greenaway, R.S. lePoole, J.D. Bregman, *OASIS: Optical Aperture Synthesis in Space, a mission concept*, ESA Colloquium SP-273, Granada, 1987.

Bély P-Y., C.J. Burrows, F. Roddier, G. Weigelt, *HARDI: High angular resolution deployable interferometer*, "New technologies for Astronomy", ECO Proc 1130, 1989.

Nein, M.E., S.H. Morgan, *Considerations for a Next Generation UV/Optical Space Telescope*, Next Generation Space Telescope Proc. Space Tel Sc. Inst., 1989.

Labeyrie A., B. Authier, J.L. Boit, T de Graauw, E. Kibblewhite, L. Koechlin, P. Rabout, G. Weigelt, *Trio: a kilometric optical array controlled by solar sails*, Bull AAS, vol 16, 3, 1984.

Stachnik R.V., K Ashlin and S. Hamilton, *SAMSI: A spacecraft array for Michelson spatial interferometry*, Bull AAS, vol 16, 3, 1984.

Kulkarni S.R., P-Y Bély, M. Colavita, A.B. DeCou, D.L. Jones, J.H. Krolik, C.D. Martin, H.A. McAlister, J.P. McGuire, T. Separated Spacecraft Interferometry, Proposal to NASA Res. Ann. 94-15, 1994.

Burns J.,*A lunar Optical Ultraviolet-Infrared Synthesis Array (LOUISA)*, NASA Conf

Publ. 3066, Burns Ed. 1989.

Labeyrie A., *Interferometry from the moon*, Targets for space-based interferometry, ESA Coll. SP-354, 1992.

Kilometric baseline space interferometry, ESA Report. SCI-96-7, June 1996.

Cornwell, T.J., 1988, "A novel principle for optimization of the instantaneous Fourier plane coverage of correlation arrays", IEEE Tr., AP-8,1165.

Gonzales J., G Saccoccia, H. von Rohden, Field emission electric propulsion: experimental investigations on microthrust FEEP thrusters, AIAA 23nd International Electric Propulsion Conf, IEPC 93-157, 1993

OBSERVING WITH OPTICAL/INFRARED LONG BASELINE INTERFEROMETERS

JOHN DAVIS
Chatterton Astronomy Department, School of Physics,
University of Sydney, NSW 2006, Australia

Abstract. A brief introduction to the basic concepts involved in long baseline optical/infrared interferometry is followed by a discussion of observational, measurement and calibration techniques. Since observational experience with long baseline optical/infrared interferometers has so far been confined almost entirely to instruments operating with a single baseline at a time, examples from this experience are used as illustrations. However, the general principles for multi-aperture observations, necessary for imaging, are generally the same. A brief introduction to the additional requirements for imaging is given to provide an introduction to following presentations.

1. Introduction

A long baseline optical/infrared interferometer is defined, for the purposes of this contribution, as an interferometer in which independently steerable input apertures are used. Observing with this type of interferometer from the ground is not easy—not only does such an instrument have to meet stringent optical and mechanical tolerances but the randomly variable temporal and spatial coherence effects introduced by turbulence in the Earth's atmosphere have to be overcome. In this contribution the observational aspects of these problems will be addressed with discussions of measurement and observing techniques including the problems of calibration and elimination of systematic errors.

Since essentially all experience to date has been obtained with interferometers combining light from two apertures, the discussion will be presented primarily in terms of two aperture instruments. It is noted, however, that multi-aperture or imaging instruments have to meet the same criteria

A-M. Lagrange et al. (eds.), High Angular Resolution in Astrophysics, 49–79.
© 1997 *Kluwer Academic Publishers.*

for each of several baselines. The discussion will be limited to amplitude interferometers since, with only one known exception, interferometers currently operating or under construction are of this type. However, much of the discussion is relevant to heterodyne and intensity interferometry.

An introduction to the basic concepts involved in long baseline optical/infrared interferometry is given in the following section in order to define the terminology used in this and subsequent contributions. In addition, to set the scene for contributions on imaging with interferometers, an introduction to the additional requirements for imaging is given in the penultimate section.

2. Background

It will be useful, before commencing a discussion of observing with interferometers, to outline the basic concepts and to define the relevant terminology. In addition, since it will be necessary to consider the coherence properties of the light from the source and the effects of atmospheric turbulence on the observational data, it is appropriate to give a brief summary of the relevant coherence and atmospheric turbulence parameters.

2.1. BASIC CONCEPTS

The basic concepts can be readily understood starting from the simplest two aperture interference experiment—Young's slits interference for a point source and monochromatic radiation. This is shown in Fig. 1. The light passed by the apertures at P_1 and P_2, from the source S, form interference fringes centred at O on the screen. The fringes are modulated by a diffraction envelope whose width is set by the size of the apertures. The angular separation of adjacent interference fringes, as seen from the plane of the apertures, is $\sim \lambda/d$ where d is the separation of the two apertures.

The fringes for a monochromatic point source have maximum contrast with zero intensity between the maxima in intensity. Consider the situation when the source has a finite angular extent in the direction parallel to the line joining the apertures, as shown in Fig. 2(a). Each point on the source produces its own set of fringes and these will be displaced relative to one another. Fringes for representative points on the source are shown in Fig. 2(b) and the result of combining the fringes from the entire source is also shown - namely fringes of lower contrast. In the figure, modulation by the diffraction envelope has been omitted—corresponding to the situation near the centre of the interference pattern for small apertures.

A measure of the fringe contrast is Michelson's fringe visibility V which is defined by

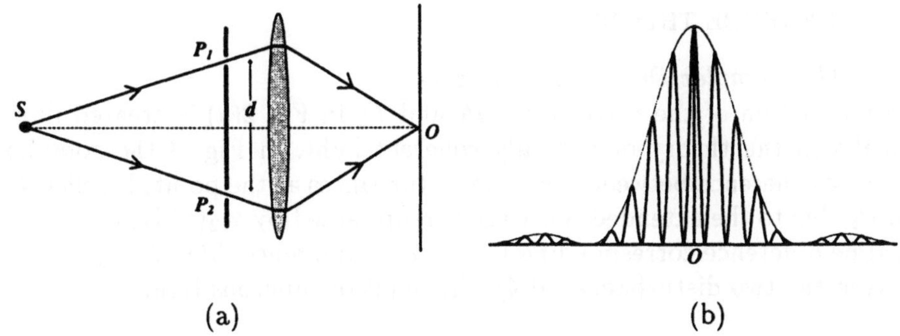

Figure 1. Young's slits experiment for a point source and monochromatic radiation. (a) S represents the point source, P_1 and P_2 two apertures and O the centre of the screen on which the interference pattern is observed. (b) The intensity distribution in the interference fringe pattern.

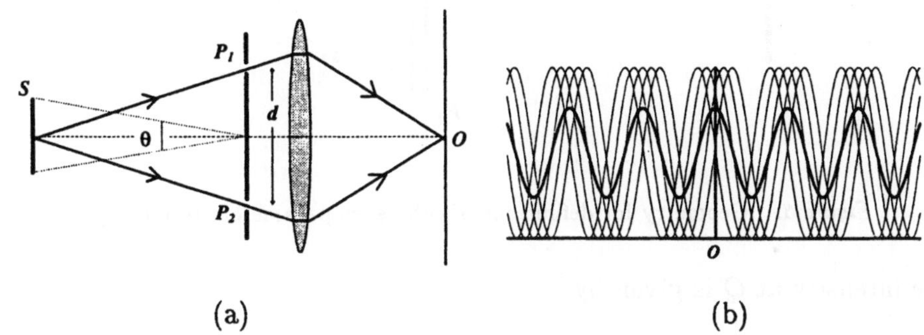

Figure 2. Young's slits experiment for an extended monochromatic source. (a) S represents the source, P_1 and P_2 two apertures and O the centre of the screen on which the interference pattern is observed. (b) The intensity distribution at the centre of the interference fringe pattern: the fine curves represent sample fringe distributions for five individual points on the source; the thick curve represents the resultant fringe distribution for the whole source. For this illustration it has been assumed that the width of the diffraction envelope is much greater than the fringe spacing.

$$V = \frac{I_{max} - I_{min}}{I_{max} + I_{min}} \tag{1}$$

where I_{max} and I_{min} represent the maximum and minimum intensities in the fringe pattern.

2.2. COHERENCE THEORY

2.2.1. *The Complex Degree of Coherence*

The interaction of the waves from P_1 and P_2 in Fig 2(a) is treated more formally in the theory of partially coherent light. In Fig. 3 the complex degree of coherence between the wave disturbances at the points labelled P_1 and P_2, due to the extended source S, is represented by $\gamma_{12}(\tau)$ [1] where τ is the time difference corresponding to the path difference $\Delta OPL = (s_1 - s_2)$ between the two disturbances at Q, the point of superposition.

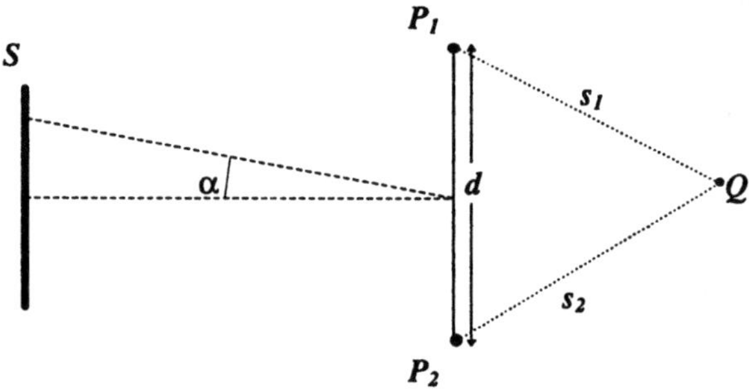

Figure 3. Geometry for definitions of coherence parameters (see text).

The intensity at Q is given by

$$I = I_1 + I_2 + 2\sqrt{I_1 I_2}\, Re[\gamma_{12}(\tau)] \tag{2}$$

where I_1 and I_2 are the intensities at Q from P_1 and P_2 respectively when each is present alone. The complex degree of coherence can be expressed in terms of a modulus $|\gamma_{12}(\tau)|$ and a phase $\phi_{12}(\tau)$:

$$\gamma_{12}(\tau) = |\gamma_{12}(\tau)| \exp i\phi_{12}(\tau) \tag{3}$$

Thus, the real part of the complex degree of coherence in equation 2 can be written:

$$Re[\gamma_{12}(\tau)] = |\gamma_{12}(\tau)| \cos\phi_{12}(\tau) \tag{4}$$

The modulus of the complex degree of coherence is known as the degree of coherence and, when I_1 is equal to I_2, it is equal to the fringe visibility given by equation 1:

$$V_{12}(0) = |\gamma_{12}(\tau)| \qquad \text{for } I_1 \text{ and } I_2 \text{ equal.} \tag{5}$$

The complex degree of *spatial* coherence $\gamma_{12}(0)$ (i.e. with $\tau = 0$) and the angular distribution of intensity across the source I_α are a Fourier transform pair - the van Cittert-Zernike theorem.

$$\gamma_{12}(0) = \int_{-\infty}^{\infty} I_\alpha \exp\left(\frac{-2\pi i\alpha}{\lambda}\right) d\alpha \tag{6}$$

Thus, if $|\gamma_{12}(0)|$ and $\phi_{12}(0)$ are measured over an appropriate range of baseline lengths and orientations, the angular intensity distribution (I_α) of the source can be found. If measurements are made with only a single baseline orientation, the Fourier transform yields the intensity distribution for the equivalent strip source (i.e. the source intensity distribution reduced to the equivalent strip source parallel to the baseline direction).

The fringe phase for an aperture separation d is the displacement of the fringes relative to the fringes for $d \sim 0$, expressed as a fraction of the fringe spacing s. Thus, if the fringes are displaced by a distance Δs, $\phi_{12}(0) = 2\pi\Delta s/s$. The fringe phase, as a function of the aperture separation, provides information on asymmetries in the angular intensity distribution—for a symmetrical source the fringe phase will be zero for all aperture separations.

In practice, the earth's turbulent atmosphere disrupts the phase and, for observations made with a single baseline at a time, only the degree of coherence $|\gamma_{12}(0)|$ is measurable. In the absence of the phase, departures from symmetry in the intensity distribution cannot be detected. Thus, measurements with a single baseline yield only the visibility and these are interpreted as the angular size of an equivalent strip source or the angular diameter of a radially symmetrical disk source. To obtain images it is necessary to measure both the modulus and phase for at least three baselines simultaneously (see Section 6).

In practice, many interferometers measure the square of the fringe visibility, which will be referred to as the correlation $C_{12}(\tau)$. Thus

$$C_{12}(\tau) = [V_{12}(\tau)]^2 = |\gamma_{12}(\tau)|^2 \tag{7}$$

As an example of the variation of visibility and correlation with aperture separation (the baseline of the interferometer), Fig. 4 shows the relationships for a uniformly bright circular source. The visibility and correlation are plotted against $x = \pi\, d\,\theta/\lambda$ where d is the baseline, θ is the angular diameter of the source, and λ is the mean wavelength of observation. The visibility is given by

$$V = \left|\frac{2J_1(x)}{x}\right| \tag{8}$$

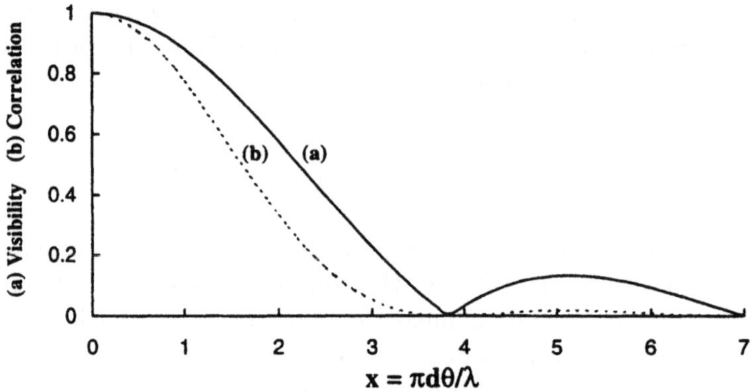

Figure 4. The variation of visibility and correlation for a uniformly bright circular source. Details of the form of the curves and definitions of the parameters are given in the text.

2.2.2. *Coherence Time and Coherence Length*

So far we have considered only monochromatic sources. The fringes for a monochromatic point source have maximum contrast with zero intensity between the maxima in intensity. If the point source emits a range of wavelengths the fringe visibility decreases for positions away from the centre of the screen. This is illustrated in Fig. 5 for a band of wavelengths $\Delta\lambda$, where $\Delta\lambda/\lambda_{mean} = 0.22$. As for Fig. 2, modulation by the diffraction envelope has been omitted.

The reduction in fringe visibility for a finite spectral bandwidth as the path difference between the beams increases can be expressed in terms of the coherence time of the light $\Delta\tau$ which is given by

$$\Delta\tau \sim \frac{1}{\Delta\nu} \quad \text{where } \Delta\nu \text{ is the spectral bandwidth} \tag{9}$$

The coherence length Δl is given by

$$\Delta l = c\,\Delta\tau = \frac{\bar{\lambda}^2}{\Delta\lambda} \tag{10}$$

where $\Delta\lambda$ is the spectral bandwidth.

The Fourier transform of the complex degree of coherence is the spectral intensity distribution and so, as an example, the fringe visibility for a flat-topped spectral bandpass of width $\Delta\lambda$ is given by

$$V_{12}(\tau) = |\gamma_{12}(\tau)| = \left|\frac{\sin x}{x}\right| \quad \text{where } x = \frac{\pi\,\Delta\mathrm{OPL}\,\Delta\lambda}{\lambda^2_{mean}} \tag{11}$$

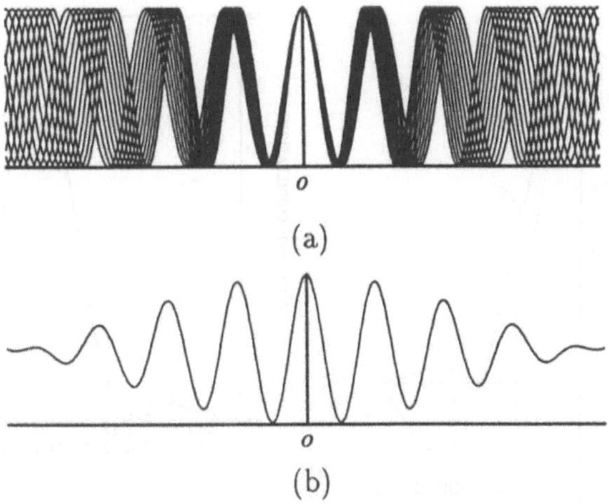

Figure 5. (a) Interference fringes for a range of wavelengths with $\lambda_{max}/\lambda_{min} = 1.25$. (b) The resultant intensity pattern for the sum of the fringes shown in part (a).

The correlation $C_{12}(\tau)$, defined by equation 7, is given by

$$C_{12}(\tau) = [V_{12}(\tau)]^2 = \left|\frac{\sin x}{x}\right|^2 \tag{12}$$

Figure 6 shows the variation of fringe visibility and correlation for a flat-topped spectral bandpass. Also shown is an empirical curve of correlation versus optical path difference obtained with the Sydney University Stellar Interferometer (SUSI) [2] for a spectral bandwidth of 0.8 nm centred on a wavelength of 442 nm. Curves of visibility or correlation versus optical path difference will be referred to as 'delay curves'.

In order to obtain accurate measurements of fringe visibility certain criteria must be met. Since it is the spatial coherence of the light that gives the intensity distribution across the source, it is $\gamma_{12}(0)$ (i.e. with $\tau \simeq 0$) that must be measured. This requires the optical paths from the source to the point of combination of the interfering beams to be equal to within a small fraction of the coherence length (i.e. $c\tau \ll \Delta l$) or, in other words, observations should be made close to the peak of the delay curve. For example, if the loss in V due to a path inequality (ΔOPL) is to be less than 1%, then ΔOPL must be less than $\sim 0.07\Delta l$. For the loss to be less than 2%, ΔOPL must be less than $\sim 0.1\Delta l$. Thus, for an optical bandwidth of 10 nm at a wavelength of 550 nm, the optical paths must be matched to within $\sim \pm 2$ μm for <1% loss and to within $\sim \pm 3$ μm for <2% loss in V.

So far it has been assumed implicitly that the interfering wavefronts are plane and are combined with a constant tilt (which can be zero) between them, and also that the relative phase of the wavefronts does not change

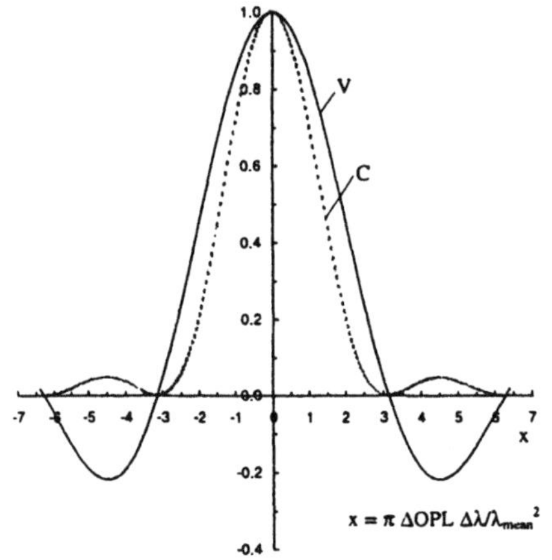

(a)

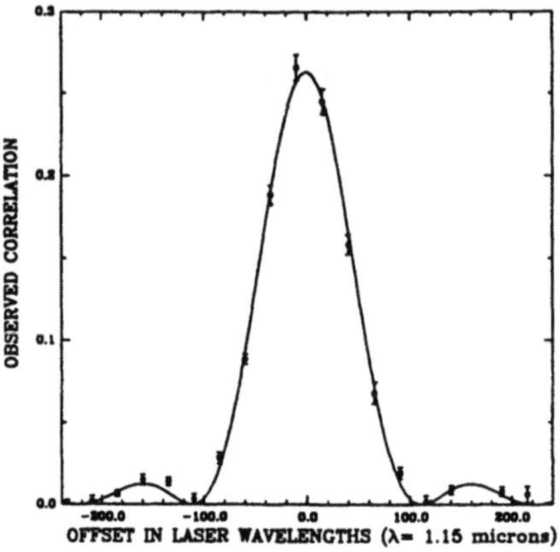

(b)

Figure 6. (a) The variation of fringe visibility V and correlation C as a function of optical path difference ΔOPL for a flat-topped spectral bandpass of width $\Delta\lambda$. (b) The observed correlation as a function of optical path difference observed for Sirius (α CMa) with SUSI. A $(\sin x/x)^2$ curve has been fitted to the observational data. The width of the 'delay' curve corresponds to the 0.8 nm spectral bandwidth used.

significantly during an observation. In practice, turbulence in the Earth's atmosphere distorts the shape of the wavefronts and introduces randomly varying phase shifts, wavefront tilts, and wavefront curvatures.

2.3. ATMOSPHERIC TURBULENCE

The wavefronts arriving at the ground from an astronomical source can be visualised as surfaces randomly corrugated in two dimensions due to the effects of turbulence in the earth's atmosphere. The corrugations have a spatial scale of the order of a few centimetres and translate across an aperture at some representative mean wind speed of the order of a few metres per second. The result is that the wavefront shape at a given location changes significantly with a timescale of a few milliseconds.

2.3.1. *The Spatial and Temporal Scales of Wavefront Distortion*

The spatial scale of the wavefront distortions is represented by the Fried parameter r_0 [3]. It is inversely proportional to the diameter of the seeing disk observed with a large telescope and has a value of $\sim$ 10 cm at a wavelength of 550 nm for 1 arc second seeing. It is also the diameter of a circle on the wavefront for which the phase deviations from a plane surface are equal to 1 radian r.m.s.

The temporal scale is characterised by the parameter t_0 which is of the order of a few milliseconds. It is related to r_0 by the relationship $t_0 = \mathbf{c} \times r_0/v$ where $\mathbf{c}$ is a constant and v is some representative wind speed. Several definitions of t_0 have appeared in the literature with $\mathbf{c}$ having values ranging from 0.314 [4] to 0.81 [5]. Following the discussion by Buscher [4] a value of 0.314 is gaining general acceptance.

It is important to note the following properties of r_0 and t_0 for interferometry:

- r_0 and t_0 both vary as $\lambda^{6/5}$
- r_0 and t_0 both vary as $\sec^{-3/5} z$ where z is the zenith angle. Shao *et al.* [6] conclude that the zenith angle dependence of t_0 is more complex and that, approximately, the dependence is $t_0 \propto 2 \cos^{3/5} z/(1 + \cos z)$.
- r_0 and t_0 vary irregularly with time and with location

Table 1 contains values for r_0 and t_0 as a function of wavelength extrapolated from values of $r_0 = 10$ cm and $t_0 = 5$ msec at λ500 nm.

TABLE 1. Spatial and temporal coherence parameters r_0 and t_0 as a function of wavelength.

λ (μm)	r_o (m)	t_o (s)
0.5	0.10	0.005
1.0	0.23	0.011
2.0	0.53	0.026
5.0	1.6	0.08
10	3.6	0.18
20	8.4	0.42

2.3.2. *The Implications for Interferometry*

The effect of atmospherically induced wavefront distortion on the measurement of fringe visibility or correlation with long-baseline optical/infrared interferometers is disastrous unless steps are taken to minimise it. For this reason, in most instruments, the aperture size is chosen to be of the order of, or less than r_0 to limit the effects of wavefront curvature and wavefront tilts are sensed and corrected in real time by piezo-electrically controlled mirrors. The signals are generally sampled with times short compared with the atmospheric coherence time to reduce the effects of phase smearing during each measurement and, finally, some form of fringe or delay tracking is used to limit the effects of differential path variations introduced by atmospheric turbulence between the separate beams of starlight. The data in Table 1 show clearly the advantages of working at longer wavelengths. Not only can larger apertures be used but also longer sampling times. In practice, even with the use of aperture sizes less than r_0 and sampling times less than t_0, there will be some residual effects of atmospheric turbulence on the observational data and these are discussed in Section 4.

3. Basic Measurement Techniques

One of the most important features of a long baseline optical/infrared interferometer is the optical path length equalising system (OPLE). Since it will be necessary to refer to OPLEs in discussing measurement techniques, a brief discussion of them is given here.

3.1. OPTICAL PATH LENGTH EQUALISATION

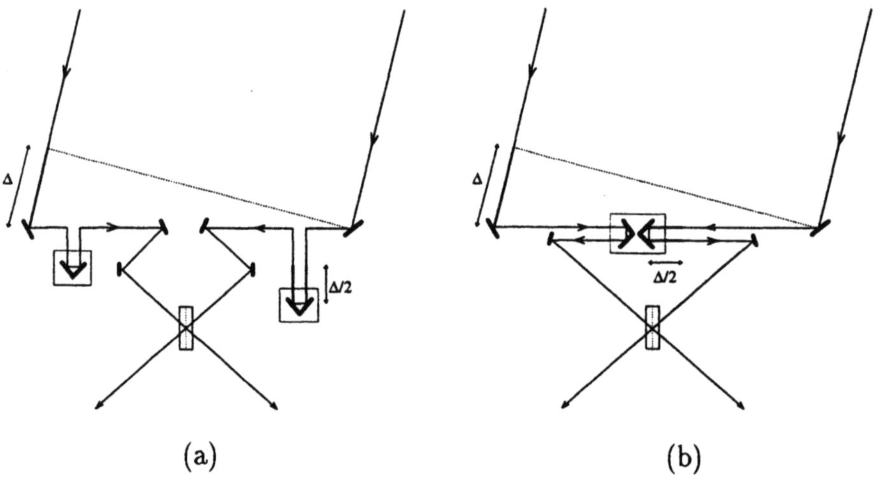

Figure 7. Two basic optical delay line configurations. In each diagram Δ is the external path difference, shown by a thickening of the line representing the incoming beam of starlight. The internal compensating path comprises two equal paths of length $\Delta/2$, also represented by thickened lines. (a) A dual delay line system with separate retroreflectors, mounted on independently controlled carriages, one in each arm of the interferometer. (b) A differential delay line in which back-to-back retroreflectors, mounted on the same carriage, move between the two beams.

The path differences from the source to the apertures of an interferometer range from zero to ~ 0.9 times the separation of the apertures, depending on the declination and hour angle. To compensate for these path differences an OPLE includes optical delay lines for the coarse correction, and piezoelectrically controlled mirrors to provide the fast, fine correction needed to meet the severe tolerances on the matching of the optical paths discussed in Section 2.2.2. Fig. 7 shows the basic form of two delay line configurations—a dual system as used in the Mark III stellar interferometer [7] and the SUSI prototype [8] and a differential system as used in SUSI [2, 9].

3.2. BEAM COMBINATION AND MEASUREMENT METHODS

Several methods have been developed for measuring fringe parameters, even for the same wavelength regime, and it is not appropriate to describe them all in detail. In most instruments the beams from the input apertures are combined in the pupil plane and three examples will be outlined with references to some of the instruments using them. For the discussion of these methods it will be assumed that wavefront tilts have been corrected, that the sampling times are short compared with phase variations and, where the fringes are not being scanned, that the optical paths are matched to within the tolerances set by the coherence length of the light (Section 2.2.2). The effects of departures from these conditions are discussed in Section 4. An alternative approach is to combine the beams in the image plane, as was done by Michelson, and an example of this method will also be outlined. Finally, the combination of beams using a single-mode fiber coupler in the near infrared is another approach being explored and this too will be outlined.

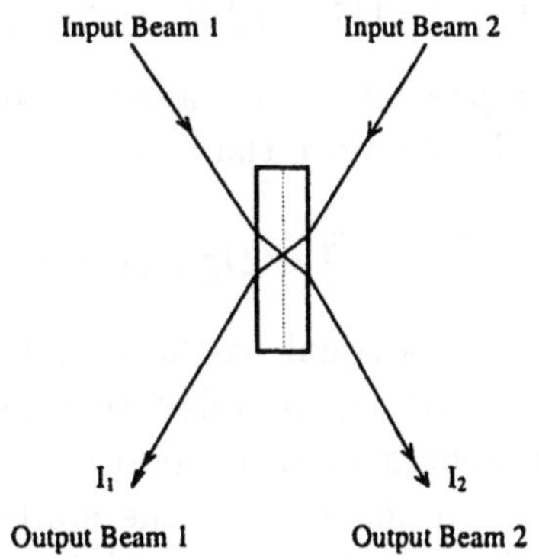

Figure 8. Beam combination at a 50:50 beamsplitter. Half of each input beam appears in each of the output beams. I_1 and I_2 represent the intensities of the output beams.

Method 1 (for V) In this method, due to Tango and Twiss [10], sub-r_0 diameter apertures are used. As shown in Fig. 8, the two starlight beams are combined with a beamsplitter which ideally transmits 50% and reflects 50% of the light from each beam. The incoming beams are aligned so that their wavefronts, after combination, are plane parallel with the result that the output beams are uniformly illuminated. Because of the phase change that occurs on reflection on one side of the beamsplitter, the combined beams are complementary in intensity. Referring to Fig. 8, the intensities in the two combined beams, assuming ideal conditions with no losses, can be written as

$$I_1 = I_0(1 + |\gamma_{12}(0)| \cos\phi(t)) \tag{13}$$
$$I_2 = I_0(1 - |\gamma_{12}(0)| \cos\phi(t)) \tag{14}$$

where I_0 is the intensity of each of the incident beams and $\phi(t)$ is a randomly varying phase dominated by the effects of atmospheric turbulence.

The two combined beams are focussed onto single pixel photon counting detectors which sample the intensities I_1 and I_2 to provide a measure of the correlation $C_{12}(0)$. If the photon counts in a sample time are n_1 and n_2, corresponding to I_1 and I_2, we have for each sample time

$$n_1 - n_2 = 2I_0 |\gamma_{12}(0)| \cos\phi(t) \tag{15}$$

The count difference is squared for each sample time and an average value for a large number of sample times is obtained.

$$\overline{(n_1 - n_2)^2} = 4I_0^2 |\gamma_{12}(0)|^2 < \cos^2\phi(t) > \tag{16}$$

where $< \cos^2\phi(t) >$ represents a time average which, if $\phi(t)$ is uniformly and randomly distributed over more than 2π, averages to 0.5 so that

$$\overline{(n_1 - n_2)^2} = 2I_0^2 |\gamma_{12}(0)|^2 \tag{17}$$

In equation 17, $|\gamma_{12}(0)|^2$ is equal to the correlation (see equation 12). In practice, equation 17 must be modified to correct for photon bias at low count rates, for differences in transmission through the two arms of the interferometer, for any imbalance in the splitting of the light at the beamsplitter and for different detector sensitivities, etc. As it stands, the equation has not been normalised by the photon flux nor does it include loss factors. These details have been omitted for simplicity but they have been discussed by Tango and Twiss [10].

This technique for the measurement of the correlation has been used in the blue wavelength region of the spectrum in the SUSI Prototype [8] and in SUSI [2, 9].

While differential path variations through the atmosphere provide satisfactory averaging, as required by equation 17, an alternative approach is to introduce a path modulation into one arm of the interferometer as proposed by Tango and Twiss [10]. For example, if each sample period is divided into two halves with a $\lambda/4$ extra path introduced into one arm for the second half of each sample period, equation 15 will contain $\sin\phi$ in place of $\cos\phi$. Summing the squares of the equations for the two halves of each sample period eliminates ϕ and gives $C_{12}(0)$ directly. However, it is still necessary to average a large number of cycles to achieve a satisfactory signal to noise ratio.

Method 2 (for V and ϕ) The approach adopted by Shao and his collaborators in the Mark III Interferometer [7] has much in common with Method 1. The two sub-r_0 beams are combined, as shown in Fig. 8, with the emerging wavefronts plane parallel on each side of the beamsplitter and the intensities of the emergent beams are complementary.

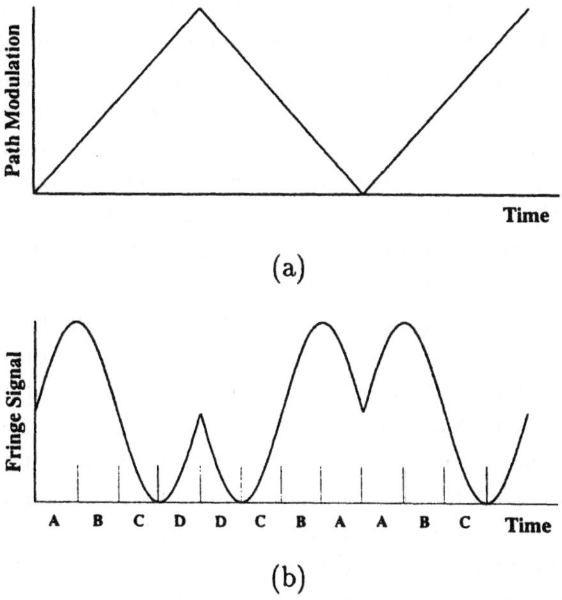

Figure 9. (a) The fringe modulation waveform used in the Mark III stellar interferometer. (b) The binning of the fringe signal used for synchronous demodulation.

The main difference between this method and Method 1 is that the path in one arm is modulated to enable fringes to be acquired and tracked automatically using a fringe detection algorithm described by Shao and Staelin [11] and which was used in the Mark I and Mark II instruments. A piezoelectrically (PZT) controlled mirror modulates the path with a 500Hz triangular waveform with an amplitude of one wavelength. The modulation and synchronous demodulation of the fringes is illustrated in Fig: 9. During each sweep of the PZT controlled mirror the photons are grouped into four time bins denoted by A, B, C and D. The amplitude and phase of the fringe are given by

$$\phi = \tan^{-1}\left(\frac{D-B}{C-A}\right) - \frac{\pi}{4} \tag{18}$$

$$Amplitude = \sqrt{(C-A)^2 + (D-B)^2} \tag{19}$$

Details of how these parameters are used to locate and track the central fringe, and the modifications to equation 19 to obtain an unbiassed estimator for the correlation are given by Shao *et al.* [7]. While the Mark III Interferometer was initially developed for wide angle astrometry [11, 12] it has proved to be a powerful instrument for measuring the angular diameters of stars and for studying binary star systems (see for example references [13, 14]). Fringe tracking is carried out at $\sim$ 800 nm and correlation measurements have been made in the wavelength range 450-800 nm [13]. It should be noted that there is a complication for simultaneous measurements of correlation at widely separated wavelengths. The phase sweeping is done for one specific wavelength, the amplitude of the sweep, and it is necessary for shorter wavelengths to make the bins narrower in proportion to the wavelength with a consequent reduction in effective observing time. Method 3 (for V and ϕ) A method, which has been used at longer wavelengths, is to scan through the fringes. This can be done by setting the OPLE to give equal paths at a later instant, keeping it stationary, and letting the Earth's rotation scan the delay through the matched path position [15, 16]. A better approach is to use the path compensator to scan the delay through the matched path position—since this allows the fringe frequency to be controlled [17, 18]. Either method results in a fringe pattern of the form illustrated in Fig. 10. As can be seen, the fringe amplitude is modulated by the visibility curve shown in Fig. 6.

The beam combination is similar to the first two methods and the combined beams are also focussed onto single pixel detectors. In the case of the Infrared Optical Telescope Array (IOTA) [18], for observations at a wavelength of 2.2 μm, the delay line has been scanned repeatedly in the same direction with a rapid slew back to the starting position between each scan. The 8 second duration scans have been made at a velocity chosen to give a nominal fringe frequency of 100 Hz, which is above most of the atmospheric noise. The fringe signals are filtered in the Fourier domain with a 50-150 bandpass filter and then inverse transformed to obtain fringes of improved signal to noise ratio. At this stage the raw, observed visibility has been

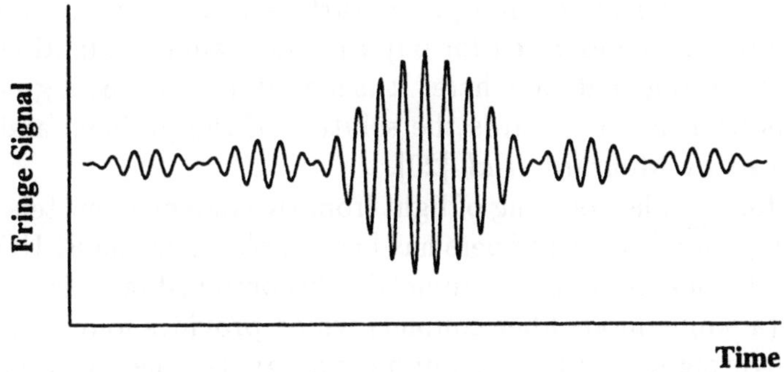

Figure 10. The fringe pattern observed by scanning through the matched optical path position.

determined from the four largest fringe peaks and normalised by similar data from unresolved sources.

The approach for the Cambridge Optical Aperture Synthesis Telescope (COAST) [17] is similar but, since the aim has been to develop a multi-aperture instrument for high resolution imaging, the beam combination system is more elaborate (see Fig. 7 of Baldwin *et al.* [17]). Four beams are involved but the same basic method is used so that all the combined wavefronts are plane parallel to each other. Again, each of the combined beams, composed of nominally 25% of each of the four input beams, is focussed onto a single pixel detector. Each incoming beam has its own optical delay line and the positions of the retroreflectors can be modulated by a sawtooth waveform to provide a linear sweep through the fringe pattern. Different sweep rates are used for each delay line so that each of the six baselines has its own unique fringe frequency in the range 100-1500 Hz. The resulting fringe pattern is analysed to obtain fringe amplitudes and phases for each baseline [17]. COAST has been designed to work in the optical and near infrared regions of the spectrum and is currently working in the wavelength range 700-900 nm. Some imaging aspects of the COAST program are discussed in Section 6.

Method 4 (for V) The Grand Interféromètre à 2 Télescopes (GI2T) [19] differs significantly from the instruments mentioned so far. It employs apertures much larger than r_0 and combines the beams from the input telescopes in the image plane. In other words, it works in the multi-speckle regime and the interference pattern in the common image plane is spread over a large number of speckles. The fringed speckles are dispersed and short exposures of the resulting spectra are taken with a photon counting detector. The problem faced in this approach is to derive a visibility estimate that is independent of the effects of atmospheric turbulence. It is more complicated than for the methods described for pupil plane beam combination and it is not appropriate to give details here. It is simply noted that a great deal of progress has been made towards the solution of the problem and this has been described by Mourard *et al.* [20].

Method 5 (for V) The coupling of light from two independent telescopes by means of single-mode optical fibers has been explored in the FLUOR experiment [21]. The beams are interferometrically combined in an integrated optics device (a single-mode fiber coupler) which provides four outputs—two interferometric signals and two photometric signals. The latter enable coupling fluctuations to be monitored and they are used as calibration signals to yield 'turbulence free' interferograms. This method of beam combination shows great promise for near-infrared long baseline interferometry.

4. Potential Sources of Systematic Error

4.1. POTENTIAL CAUSES OF VISIBILITY AND CORRELATION LOSS

The measured fringe visibility or correlation will generally be less than the true visibility or correlation although the losses may be partially masked by spurious correlation due to the detectors employed. Visibility losses result from a number of causes including:

1. Residual phase smearing
2. Residual wavefront distortion
3. Residual effects of scintillation
4. Incorrect optical path matching
5. Differential diffraction effects between the beams
6. Polarisation effects.

The first four of these potential sources of fringe visibility loss result primarily from the effects of turbulence in the Earth's atmosphere. Each item will be discussed in more detail.

(1) Residual phase smearing—If the relative phase of the interfering waves changes during a measurement sample time, the measured visibility or correlation will be reduced. For accurate results it is therefore essential to limit the relative phase variation during a measurement. It is reasonable to assume that instrumental sources of phase smearing will be negligible during a sample time through careful design and that the only concern is with the effects of the atmosphere. The effect of the length of the sample time, in terms of the atmospheric coherence time t_0, is shown in Buscher's Fig. 1 [22] and it is plotted for a range of values of the coherence time in Fig. 11. It is clear that even for sampling times of a very few milliseconds the loss of correlation can be large and that it critically depends on the prevailing

value of t_0. It is highly desirable to measure t_0 in parallel with measurements of correlation so that the effects of phase smearing can be corrected and Davis and Tango [23] have developed a technique for achieving this as part of the SUSI program.

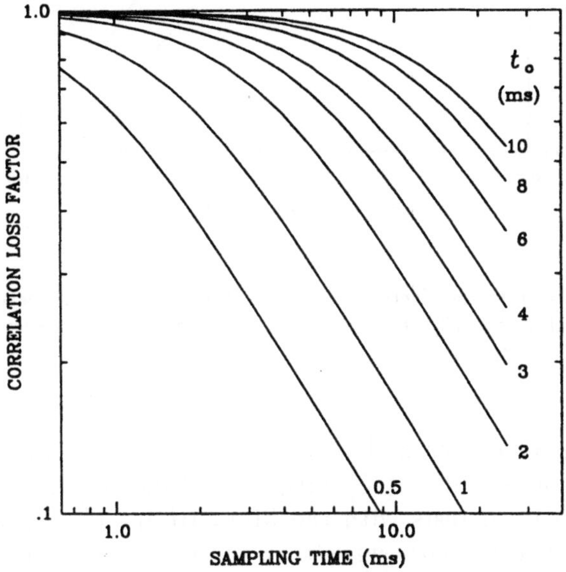

Figure 11. The correlation loss factor as a function of sample time for a range of values of the atmospheric coherence time t_0.

(2) Residual wavefront distortion Fried [24] has pointed out that successive correction of atmospherically induced aberrations of increasing order with an adaptive optical system steadily reduces the mean square residual wavefront distortion $\overline{\Delta^2}$. $\overline{\Delta^2}$ is proportional to $(D/r_0)^2$, where D is the diameter of the aperture. By far the largest improvement comes from the correction of wavefront tilts. Fig. 12 compares the loss in visibility as a function of the ratio of aperture size to r_0 (a) for no wavefront tilt correction and (b) for the case when the tilts are corrected by an active mirror. The curves are taken from Tango and Twiss [10], who noted that, with an aperture diameter equal to r_0, the loss in visibility for no correction is $\sim 50\%$ but only $\sim 13\%$ with tilt correction.

It is clear that, even with complete tilt correction, there are still significant losses due to higher order wavefront aberrations unless apertures much smaller than r_0 are employed. Adaptive optics will reduce these aberrations at the cost of increased complexity. An alternative and promising approach is the use of single-mode optical fibers as spatial filters to reduce the higher order aberrations. The fibers can be placed in each beam either

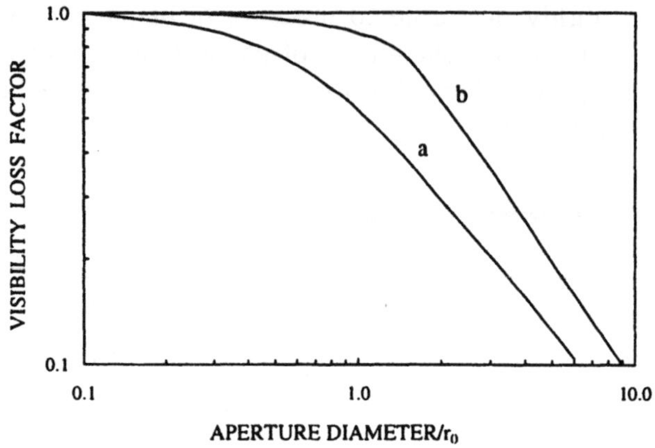

Figure 12. The visibility loss factor as a function of the ratio of aperture diameter to r_0 (After Tango and Twiss [10]). (a) Loss for no wavefront-tilt correction. (b) Loss with a wavefront tilt correcting servo.

before the beam combiner or between the beam combiner and detector. Shaklan, Colavita and Shao [25] have carried out an investigation which included observational tests with the Mark III Interferometer. While the fibers remove the detrimental wavefront aberrations they do introduce significant scintillation. However, Shaklan and his colleagues have shown that measurements using fibers result in correlation values that are easily calibrated and which have a greatly reduced seeing dependence. They point out that fibers allow larger apertures to be used and they have also shown that the signal to noise ratio is similar to the case when fibers are not used. For initial Mark III observations with a blue channel the use of a fiber increased the observed correlation by a factor of ~ 3 although some residual dependence on seeing remained.

It is inevitable that, whatever technique is used to minimise the effects of atmospheric turbulence, there will always be residual wavefront distortion at some level. It will be a combination of that due to the atmosphere and the instrumental aberration due to the numerous optical components. The calibration and correction for the effects of residual wavefront distortion is discussed in Section 5.4.

(3) Residual effects of scintillation

Tango and Twiss [10] have shown that the effect of scintillation on the measurement of correlation is negligible using measurement methods in which the beams from both sides of the beamsplitter are combined as, for example, in Method 1 of Section 3. This is not true for measurements made using a detection system involving only a single combined beam. The use of optical fibers as spatial filters introduces significant scintillation but measurement of the average total intensity and average intensity ratio of the two beams allows calibration of an unbiassed estimator of the correlation [25].

(4) Incorrect optical path matching The importance of matching the optical paths and the tolerances involved were discussed in Section 2.2.2. There are several sources of uncertainty in determining the correct configuration for the OPLE at any given instant. These include fluctuating optical path differences introduced by the atmosphere, uncertainties in the baseline orientation and length, and inaccuracies in stellar positions.

The fluctuating optical path difference introduced by the atmosphere, between the light received at separate apertures, has rms values ranging up to ~ tens of micrometres and time scales measured in seconds. Roddier [26] has shown that, for apertures separated by a baseline of length B, the rms optical path length fluctuation σ_{OPL} is given by

$$\sigma_{OPL} = 0.42\,\lambda_o \left[\frac{B}{r_0}\right]^{5/6} \tag{20}$$

Since r_0 varies as $\lambda^{6/5}$ it follows that σ_{OPL} is independent of wavelength. Equation 20 is only valid if the atmospheric turbulence follows Kolmogorov statistics and the frozen turbulence assumption holds. For baselines of the order of and greater than the outer scale of the turbulence, σ_{OPL} increases less rapidly and is expected to approach a limiting value which will depend on the site and meteorological conditions. Davis *et al.* [27] have measured the rms optical path length variations for baselines up to 80 m and, from a limited set of data, showed that they departed from equation 20 for baselines of a very few metres and approached a maximum rms value of the order of ~15-20 μm.

The uncertainties introduced by the atmosphere, baseline parameters, and stellar positions, are overcome simultaneously if a fringe or delay tracking system is used. The fringe tracking system used in the Mark III Interferometer was outlined in Section 3.2. An alternative method for maintaining equality of the optical paths within the required tolerance is known as 'group delay tracking' (GDT) [28, 29]. In this technique a combined beam is dispersed to reveal a channelled spectrum in which the number of fringes is directly proportional to the optical path difference between the beams. Averaging of the power spectrum of short exposures of the channelled spectrum provides a delay tracking signal. Even if a fringe or delay tracking system is used, an accurate knowledge of the baseline parameters, which can be gained through stellar observations, is highly desirable for efficient observing since it assists in rapid acquisition of the matched path condition. Similarly, the importance of using the best available stellar positions corrected to the epoch of observation cannot be overemphasised. In the absence of a fringe or delay tracking system the tolerances on matching the optical paths can only be met for short baselines and relatively narrow spectral bandwidths [30].

One complicating aspect of optical path equalisation has not been mentioned. Since both incoming beams traverse the same atmospheric path length, at least to first order, the path difference between the beams is a vacuum path difference. The OPLE should therefore use optical delay lines in vacuum, as has been done in the Mark III interferometer [7]. Because of the cost and complexity of vacuum delay lines, many instruments have delay lines operating in air. Since air is a dispersive medium, the resulting longitudinal dispersion will differ in the two arms of the instrument. The difference in dispersion, or differential longitudinal dispersion, will be a function of time, source coordinates, and baseline length. For other than narrow spectral bandwidths and short baselines, it is necessary to use a dispersion compensation system involving varying thicknesses of one or more glasses [31, 32, 33]. Dispersion also complicates the group delay tracking technique. The angular dispersion of the spectrograph used to produce the channelled spectrum for GDT will be non-linear for either a prism or a diffraction grating as the dispersing element. This is undesirable since it will broaden the fringe peak in the Fourier transform thereby reducing the signal to noise ratio of the tracking signal. Traub [34] has proposed a constant-dispersion grism spectrometer to linearise the channelled spectrum in the absence of differential longitudinal dispersion. However, if differential longitudinal dispersion is present, the non-linearity of air dispersion can add to, or subtract from, the non-linearity introduced by the dispersing element depending on which side of the zero group delay position the channelled fringes are formed—the prism and grating cases add to the non-linearity on opposite sides of the zero group delay position. By adjusting the criteria for dispersion correction and modifying the glass thicknesses in the dispersion compensation system it is possible to linearise the fringes in the channelled spectrum and, at the same time, match the group delays in a separate visibility/correlation measurement bandwidth [35].

(5) Differential diffraction effects The beams within an interferometer are necessarily restricted in size and this gives rise to differential diffraction effects which reduce the fringe visibility. This problem has been studied by several workers [36, 37, 38, 39] and while, in principle, the differential effects can be computed, they are complicated by the presence of aberrations. The existence of diffraction effects needs to be considered in the context of the calibration of visibility data and we return to it in Section 5.4.

(6) Polarisation effects In the design of an interferometer it is clearly essential to ensure that the orientation of the polarisation planes is the same for

both beams after they have traversed the instrument and that one beam is not rotated with respect to the other. However, by itself, this is not sufficient to eliminate the loss of visibility due to polarisation effects. Traub [40] has discussed the difference in the phase changes that occur for the *s* and *p* polarisations on reflection and shown that significant losses in visibility will occur if, for a given plane of polarisation entering the instrument, *s* and *p* reflections do not take place in the same sequence for each beam.

Experience with SUSI has shown that there can still be a loss of visibility due to polarisation effects even if the paths are matched in terms of the geometry of the reflections. In SUSI the *s* polarisation on the beam-combining table is used for wavefront-tilt detection and the *p* polarisation is used for the correlation measurements. A small amount of the *s* component is leaked by the polarising beamsplitters and it reduces the observed fringe visibility—the introduction of linear polarisers to ensure that only the *p* component was present for correlation measurements increased the observed correlation and improved the signal to noise ratio.

4.2. A POTENTIAL CAUSE OF SPURIOUS CORRELATION

Detectors which suffer from after-pulsing, such as photomultipliers, will produce spurious correlation which is a function of the photon count rate. At high count rates the spurious correlation may be insignificant but, since it increases with decreasing count rate, it is essential that observed values of correlation for low count rates are corrected for its effect. The magnitude of the spurious correlation as a function of count rate can be established by observing stars with an optical path difference much greater than the coherence length of the light or, alternatively, by illuminating the detectors with incoherent light from a laboratory source.

5. Observing Technique

The technique of making interferometric observations depends on the degree to which the instrument is automated. Mature instruments are generally highly automated with many of the factors influencing the quality of the data or the efficiency of operation taken care of automatically. For such instruments many of the factors described will be 'transparent' and the user unaware of them. Nevertheless, the observer should be conscious of all the factors which could influence their data.

5.1. ALIGNMENT

At the commencement of each night of observations a check on the internal alignment of an interferometer is essential. This is performed with the aid of internal reference light sources—lasers for optical alignment and a 'white light source' for establishing the zero point of the laser metrology system which monitors the position and velocity of the OPLE delay lines. The optical alignment is generally carried out by autocollimating the reference beams from the input apertures. This allows the siderostat or telescope encoder values corresponding to the optical axes of the arms of the instrument to be checked. Ideally the alignment procedures are fully automated and can be performed without a detailed knowledge of the optical system but, nevertheless, they are vital if reliable data are to be obtained.

5.2. POINTING AND BASELINE SOLUTIONS

Important factors for the efficient operation of an interferometer are the pointing of the input siderostats or telescopes and an accurate knowledge of the baseline coordinates.

Accurate pointing of the siderostats or telescopes is important for the rapid acquisition of the source to be observed. The parameters of the siderostat or telescope are determined from observations of stars for which the pointing errors have been recorded. These parameters include the orientation of the axes of rotation with respect to each other and relative to the zenith or pole, encoder offsets, and the orientation of the mirror with repect to the mechanical axes. Accurate knowledge of the baseline parameters (i.e. the length plus two angles to define the orientation) is essential for observations made without the benefit of fringe or delay tracking, and important for rapid acquisition of lock if either fringe or delay tracking is employed. A solution for the baseline parameters is found through observations of stars for which the delay errors have been established (i.e. the offsets of the observed zero group delay positions from the computed positions). Both pointing and baseline solutions will change slowly with time and ideally should be continuously upgraded and the process should be transparent to the observer.

5.3. CHOICE OF OBSERVATIONAL PARAMETERS

The observational parameters of importance are: wavelength, baseline, aperture size, and sampling time.

<u>Wavelength</u>—if a range of observing wavelengths is available, the choice of wavelength will generally be governed by the science problem. However, the advantages of working at longer wavelengths to reduce the magnitude of the deleterious effects of the atmosphere should be borne in mind.

<u>Baseline</u>—again this may be restricted by the instrument but, where there is a choice, the following points should be considered. First, the resolution for a given baseline varies inversely with the wavelength. Secondly, for a

simple angular diameter measurement, there is an optimum baseline which minimises the observing time required for a given accuracy (at the point where $C \sim 0.3$). This of course implies knowing, or at least estimating the angular size in advance but, nevertheless, it should be recognised that the choice of baseline influences the accuracy that can be achieved in a given amount of observing time.

For imaging, as will be discussed in following contributions, it is important to choose baselines that will give adequate two-dimensional cover.

Aperture size—the internal aperture size used for visibility or correlation measurements should be comparable with or less then r_0 unless adaptive optics are employed. If the selected aperture is much less than r_0, the effects of residual tilts and wavefront curvature will be minimised. Experience with SUSI has shown that it is highly desirable to measure r_0 in parallel with measurements of visibility or correlation and to calibrate the residual effects by observations of calibration sources. It is noted that there will be an optimum aperture for maximum signal to noise ratio but that it depends on both the variable r_0 and the internal optical aberrations. In any case it is important for each complete series of observations, including both target and calibration sources, to choose an aperture size and stick with it.

Sampling time—the choice of sampling time depends on whether the atmospheric coherence time t_0 is being measured in parallel with the measurement of visibility or correlation. If it is, the choice of sample time is not so critical since correction of the data to zero sample time is possible [23]. If t_0 is not being measured, a sample time short compared with the expected values of t_0 is highly desirable as shown by Fig. 11.

5.4. CALIBRATION

In order to correct observed visibilities or correlation measurements for residual atmospheric and instrumental effects it is essential to include unresolved sources in the observational program. Calibration sources will generally be early-type stars chosen because the expected visibility is close to unity at the observing baseline. Because the visibility curve is close to horizontal for 'short' baselines, as can be seen in Fig. 13, it is not necessary to know the angular sizes of the calibrating stars with great accuracy. For example, if the calibration star is estimated to give $V = 0.98$ at the observing baseline, a 20% uncertainty in the assumed angular diameter only results in an error of $\pm 0.8\%$ in V. Generally the angular diameters of early type stars can be estimated to much better accuracy than 20% based on the results of the Narrabri Stellar Intensity Interferometer program [41].

In choosing calibration stars, care should be taken to choose single stars, to minimise the angular distance from the program stars and, during obser-

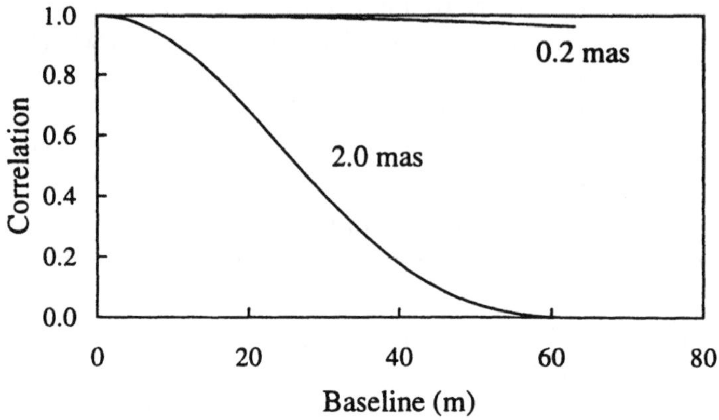

Figure 13. The variation of correlation with baseline for two stars with angular diameters differing by a factor of ten.

vations, to minimise the time difference between observations of calibration and program stars. Choosing calibration stars close to the program stars not only minimises the time involved in switching between them, but it also ensures that the effects of differential diffraction within the instrument (Section 4.1) will be similar and the loss of fringe visibility due to this cause will be calibrated along with the loss from other causes. Ideally, the calibration stars will be of comparable brightness to the program stars although, in practice, some compromise will generally be necessary. These precautions are aimed at minimising calibration errors due to changing seeing conditions with zenith angle and time. Instruments employing real time wavefront-tilt correction, and which provide measurements of t_0 and r_0 in parallel with visibility or correlation measurements, have an advantage in minimising seeing dependent calibration errors.

5.5. OBSERVATIONS

In carrying out observations, unless the conditions are particularly stable, it is advisable to switch frequently and rapidly between program and calibration sources to minimise calibration errors due to changing seing conditions with zenith angle and time. The advantages of a fully automated instrument, like the Mark III interferometer [7], for this are obvious. It is also desirable to use more than one calibration source to guard against the possibility that a calibration source is an unknown double source - one has to guard against the well known phenomenon that stars chosen as standards very often turn out to be unsuitable as standards. Regular measurements of dark count and sky background should be included in the observing sequence unless, when observing bright sources, they are known from experience to be negligible.

Although individual sample measurements have a duration of only a very few milliseconds in order to avoid visibility loss due to phase smearing, it is necessary to accumulate and average the results of a large number of samples to reduce the uncertainty in a measurement. Integration periods of the order of one to a few minutes are appropriate and a number of short 1-2 minute integrations are better than a single long one since they allow a direct computation of the uncertainty of the measurement from the scatter of the individual integrations. It also allows for easier recognition of poor data points due to cloud etc.

The total integration time on a program or calibration object between switches should be limited because of the possibility of seeing changes - regular switching between program and calibration objects is recommended unless one can rely on the stability of the seeing. In the SUSI program the switch between program and calibration sources is made every 5-20 minutes, depending on the brightness of the sources and the stability of the seeing which is monitored continuously during observations.

Observing with an instrument that allows observations with only a single baseline at a time requires caution. A single measurement can be misleading if the source, assumed to be single, is in fact multiple. It is essential to observe the source at different hour angles so that the brightness distribution is projected onto the baseline with different orientations. A multiple system will give different visibility values as the projected angular separation of its components onto the baseline changes. This of course requires consideration of the change in projection angle as a function of declination for the baseline orientation. This problem does not exist for an imaging array that provides simultaneous data for a range of baseline orientations.

There is no standard technique for analysing the observational data since every instrument built has included different features and monitored the seeing by different means. For SUSI data, the values of observed correlation are corrected to zero sample time and for background correlation. Calibration curves of the corrected correlation versus aperture size/r_0 are prepared for each calibration source and similar curves are made for the target stars. The curves for the target stars are then normalised by the curves for the calibration sources to obtain the true correlation freed from instrumental and residual atmospheric effects. The methods of analysis for data from the Mark III Interferometer, COAST, IOTA, and the GI2T are discussed in references [13, 47, 18, 20] respectively.

6. Closure Phase and Imaging

It was pointed out in Section 2.2.1 that more than two apertures must be used simultaneously if images are to be constructed. For each *pair* of apertures of an imaging array it is necessary to meet the same contraints that apply to a single baseline, two aperture interferometer. This includes matching the optical paths to within a small fraction of the coherence length, dealing with the effects of wavefront distortion, and using rapid signal sampling to limit phase smearing during each observational sample period.

Even though the phase measurement for a single baseline is corrupted by the atmosphere, it is possible to determine a 'closure phase', free of corruption, for each independent triangle of baselines formed by three apertures of an array.

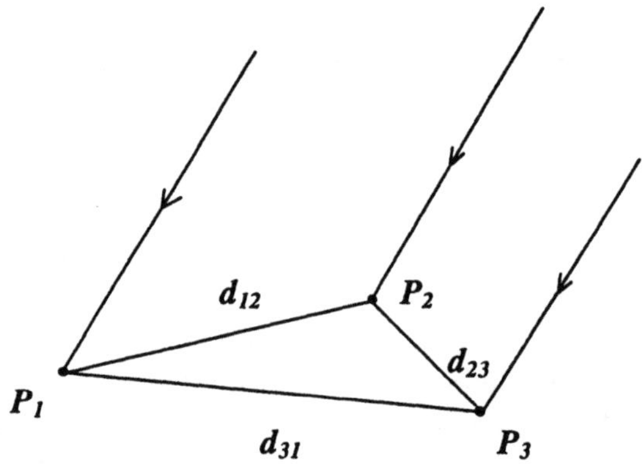

Figure 14. Geometry for a three aperture interferometer. Details are given in the text

In Fig. 14 three input apertures located at P_1, P_2 and P_3 form baselines d_{12}, d_{23} and d_{31}. Let the observed phases for the three baselines be θ_{12}, θ_{23} and θ_{31} respectively. Let the true phases for the three baselines, due to the source alone, be represented by ϕ_{12}, ϕ_{23} and ϕ_{31} respectively and the phase errors for each aperture, due to the atmosphere and the instrument, be Δ_1, Δ_2 and Δ_3. It follows that

$$\theta_{12} = \phi_{12} + \Delta_1 - \Delta_2 \tag{21}$$

$$\theta_{23} = \phi_{23} + \Delta_2 - \Delta_3 \tag{22}$$

$$\theta_{31} = \phi_{31} + \Delta_3 - \Delta_1 \tag{23}$$

The closure phase Ψ_{123} is defined as the sum of the observed phases

$$\Psi_{123} = \theta_{12} + \theta_{23} + \theta_{31} \tag{24}$$

Substitution for the θs from equations 21-23 in equation 24 leads to

$$\Psi_{123} = \phi_{12} + \phi_{23} + \phi_{31} \tag{25}$$

The closure phase given by equation 25 depends only on the true phases due to the source and it is therefore a property of the angular intensity distribution across the source. Thus measurements of $|\gamma(0)|$ and $\phi(0)$, for three or more baselines simultaneously, enable images to be constructed using the techniques developed and successfully exploited by radio astronomers [42, 43]. These techniques have been demonstrated at optical wavelengths using masks containing small apertures on conventional telescopes [45, 46] and recently the first image using separate apertures has been obtained with COAST by Baldwin *et al.* [47].

For three input apertures only one closure phase is obtained with three values for the fringe visibility, one for each baseline. For N baselines, $(N/2-1)(N-1)$ independent closure phases and $N(N-1)/2$ values for the fringe visibility are obtained. This suggests that increasing the number of apertures will improve the imaging capabilty of an interferometric array. In practice this is true for a small number of apertures but, because beam combination techniques involve beamsplitters and the division of the light between detectors with associated losses of light, in the absence of coherent amplifiers there is a practical limit to the number of apertures that can be usefully employed. There is no clear consensus on the optimum number of apertures but it is within the range 4-10.

In Section 2 the Fourier transform relationship between the complex degree of spatial coherence and the angular distribution of intensity across the source was discussed and it was noted that if $|\gamma_{12}(0)|$ and $\phi_{12}(0)$ were measured over an appropriate range of baseline lengths and orientations, the angular intensity distribution of the source can be found. For the purposes of the transformation the components of each baseline vector are expressed in wavelength units for the mean wavelength of the spectral band. Observations are effectively taken in a plane perpendicular to the source direction and the components of each baseline vector are given for the orthogonal axes, u and v. The components u and v are known as spatial frequencies and the plane containing u and v is referred to as the (u,v) plane. For a detailed discussion and more formal definitions see Chapter 4 of reference [43]. Images may be reconstructed from observations that only partially fill the (u,v) plane using techniques developed in radio astronomy [42, 43, 44], the resolution in two dimensions being set by the maximum values of u

and v for which there are observations. The quality of images depends on the (u,v) coverage of the observations and, as will be discussed in other contributions, the design of interferometric arrays is largely governed by the need to provide good observational cover of the (u,v) plane.

7. Summary

Initially this contribution was intended to discuss solely the topic of "Observing with Interferometers" but, during the development of the program, the author was asked to include a basic introduction to interferometers and an introduction to the concept of closure phase. The result is a contribution that is only 'partially coherent' since the different topics do not blend easily into a single contribution. Nevertheless, an attempt has been made to present the material in a logical sequence. Much of the material presented will not concern users of future interferometers that are built as 'user instruments'. However, such instruments are only now under development and users should always be aware of the underlying problems in case something goes wrong with an instrument that affects their observational data.

The field of optical/infrared long baseline interferometry has gone through a long period of development. The record of this development is largely contained in the proceedings of a number of conferences held during the last decade or so [48, 49, 50, 51]. The emphasis has been primarily on the technical developments necessary to achieve very high angular resolution from the ground and the building of prototype instruments to demonstrate and evaluate the potential of the technique. To date the scientific output has not been commensurate with the effort that has been expended in developing the field but the stage has been reached where several major instruments are under development with the confident expectation that they will open up exciting new scientific possibilities. The announcement of the first image obtained with separate input apertures [47] reinforces this confidence.

There are a number of long baseline optical and infrared interferometers and interferometric imaging arrays at various stages of development. These include the Grand Interféromètre à 2 Télescopes (GI2T) [19], the Infrared Spatial Interferometer (ISI) [52], SUSI [2], IOTA [18, 53], COAST [17], the Palomar Interferometer [54] which is a testbed for a planned Keck Interferometer, the US Navy Prototype Optical Interferometer (NPOI) [55, 56], Georgia State University's CHARA Array [57], and ESO's Very Large Telescope Interferometer (VLTI) [58]. The development of these major facilities will see a rapid increase in the science output of high angular resolution optical and infrared interferometers and interferometric arrays over the next decade.

References

1. Born, M. and Wolf, E. (1964) *Principles of Optics*, Pergamon Press.
2. Davis, J. (1994) The Sydney University Stellar Interferometer (SUSI), in J.G. Robertson and W.J. Tango (eds.), *Proc. IAU Symp. 158, Very High Angular Resolution Imaging*, Kluwer, Dordrecht, pp. 135–142.
3. Fried, D.L. (1965) Statistics of a geometric representation of wavefront distortion, *J. Opt. Soc. Am.* **55**, 427–1435.
4. Buscher, D.F. (1994) A thousand and one nights of seeing on Mount Wilson, in J.B. Breckinridge (ed.), *SPIE Proc.* **2200**, SPIE, Washington, pp. 260–271.
5. Colavita, M.M., Shao, M. and Staelin, D.H. (1987) Atmospheric phase measurements with the Mark III stellar interferometer, *Applied Optics* **26**, 4106–4112.
6. Shao, M. *et al.* (1988) Initial stellar diameter measurements with the Mark III Interferometer, *Ap. J.* **327**, 905–910.
7. Shao, M. *et al.* (1988) The Mark III stellar interferometer, *Astron. Astrophys.* **193**, 357–371.
8. Davis, J. and Tango, W.J. (1985) The Sydney University 11.4 m prototype stellar interferometer, *Proc. Astron. Soc. Australia* **6**, 34–38.
9. Davis, J. and Tango, W.J. (1985) A new very high angular resolution stellar interferometer, *Proc. Astron. Soc. Australia* **6**, 38–43.
10. Tango, W.J. and Twiss, R.Q. (1980) Michelson stellar interferometry, *Progress in Optics* **XVII**, 239–277.
11. Shao, M. and Staelin, D.H. (1977) Long-baseline optical imterferometer for astrometry, *J. Opt. Soc. Am.* **67**, 81–86.
12. Shao, M. *et al.* (1990) Wide-angle astrometry with the Mark III stellar interferometer, *Astron. J.* **100**, 1710–1711.
13. Mozurkewich, D. *et al.* (1991) Angular diameter measurements of stars, *Astron. J.* **101**, 2207–2219.
14. Armstrong, J.T. *et al.* (1992) The orbit of α Equulei measured with long-baseline interferometry: component masses, spectral types, and evolutionary state, *Astron. J.* **104**, 241–252.
15. Gay, J. and Mekarnia, D. (1988) Infrared interferometry at CERGA, in F. Merkle (ed.), *High-Resolution Imaging by Interferometry*, ESO Conference and Workshop Proceedings No. 29, ESO, Garching bei München, pp. 811–816.
16. Benson, J.A. *et al.* (1991) The infrared angular diameter of Alpha Herculis measured with a Michelson interferometer, *Astron. J.* **102**, 2091–2097.
17. Baldwin, J.E. *et al.* (1994) Design and performance of COAST, in J.B. Breckinridge (ed.), *SPIE Proc.* **2200**, SPIE, Washington, pp. 118–128.
18. Dyck, H.M. *et al.* (1995) First 2.2 μm results from the IOTA interferometer, *Astron. J.* **109**, 378–382.
19. Mourard, D. *et al.* (1994) The GI2T interferometer on Plateau de Calern, *Astron. Astrophys.* **283**, 705–713.
20. Mourard, D. *et al.* (1994) Estimation of visibility amplitude by optical long-baseline Michelson interferometry with large apertures, *Astron. Astrophys.* **288**, 675–682.
21. Coudé du Foresto, V. and Ridgway, S. (1992) FLUOR: a stellar interferometer using single-mode infrared fibers, in J.M. Beckers and F. Merkle (eds.), *High-Resolution Imaging by Interferometry II*, ESO Conference and Workshop Proceedings No. 39, ESO, Garching bei München, pp. 731–740.
22. Buscher, D.F. (1990) Optimizing a ground-based optical interferometer for sensitivity at low light levels, *Mon. Not. R. Astron. Soc.* **235**, 1203–1226.
23. Davis, J. and Tango, W.J. (1996) Measurement of the atmospheric coherence time, *Pub. Astron. Soc. Pacific* **108**, *in press*.
24. Fried, D.L. (1978) The nature of atmospheric turbulence effects on imaging and pseudo-imaging systems and its quantification, in J. Davis and W.J. Tango (eds.), *Proc. IAU Colloquium No. 50: High Angular Resolution Stellar Interferometry*,

Chatterton Astronomy Department, University of Sydney, pp. 4-1–4-44.
25. Shaklan, S.B., Colavita, M.M. and Shao, M. (1991) Visibility calibration using single mode fibers in a long-baseline interferometer, in J.M. Beckers and F. Merkle (eds.), *High-Resolution Imaging by Interferometry II*, ESO Conference and Workshop Proceedings No. 39, ESO, Garching bei München, pp. 1271-1283.
26. Roddier, F. (1981) The effects of atmospheric turbulence in optical astronomy, in E. Wolf (ed.), *Progress in Optics* **XIX**, North-Holland, Amsterdam, pp. 283–376.
27. Davis, J. *et al.* (1995) Atmospheric path variations for baselines up to 80 m measured with the Sydney University Stellar Interferometer, *Mon. Not. R. Astron. Soc.* **273**, L53–L58.
28. Nisenson, P. and Traub, W. (1987) Magnitude limit of the group delay fringe tracking method for long baseline interferometry, in J.W. Goad (ed.), *Interferometric Imaging in Astronomy*, Proceedings of a Joint Workshop on High-Resolution Imaging from the Ground using Interferometric Techniques, ESO-NOAO, pp.129–133.
29. Lawson, P.R. (1995) Group-delay tracking in optical stellar interferometry with the fast Fourier transform, *J. Opt. Soc. Am. A* **12**, 366–374.
30. Davis, J. (1984) Long baseline optical interferometry, in J.A. Roberts (ed.), *Indirect Imaging*, Cambridge University Press, Cambridge, pp. 125–141.
31. Lacasse, M.G. and Traub, W.A. (1988) Glass compensation for an air filled delay line, in F. Merkle (ed.), *High-Resolution Imaging by Interferometry*, ESO Conference and Workshop Proceedings No. 29, ESO, Garching bei München, pp. 959–970.
32. Tango, W.J. (1990) Dispersion in stellar interferometry, *Appl. Opt.* **29**, 516–521.
33. ten Brummelaar, T.A. (1995) Differential path considerations in optical stellar interferometry, *Appl. Opt.* **34**, 2214–2219.
34. Traub, W.A. (1990) Constant-dispersion grism spectrometer for channeled spectra, *J. Opt. Soc. Am. A* **7**, 1779–1791.
35. Lawson, P.R. and Davis, J. (1996) Dispersion compensation in stellar interferometry, *Appl. Opt.* **35**, 612–620.
36. Tango, W.J. and Twiss, R.Q. (1974) Diffraction effects in long path interferometers, *appl. Opt.* **13**, 1814–1819.
37. Bagnuolo, W.G. (1988) Simulations for the CHARA/GSU interferometer and binary star speckle photometry, in F. Merkle (ed.), *High-Resolution Imaging by Interferometry*, ESO Conference and Workshop Proceedings No. 29, ESO, Garching bei München, pp. 981–994.
38. Mekarnia, D. and Gay, J. (1989) Alteration du visibilite par diffraction de Fresnel en synthese d'ouverture, *J. Optics* **20**, 131–140.
39. Hrynevych, M. (1994) Differential diffraction in Michelson stellar interferometry, in J.G. Robertson and W.J. Tango (eds.), *Proc. IAU Symp. 158, Very High Angular Resolution Imaging*, Kluwer, Dordrecht, pp. 224–226.
40. Traub, W.A. (1988) Polarization effects in stellar interferometers, in F. Merkle (ed.), *High-Resolution Imaging by Interferometry*, ESO Conference and Workshop Proceedings No. 29, ESO, Garching bei München, pp. 1029–1038.
41. Hanbury Brown, R., Davis, J. and Allen, L.R., (1974) The angular diameters of 32 stars, *Mon. Not. R. astr. Soc.* **167**, 121–136.
42. Pearson, T.J. and Readhead, A.C.S. (1984) Image formation by self-calibration in radio astronomy, *Ann. Rev. Astron. Astrophys.* **22**, 97–130.
43. Thompson, A.R., Moran, J.M. and Swenson, G.W. (1986) *Interferometry and Synthesis in Radio Astronomy*, John Wiley & Sons, New York.
44. Clark, B.G. *et al.* (1989) Synthesis Imaging in Radio Astronomy, R.A. Perley, F.R. Schwab and A.H. Bridle (eds.), *Astron Soc. Pacific Conf. Series* **6**, Astron. Soc. Pacific., San Francisco.
45. Baldwin, J.E. *et al.* (1986) Closure phase in high-resolution optical imaging, *Nature* **328**, 595–597.
46. Haniff, C.A. et al. (1987) The first images from optical aperture synthesis, *Nature* **328**, 694-696.

47. Baldwin, J.E. *et al.*, (1996) The first images from an optical aperture synthesis array: mapping of Capella with COAST at two epochs, *Astron. Astrop.* **306**, L13–L16.
48. NOAO-ESO Conference (1987) J. Goad (ed.), *Interferometric Imaging in Astronomy*, NOAO, Tucson.
49. NOAO-ESO Conference (1988) F. Merkle (ed.), *High-Resolution Imaging by Interferometry*, ESO Conference and Workshop Proceedings No. 29, ESO, Garching bei München.
50. ESO Conference (1991) J.M. Beckers and F. Merkle (eds.), *High-Resolution Imaging by Interferometry II*, ESO Conference and Workshop Proceedings No. 39, ESO, Garching bei München.
51. SPIE Conference (1994) Amplitude and Intensity Spatial Interferometry II, J.B. Breckinridge (ed.), *SPIE Proc.* **2200**, SPIE, Washington.
52. Bester, M. *et al.* (1994) The U.C. Berkeley Infrared Spatial Interferometer—recent system upgrades and analysis of atmospheric fluctuations, in J.B. Breckinride (ed.), *SPIE Proc.* **2200**, SPIE, Washington, pp. 274–285.
53. Carleton, N.P. *et al.* (1994) Current status of the IOTA interferometer, in J.B. Breckinridge (ed.), *SPIE Proc.* **2200**, SPIE, Washington, pp. 152–165.
54. Colavita, M.M. *et al.* (1994) ASEPS-0 Testbed Interferometer, in J.B. Breckinridge (ed.), *SPIE Proc.* **2200**, Washington, pp. 89–97.
55. Armstrong, J.T. (1994) Progress on the Big Optical Array (BOA), in J.B. Breckinridge (ed.), *SPIE Proc.* **2200**, SPIE, Washington, pp. 62–70.
56. Hutter, D.J. (1994) U.S. Naval Observatory Astrometric Interferometer, in J.B. Breckinridge (ed.), *SPIE Proc.* **2200**, SPIE, Washington, pp. 81–88.
57. McAlister, H.A. *et al.* (1994) The CHARA Array, in J.B. Breckinridge (ed.), *SPIE Proc.* **2200**, SPIE, Washington, pp. 129–139.
58. von der Lühe, O. (1994) The interferometric mode of the ESO Very Large Telescope, in J.B. Breckinridge (ed.), *SPIE Proc.* **2200**, SPIE, Washington, pp. 168–179.

IMAGING IN INTERFEROMETRY

OSKAR VON DER LÜHE AND NANCY AGEORGES
European Southern Observatory
Karl-Schwarzschild-Str. 2, D-85748 Garching

1. Statement of Problem

The goal of imaging with interferometers is to obtain an estimate of the spectral flux $I_\lambda(\overline{\alpha})$ of the source of interest, as a function of two-dimensional angular coordinates $\overline{\alpha}$ on the celestial sphere, with an angular resolution that corresponds to the maximum baseline in the array, and which is consistent with the observations.

The distribution of the source's flux at wavelength λ is given in units of $W\ \text{Å}^{-1}\ m^{-2}\ s^{-1}$. The two-dimensional angular coordinate system $\overline{\alpha}$ has units in radians, its origin is chosen suitably close to the observed source and its axes are preferrably aligned with the directions of right ascension and declination. Since we consider the action of an interferometer, we are interested in the Fourier transform of the source's intensity distribution:

$$\hat{I}_\lambda(\overline{u}) = \int\int_{\mathcal{F}} I_\lambda(\overline{\alpha}) \exp 2\pi j \overline{\alpha} \cdot \overline{u}\ d\overline{\alpha} \tag{1}$$

where $j = \sqrt{-1}$, $\overline{u}$ denotes an *angular frequency* measured with units "line pairs per radian" $[rad^{-1}]$. The set $\mathcal{F}$ of coordinates $\overline{\alpha}$ chosen such that the detected flux vanishes outside the set. The Fourier transform has a number of interesting properties which result from the circumstance that the source's intensity distribution is a positive semi–definite and real valued function:

$$\begin{aligned} \text{Hermiticity:} \quad & I_\lambda(\overline{\alpha}) \in \Re \quad \Rightarrow \hat{I}_\lambda(-\overline{u}) = \hat{I}^*_\lambda(\overline{u}) \\ \text{Reduced spectrum:} \quad & I_\lambda(\overline{\alpha}) \geq 0 \quad \Rightarrow \hat{I}_\lambda(\overline{u}) = \hat{G}(\overline{u}) \otimes \hat{G}(\overline{u}) \end{aligned} \tag{2}$$

A-M. Lagrange et al. (eds.), High Angular Resolution in Astrophysics, 81–108.
© 1997 *Kluwer Academic Publishers.*

The latter property expresses the Fourier transform of the source distribution in the form of a convolution of the Fourier transform ("reduced spectrum") of another function, $G(\overline{\alpha})$, which in general is complex valued. In the case of monochromatic light, G could be the complex electromagnetic wave which gives rise to the detected intensity.

The interferometer measures with the complex degree of coherence of the light waves which are received by the array elements i and k, the Fourier component of the flux distribution of an *incoherent* source at discrete angular frequencies $\overline{u}_{ik}$ (theorem of van Cittert – Zernike). The frequencies are given by the *projected baselines* in the array and the observing wavelength λ:

$$\overline{u} = \frac{1}{\lambda}\overline{B\prime}_{ik}, \quad i,k \in \{1,\cdots,N\} \tag{3}$$

$\overline{B\prime}_{ik}$ is the projected baseline (see below) between elements i and k of the interferometric array with N elements.

The value of the Fourier transform of the source's flux distribution at the frequency which corresponds to the projected baseline between array elements i and k is called the "correlated flux" between the two telescopes. The normalised quantity

$$V_{ik} = \frac{\hat{I}_\lambda(\overline{u}_{ik})}{\hat{I}_\lambda(\overline{0})} = \frac{\int\int_{\mathcal{F}} I_\lambda(\overline{\alpha}) \exp 2\pi j\overline{\alpha}\cdot\overline{u}_{ik}\, d\overline{\alpha}}{\int\int_{\mathcal{F}} I_\lambda(\overline{\alpha})\, d\overline{\alpha}} \tag{4}$$

is called the *complex visibility*. The denominator in eqn. 4, the Fourier transform at the origin of frequency space, is equal to the total flux of the source. The complex visibility is therefore the correlated flux divided by the total flux, its magnitude is always less or equal to unity.

2. Sample Source Fourier Transforms

Two-dimensional Fourier transforms have a number of interesting properties which are important to remember when inspecting visibility data. A good introduction into the Fourier transform and its properties is given in Bracewell [1]. We shall review some of them here.

One of the most prominent properties of the Fourier transform is the reciprocity of scales. Sources with small angular scales contain information at high frequencies, and therefore contribute to large values of angular frequencies. Conversely, the spectrum of extended, featureless sources is very limited in extent. For example, the Fourier transform of a two-dimensional Gaussian distribution with a width of σ is itself a Gaussian whose width is proportional to $\frac{1}{\sigma}$. This property reflects the fact that in order to resolve something extended, only low angular frequencies and therefore short baselines between telescopes are necessary, while long baselines are needed in order to resolve a source of very limited angular extent.

A source which consists of a single, point-like object has a Fourier transform which is constant, i. e., the complex visibility is unity everywhere. A source which consists of two point-like objects has a Fourier transform of the form of a two-dimensional cosine function which extends to infinity, and whose ridges are oriented perpendicular to the direction of separation of the double star. Owing to the law of reciprocity of scales, the period of the cosine is inversely proportional to the separation of the two sources. For a source with N point-like objects, there is a cosine contribution with a appropriate period for each pair of sources, there are $N[N-1]/2$ such pairs. If the N sources are arranged in a random fashion, the mean visibility level will be of order $\frac{1}{N}$ except close to the frequency origin.

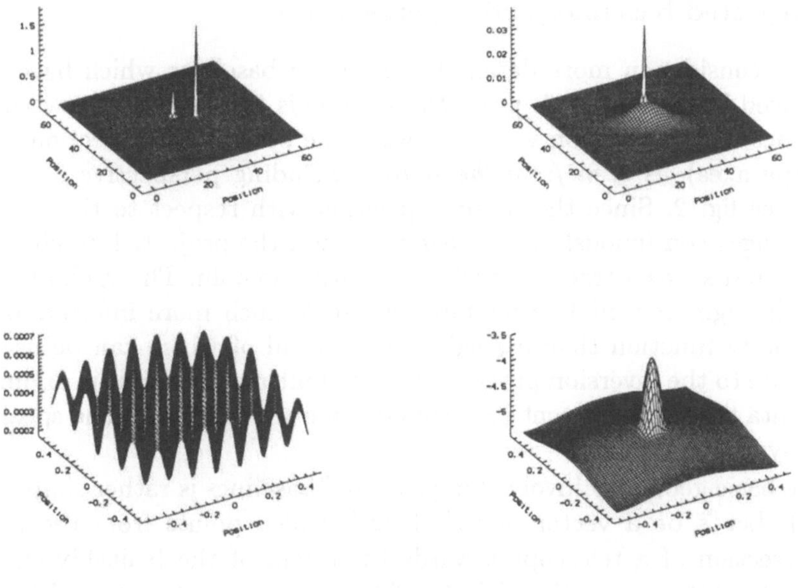

Figure 1. A double source (top left) and its Fourier transform magnitude (bottom left). A source which consists of an extended disk with an embedded point-like source (top right) and the logarithm of its Fourier transform (bottom right). All units are arbitrary.

An important characteristic of the visibility function is its "randomness'. The more complex the structure of the visibility function is, the more statistically independent samples will be needed in order to afford a reliable inversion of the data. An extended object which contains a lot of quasi-random, small scale structure will have a Fourier transform which varies at scales that correspond to the inverse of the full extent of the source, while it extends to the cutoff scale of the small scale structure. The number of independent samples of the Fourier transform which will be required for a reliable image of the source is roughly $\frac{(\textit{extent of source})^2}{(\textit{small-scale structure cutoff scale})^2}$. Conversely, a collection of N point sources will, in most cases, only require sufficient samples to identify the $N(N-1)/2$ cosine patterns, regardless of the extent of the source as a whole.

Figure 1 shows two sample sources, a double star with unequal brightness and an extended disk-like distribution with an embedded point-like star. The Fourier transform of the double star shows the already described cosine pattern. The minima of the cosine do not extend to zero; they would be if the sources were of equal brightness. The contribution of the extended disk to the Fourier transform is the peak at the center, while the nearly unresolved star's contribution is the, rather faint, extended background in the Fourier transform.

3. Projected baselines and supersynthesis

We now consider in more detail the projected baselines which have been introduced in Section 1. A *projected baseline* is the baseline between two telescopes (the separation vector between the points of intersection of the telescope axes) *as seen from the source*, including perspective foreshortening (see fig. 2. Since the source's position with respect to the observatory changes continuously as the earth rotates, the projected baseline also changes and scans a trace through the Fourier domain. The visibility data taken throughout a night therefore represent much more information on the visbility function than a single snapshot, all of which can be used to contribute to the inversion process which will ultimately result in an image. Since data taken at different moments in time are combined, one speaks of "supersynthesis".

The basic geometry involved in projected baselines is rather simple (see Fig. 2). Let $\overline{s}$ be a vector of unit length which points from the points of intersection of a telescope towards the origin of the (suitably chosen) coordinate system α in the vicinity of the source of interest, and let $\overline{B}_{ik}$ be the vector separation between the telescopes i and k. The projected baseline is

$$\overline{B\prime}_{ik} = \overline{s} \times (\overline{B}_{ik} \times \overline{s}) \; . \tag{5}$$

The symbol $\times$ denotes a vector product. A related quantity is the *geometric delay* λw_{ik} which is given with

$$\lambda w_{ik} = \overline{B}_{ik} \cdot \overline{s} \; . \tag{6}$$

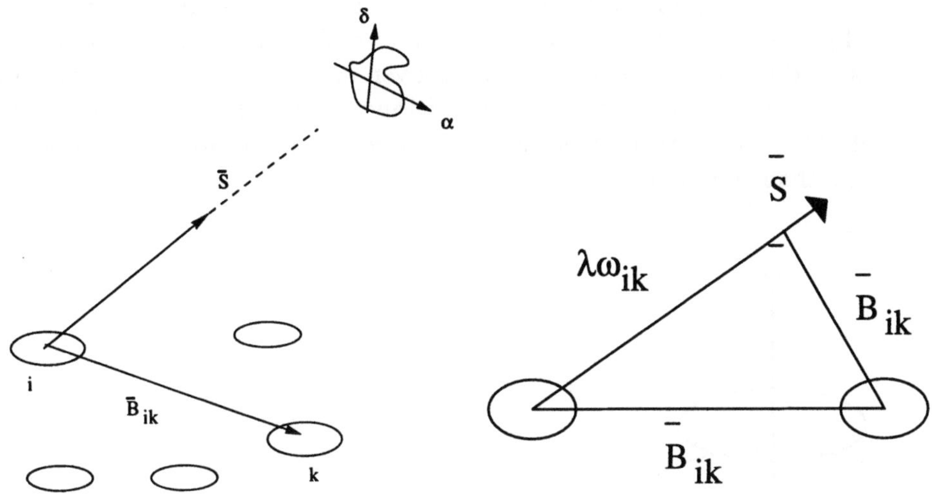

Figure 2. Left: geometry of a baseline. Right: geometry of projected baseline.

Here, the symbol "·" denotes a scalar product. It is easy to see that

$$\overline{B\prime}_{ik} = \overline{B}_{ik} - \lambda w_{ik} \cdot \overline{s} \,. \tag{7}$$

As already stated above, the projected baseline changes with time. For a given latitude ξ of the array, the region scanned by the projected baseline in Fourier space depends on the position of the source - its declination δ and its hour angle h (sidereal time minus right ascension). Both the angular frequency $\overline{u}_{ik} = \begin{pmatrix} u_{ik} \\ v_{ik} \end{pmatrix}$ and the geometric delay, expressed in wavelengths, w_{ik}, can be conveniently combined into a single equation. We express the position of the point where the two axes of telescope i intersect, in the northwards direction by X_i, in the eastward direction with Y_i, and zenithwards with Z_i. We introduce the terms

$$\begin{aligned} c_\xi &= \cos\xi & s_\xi &= \sin\xi \\ c_\delta &= \cos\delta & s_\delta &= \sin\delta \\ c_h &= \cos h & s_h &= \sin h \end{aligned}$$

The Fourier and delay coordinates are given with

$$\begin{pmatrix} u_{ik} \\ v_{ik} \\ w_{ik} \end{pmatrix} = \frac{1}{\lambda} \begin{pmatrix} -c_\xi s_h - s_\xi c_\delta c_h & s_\xi c_\delta s_h - c_\xi c_h & -s_\xi s_\delta \\ -s_\delta c_h & s_\delta s_h & c_\delta \\ -s_\xi s_h - c_\xi c_\delta c_h & c_\xi c_\delta c_h - s_\xi c_h & -c_\xi s_\delta \end{pmatrix} \begin{pmatrix} X_k - X_i \\ Y_k - Y_i \\ Z_k - Z_i \end{pmatrix} \tag{8}$$

The loci of $\overline{u}_{ik}$ in the Fourier domain as a function of hour angle lie on an ellipse with an excentricity $\epsilon = \cos\delta$, see also fig. 3. Only part of the ellipse will be actually traced during a given night, depending on dusk and dawn, and limitations of the zenith distance to which the telescopes can point. There are further constraints if the source is variable on time scales shorter than the night.

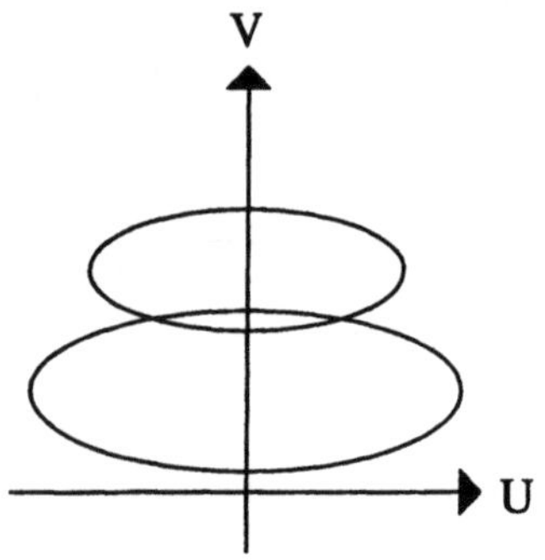

Figure 3. UV plane tracks of two baselines.

If there are more than two telescopes contributing to the array, there is a baseline which connects each pair, and a corresponding time-varying projected baseline. For an array with N elements (telescopes), there are $\frac{N}{2}(N-1)$ baselines, i. e., their count rises with the square of the number of telescopes in the array. An array is called *non-redundant* if each vector separation of telescopes is unique. In this case, each baseline will scan a different trace in Fourier space and hence increase the amount of unique information which is gained during observing. A baseline is called *redundant* if there is at least one more identical or nearly identical vector separation between another pair of telescopes. In order to use telescopes efficiently, one normally seeks to avoid redundant baselines in an array, but there are exceptions (see below).

The Very Large Telescope, which consists of four 8m telescopes on a single mountain, constitutes an interferometric array. Figure 4 shows the Fourier domain coverage which can be achieved with the VLT, assuming a source at -30 degrees declination and a full night of obervations.

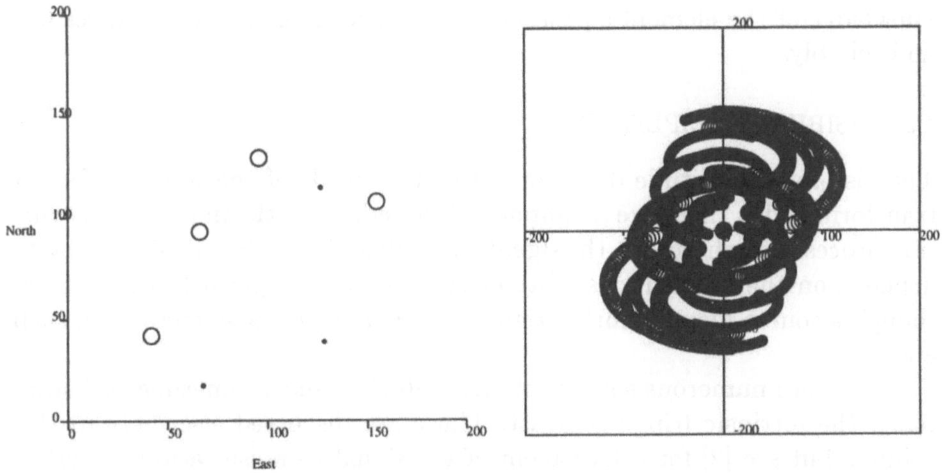

Figure 4. **Left: array layout, viewed from zenith, of the VLT 8m Unit telescopes (large circles) and three 1.8m Auxiliary telescopes (small circles). Right: the corresponding coverage of the UV plane for earth rotational synthesis of a source with a declination of $\delta = -30°$. Wide"sausages" represent baselines between 8m telescopes, medium size tracks baselines between 8m and 1.8m telescopes, narrow tracks between 1.8m telescopes.**

4. Observables

We now consider the essential observables of an interferometric experiment and their relation to imaging. These are the following:

- optical delay between the elements of the array,
- visibility amplitude $|V_{ik}|$,
- referenced visibility phase,
- closure (bispectrum) phase.

4.1. OPTICAL DELAY

Each element has an adjustable delay which is used to cancel differences in the optical path through the interferometer for a given position in the sky. The origins for optical path differences are the geometric delay already described above, any fixed path difference due to the configuration of the array, and random fluctuations which are of instrumental (vibrations) and atmospheric (seeing) origin. The source's position can be determined from the delay data provided that static and random instrumental fluctuations of the optical path are well understood and that the positions of the array elements are precisely known. This is important for astrometry, but of little consequence for imaging, where precise stabilisation of fluctuations of optical path are important. These fluctuations would otherwise disturb a precise measurement of the source's Fourier components. A fairly good knowledge of the element's positions are needed mainly to find fringes fast and reliably.

4.2. VISIBILITY AMPLITUDE

The visibility amplitude determines the magnitude of the source's Fourier transform and is therefore an important ingredient in the image reconstruction process. Furthermore, the signal strength and therefore its detectability depends on the intrinsic visibility of the source at a given baseline. Even though a source may be fairly bright, fringes may not be detectable in such a case.

There are numerous sources of error which reduce, sometimes substantially, the intrinsic fringe contrast. Those are discussed elsewhere in this volume, but see [2] for a treatment of the signal to noise ratio in interferometers. It is important that these errors are well calibrated in order to be able to do image recovery with at least some photometric fidelity.

4.3. REFERENCED PHASES

The visibility phase, the argument of the complex Fourier transform, is the second important ingredient in the image reconstruction. A direct inversion of the data would be impossible without phases. However, atmospheric and instrumental variations do not permit a direct measurement of the source's Fourier transform phase from the measured visibility phases, which therefore are no good observables. Consider the measured phase for a given baseline ik in the presence of element-dependent optical delay errors ϕ_i:

$$\psi_{ik} = \psi_{0,ik} + \phi_k - \phi_i \tag{9}$$

The element-dependent errors, which often are much larger than 2π, make it impossible to retrieve the Fourier phase of the observed source, $\psi_{0,ik}$, from the measured phase. Furthermore, the measured phase ψ_{ik} can be changed by shifting the origin of the coordinates without affecting the morphology of the reconstructed source at all. If one disregards the position of the image, there is a set of solutions for the visibility phase which do not change the source's shape. Fortunately, this dependence is linear with the image's position and therefore easily separated.

A reference source removes this ambiguity. It is necessary to compare the measured visibility phase with that of a near-by reference source whose Fourier phase is known – ideally a point-like source for which the intrinsic phase is zero for all baselines. The comparison can be done by measuring both (program and reference sources) visibilities simultaneously. In some cases, the target under investigation itself can serve as a reference at a different wavelength. This would be the case if the target was unresolved at continuum wavelengths but appears resolved in spectral lines when observing fringes with sufficient spectral resolution. The Fourier phase then varies with wavelength.

4.4. CLOSURE PHASES

Closure phases are obtained by adding measured visibility phases along a closed path along baselines between array elements. The simplest meaningful closed path involves three elements (adding the phases of a closed path between two elements would result in a phase which is always zero). It is easily seen that all element-dependent errors ϕ_i cancel within a three element closed path:

$$\begin{aligned}\psi_{ikl} &= \psi_{ik} + \psi_{kl} + \psi_{li} \\ &= \psi_{0,ik} + \phi_k - \phi_i + \\ &\quad \psi_{0,kl} + \phi_l - \phi_k + \\ &\quad \psi_{0,li} + \phi_i - \phi_l \\ &= \psi_{0,ik} + \psi_{0,kl} + \psi_{0,li}\end{aligned} \tag{10}$$

The same would be true for closed paths with more than three elements, although those are practically never used, for reasons of a reduced signal-to-noise ratio. Since element-dependent errors which are usually not directly observable are eliminated, closure phases are in general good observables. Closure phases are insensitive to the position of the source.

An array with several elements may allow for many closure relations. However, those are not independent, e. g., for a 4 - element array:

$$\psi_{123} = \psi_{124} - \psi_{134} + \psi_{234} \tag{11}$$

4.5. HOW MANY GOOD OBSERVABLES DO WE HAVE?

The number of baselines increases with the square of the number of elements N of an interferometer. The number of closure relations betwen three elements even increases with N^3. However, the number of *independent* closure relations between three elements only increases with N^2, and they are always less than the number of baselines. Therefore, the high degree of redundancy of closure phases can be exploited only to improve the robustness of a recovery technique, but not to increase the amount of independent information. The number of observables are given in Table 1.

Since the number of independent closure phases is always less than the number of baselines in a non-redundant array, there is in general no possibility to recover the visibility phase completely from closure phases.

Array elements	Baselines (vis. amplitude)	Closure phases	Independent closure phases
N	$\frac{N}{2}(N-1)$	$\frac{N}{6}(N-1)(N-2)$	$\frac{N}{2}(N-1)-(N-1)$
2	1	0	0
3	3	1	1
4	6	4	3
5	10	10	6
6	15	20	10
etc.	etc.	etc.	etc.

TABLE 1. Number of observables of an N element array.

However, if the array is redundant or the number of elements is small, the closure phase may complete the missing information. Consider an array of three elements which are spread in a two-dimensional arrangement (i. e., not on the same line, cf. fig 5, left). The observed visibility phase can then be made to zero by adjustment of the source's position in two directions. The single observed closure phase is then equal to the object phase on the third baseline. Of course, the information collected this way is very limited, and the technique would not apply to rotational synthesis mapping because the source can be positioned only once and there is no phase connection of the information between observations taken at different aspect angles.

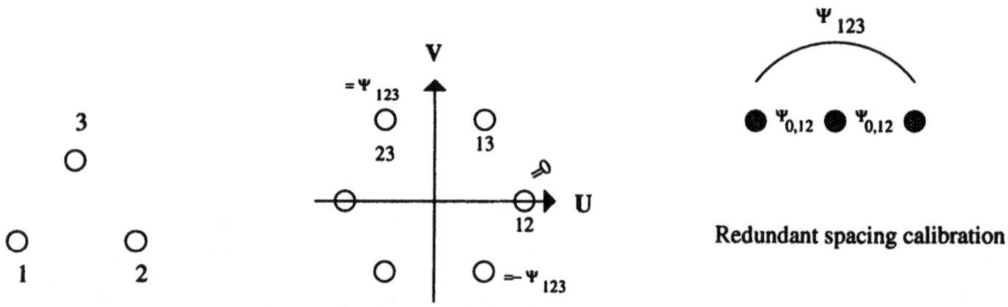

Figure 5. Left: a three-element array. Middle: the UV coverage of the three-element array showing the signal; areas of all three baselines. Right: a linear, redundant three-element array.

Another way to exploit phase closures is redundant spacings calibration. Here, the array's configuration does have some redundant baselines and beam combination must take place in such a way that the signal from different redundant baselines can be distinguished. For example, consider an array which consists again of three elements, this time arranged on a line (see fig 5, right). If the separation between the first and the second element is the same as between the second and third, the object phase on the corresponding two baselines will be equal and can be made to zero by chosing a suitable origin of the image space coordinate system. The closure phase is again equal to the object phase for the third baseline. This technique can be iterated in a succession of linear arrays. Figure 6 shows the locations of ten stations, from which Auxiliary telescopes of the Very Large Telescope Interferometer can operate, at the top. There are 17 configurations of three telescopes between the ten stations shown near the bottom of the figure. Within each configuration, two out of three baselines are equal to baselines from a configuration measured earlier (except for the first one, of course). Starting with the redundant configuration of three telescopes just described, this sequence of configurations is capable to measure all object phases for baselines equal to integer multiples of the shortest baseline, up to a factor 18. It should be noted that this is only possible for one orientation of the baseline with respect to the source. Auxiliary data taken with a single dish telescope may serve to establish the phase relation between observations with different aspect angles.

4.6. BASELINE DEPENDENT ERRORS

Phase closures provide reliable information on the object phases only if the errors depend only on the elements of the array, and therefore cancel in the closure equation. As soon as there are errors which depend on a baseline rather than on an element, closure phases may no longer be useful. It is therefore important to avoid baseline-dependent errors. An easy way to introduce baseline dependent errors is with a pair-wise beamcombination design when the interferometer has more than two elements.

Consider the beamcombiner shown in Figure 7 which shows a fiber-optical combiner with two and three elements (baseline-dependent errors occur with bulk optics beam combiners in exactly the same manner, but fibers are easier to draw). The light enters the fibers and first encounter a Y coupler where part of their light is divided off to make an intensity measurement. An X coupler then mixes the two beams of the two-element combiner. If ϕ denotes the phase between the two beams, then the visibility magnitude V is given with

$$V\cos(\phi) = \frac{C_1 - C_2}{4\sqrt{I_1 I_2}} \tag{12}$$

where I_1, I_2, C_1 and C_2 are the intensities measured at the appropriate

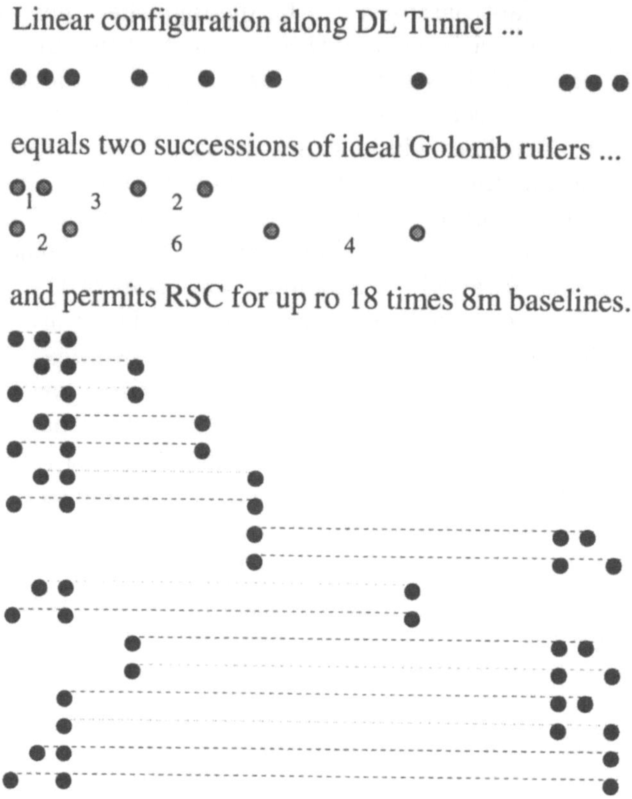

Figure 6. **Configuration of ten stations for Auxiliary telescopes along the delay line tunnel of the ESO VLT Interferometer (top). A sequence of 17 redundant-spacings calibration (RSC) configurations (bottom).**

fiber outputs. For the three-element beamcombiner, one would use another Y coupler to split the light of each element two ways to be combined with the other two elements. It is on these paths, between the splitting and the recombination, where arbitrary delays may enter the system and produce errors which would not cancel in a closure phase. Therefore, a separate metrology system would be required which continuously measures the system closure phase so the object closure phase could be calibrated.

It is important to perform beam combination at an as late a stage as possible in order to avoid closure phase errors. The combination of several parallel, collimated beams with a lens in an image plane conjugate is a proven concept (fig. 8).

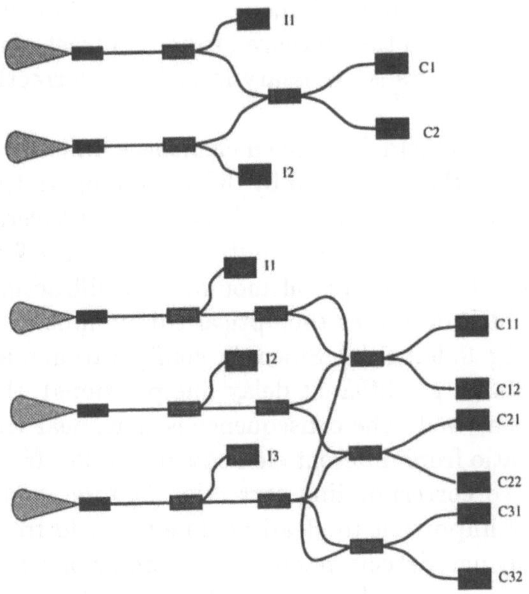

Figure 7. A two-element (top) and three-element (bottom) beamcombiner with optical fibers.

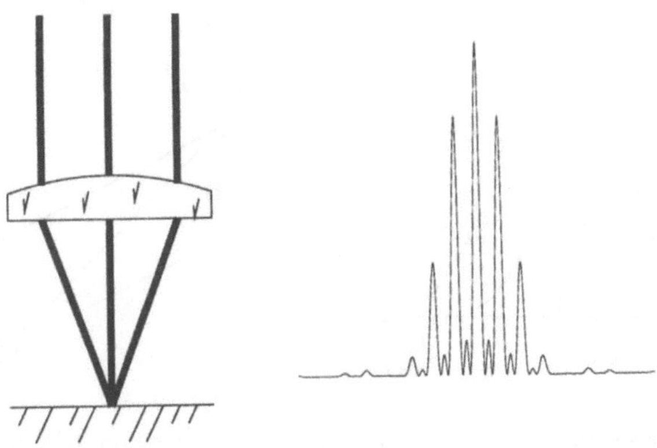

Figure 8. The collimated beams of three array elements are combined on a detector in the focal plane of a lens (left) which forms an image of the source with fringes (right).

5. Effects of an Extended Field of View

We shall now return to the set $\mathcal{F}$ of coordinates which appeared in the Fourier integral in eqn. 1 and consider the effects of an extended source. Contrary to radio interferometers, an optical interferometer can instantaneously accept the radiation from a field which may exceed substantially the field covered by the point response of a single element. Sources within this field may make different contributions to the measured fringe contrast, and a thorough understanding of their effects is necessary in order to correctly interpret the measurements.

The optical throughput across the field for a given element is limited by vignetting due to the limited size of the optics and by diffraction due to the length of the propagation which may sometimes be of the order of several hundred meters. The unvignetted field is usually quite small, i. e., a few arcsec. The consequences of vignetting by optical mounts and diffraction is a smooth "taper function" which describes the optical throughput as a function of field coordinate (see fig. 9, left side). Since the configuration may be different for different elements (e. g., different delay line positions), the taper functions may be different as well. The consequence is that, near the edges of the field, the intensity ratio from different elements may differ from that at the center of the field, so corresponding correction factors would not be applicable. It is therefore important to clearly define the field from which light is detected by the fringe detection system. A good approach would be a common field stop in an image plane conjugate where all beams are properly superimposed.

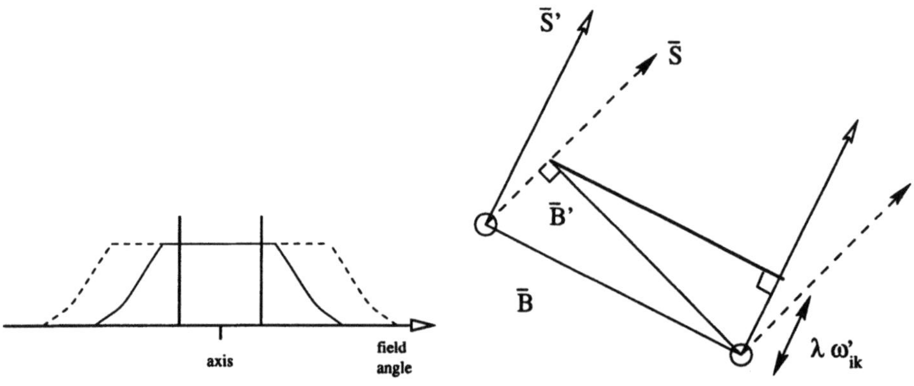

Figure 9. Left: vignetting taper function for two array elements with different positions of the delay line (solid and dashed). A field stop (vertical lines) limits the common field seen by the detector. Right: geometry of the field delay for a given baseline.

Another effect to consider is that the fringe phase depends on the position of the source in the field. Consider the right side of fig. 9 which shows the baseline $\overline{B}$ for two elements i, k, and the projected baseline $\overline{B'}$ in the direction $\overline{S}$ of the origin of the field coordinate system. Suppose that the internal and geometric delays have been canceled in the direction $\overline{S}$ by a delay line. A source at the position $\overline{S'}$ then suffers a differential geometric delay

$$\lambda\omega'_{ik} = \overline{B}' \cdot (\overline{S}' - \overline{S}). \tag{13}$$

Since the angle between the directions $\overline{S}$ and $\overline{S'}$ are quite small, one can approximate the vector angle $\overline{\alpha}$ between the directions by

$$\overline{\alpha} = \overline{S}' - \overline{S} \tag{14}$$

which yields a phase delay of

$$\omega\prime_{ik} = \frac{1}{\lambda}\overline{B}' \cdot \overline{\alpha} = \overline{u}_{ik} \cdot \overline{\alpha} \tag{15}$$

Of course, the phase delay depends on the wavelength λ of the light. We note that the fringe phase is a function of the position of the source in the field.

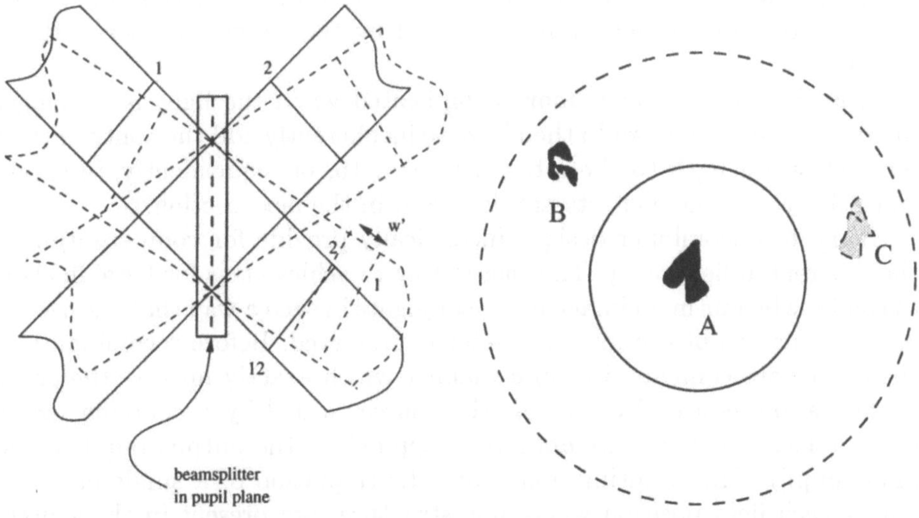

Figure 10. **Left: a beamcombiner plate located at the intersection of collimated beams from two telescopes with transferred pupils overlapping. Beams from two sources at different field angles are shown. For the on-axis source (solid lines), the optical path difference is cancelled, while the off-axis source (dashed lines) experiences a differential delay of w'. Right: field stops must assure that only the light from the sources of interest reach the fringe detector - in this case, source A only.**

To illustrate which effect the differential delay has on the measured fringe visibility, let us consider monochromatic two-element beamcombination in a pupil plane (fig. 10 left). The pupil is an axial location where the collimated beams for all field positions intersect. Superimposing the beams in a pupil therefore assures that all sources in the accepted field participate in the interference in the same way. The fringe signal is obtained by modulation of the optical path difference between the participating elements and by monitoring the resulting intensity variations after the beamcombiner. The differential delay (eqn. 15) will cause corresponding shifts of the fringe position for off-axis sources. The measured fringe contrast is proportional to the incoherent superposition of the fringes which result from all detected sources. If $I(\overline{\alpha})$ denotes the true intensity distribution within the observed field $\mathcal{F}$, we obtain

$$V_{ik} = \frac{\int\int_{\mathcal{F}} I(\overline{\alpha}) \exp \overline{u}_{ik} \cdot \overline{\alpha} d\alpha}{\int\int_{\mathcal{F}} I(\overline{\alpha}) d\alpha} = \frac{\hat{I}_\lambda(\overline{u}_{ik})}{\hat{I}_\lambda(0)} \tag{16}$$

which is exactly the normalized Fourier transform at the angular frequency $\overline{u}_{ik}$ of the intensity distribution within the accepted field - which is the output expected from the interferometer! If one is interested in imaging only a particular part of the source, care must be taken that light from other sources in the field is not detected by the fringe detectors (fig. 10 right).

The situation becomes more complicated when the light is no longer monochromatic. One would then have to incoherently add the contributions for each wavelength to the integral in eqn. 16, or, equivalently, integrate eqn. 16 over λ. These effects are discussed in the next section.

Some beamcombiner designs intrinsically provide for compensation of the differential field delay. The easiest way to achieve this for large fields is to combine beams in an image plane conjugate in such a way that the overall optical constant of the interferometer is preserved. Before recombination, the beams are configured in in a manner which exactly mirrors the array configuration as seen by the celestial source, scaled by a common factor (this is also called the "homothetic mapping" of the output pupil to the input pupil). The resulting image has the resolution of a single element, but at each field position where fine structures are present in the source, there will be fringes. This way one can map the visibility of a source over areas which correspond to the resolution element of a single element, as a function of much larger field angles.

6. Effects of Finite Bandwidth

We now consider the effect of a finite bandwidth on the detected fringe signal. To this end, let us repeat the Fourier transform in eqn. 3 in a slightly different manner in order to emphasize the wavelength dependence.

$$\hat{I}_\lambda(\overline{u}_{ik}) = \int\int_{\mathcal{F}} I_\lambda(\overline{\alpha}) \exp 2\pi j \overline{\alpha} \cdot \overline{u}_{ik} \, d\overline{\alpha} = \int\int_{\mathcal{F}} I_\lambda(\overline{\alpha}) \exp 2\pi j \overline{\alpha} \cdot \frac{\overline{B}_{ik}}{\lambda} \, d\overline{\alpha} \tag{17}$$

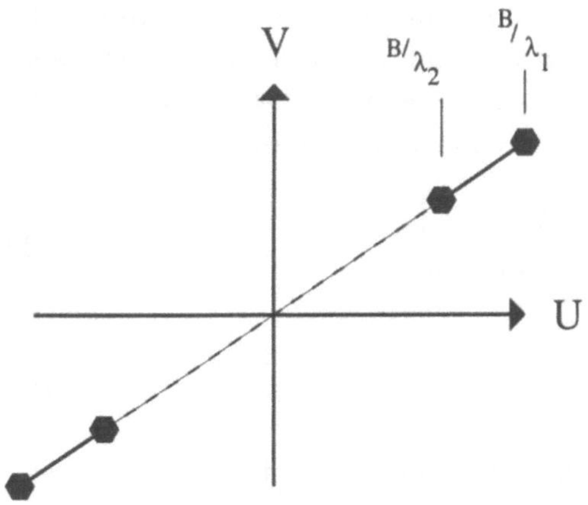

Figure 11. Loci of positions in the UV plane covered by a finite spectral range.

For a given measurement, the projected baseline $\overline{B}_{ik}$ is constant, hence the angular frequency $\overline{u}_{ik}$ varies inversely proportional to the wavelength. If the measurement covers the range from λ_1 to $\lambda_2 > \lambda_1$, the actually observed fringe contrast will be given by the incoherent sum of eqn. 3 for all wavelengths:

$$\hat{I}_{ik} = \int_{\lambda_1}^{\lambda_2} \hat{I}_\lambda(\overline{u}_{ik}) d\lambda \tag{18}$$

and

$$V_{ik} = \frac{\int_{\lambda_1}^{\lambda_2} \hat{I}_\lambda(\overline{u}_{ik}) d\lambda}{\int_{\lambda_1}^{\lambda_2} \hat{I}_\lambda(0) d\lambda}. \tag{19}$$

For any given measurement, the loci which contribute to the fringe signal lie on two radial lines centered on the origin in the UV plane (see fig. 11). In general, the object visibility along these lines will vary in phase and amplitude. The integrated visibility in eqn. 19 will be reduced and not be a good representation of the object visibility at, say, the central wavelength when the measurement is not spectrally resolved. If measurements are taken with sufficient wavelength resolution (by, e. g., using a prism or grating spectrometer), the contribution of different spectral channels to the visibility function can be distinguished and used advantageously for reconstructing an image in each channel. In order to do this, several conditions must be met. The spectral resolution must be good enough that the variation of the visibility due to the source remains essentially constant across the area in the UV plane which is covered by a spectral channel. This is the case when the source has a limited support (the source intensity vanishes outside a confined area). Let $\delta\alpha$ be the largest extent of the source in units of radians. The Fourier transform of the source distribution will show variations at scales of order $1/\delta\alpha$ line pairs per radian and above. Let $\Delta\lambda$ denote the width of the spectral channel at wavelength λ, the spectral channel then extends from B/λ to $B/(\lambda + \Delta\lambda)$. If we require that the source's Fourier transform does not significantly vary over scales of order $1/\delta\alpha$, then, obviously

$$\frac{B}{\lambda} - \frac{B}{\lambda + \Delta\lambda} \geq \frac{1}{\delta\alpha} \tag{20}$$

or, equivalently,

$$\frac{\lambda + \Delta\lambda}{\Delta\lambda} \frac{\lambda}{B} \leq \delta\alpha \ . \tag{21}$$

We introduce the *spectral resolution* of the measurement

$$\mathcal{R}_\lambda = \frac{\lambda + \cdot\lambda}{\cdot\lambda} \ . \tag{22}$$

Eqn. 21 means simply that, in order to avoid visibility loss due to spectral smearing, the spectral resolution at wavelength λ, $\mathcal{R}_\lambda$, times the fringe period of the baseline at wavelength λ must be at least equal to the extent of the source.

This result warrants some discussion. One can assume that interferometers will use the broadest spectral bands possible when observing faint sources in order to optimize the detected flux. Atmospheric transmission windows in the near infra-red spectral regime (Johnson I, J, H, and K bands) correspond to a spectral resolution of 3 ... 5. Using the full bands therefore requires that the source is only marginally resolved with the interferometer in order to measure valid visibilities. For instance, a source observed with 100m baselines in the I band should not be much larger than 3 milli-arcsec to be observed with a resolution of 1 milli-arcsec.

In general, the spectral resolution will be significantly better. The factor which then limits the reconstructed field is the coverage of the UV plane and the residuals from the inversion and mapping processes.

7. From Observables to Images

Once one has observed and calibrated a set of complex visibility data for a given source, the remaining task is to subject them to astrophysical interpretation. The subject of this lecture is imaging, so we shall discuss how to generate an image of the source which best fits the available data. The whole issue of imaging with interferometric data warrants a textbook of its own and we will be able only to address the very basic topics. This problem has been addressed since a long time by astronomers who do Radio interferometry, and very much of their results is directly applicable to

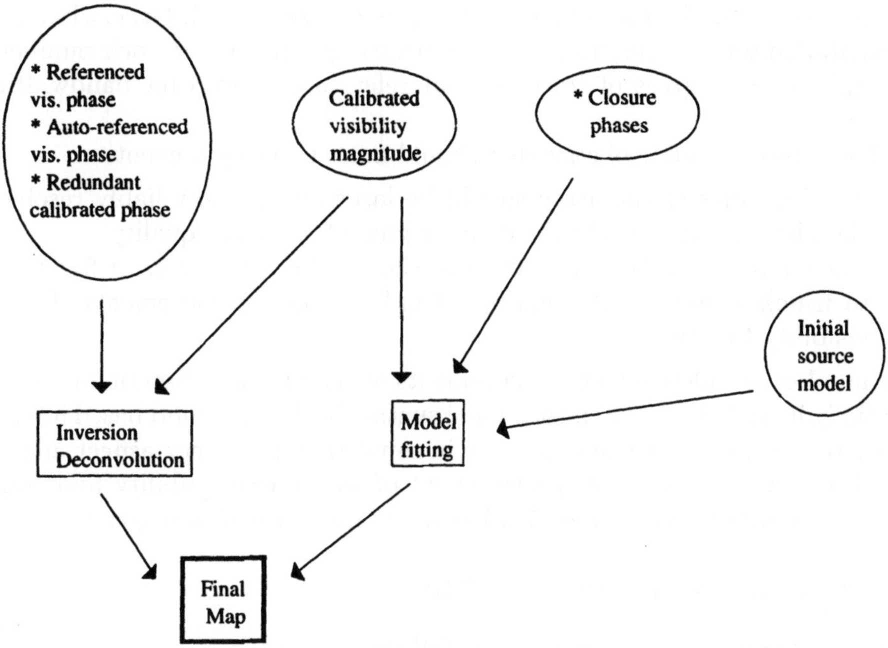

Figure 12. Relation between observables and the map.

interferometry at optical wavelengths. Textbooks on imaging with Radio interferometers are therefore excellent references [3].

We will discuss two basic techniques which permit generating images from interferometry data. These are inversion/deconvolution and model fitting. Fig. 12 shows these two techniques and their relation with the input observables. Before we set out describing them, a big caveat is in order.

One always has to bear in mind that interferometric data are limited. Sometimes, e. g., when there is only a single baseline (which is the case for most contemporary optical interferometers), the data are very limited and permit very limited information about the object to be gained. Such data permit only an estimate of the dimension of the observed source. In general, the results derived from interferometric data are easily over-interpreted and should always be taken with a lot of caution.

Any source can be characterised by the following three parameters which are essential for imaging: the scale of the phenomenon of interest (small scale structure), the overall dimension of the source (extent of source's support region), and the "number of degrees of freedom" in the image (the number of independent parameters which fully describes the source at the desired resolution). If the source is connected and compact, the number of degrees of freedom is roughly equal to the square of the dimension of the source divided by the phenomenon scale (see also section 2). Otherwise, for a distributed source consisting of a collection of point sources, their number determines the degrees of freedom (but refer to section 6 for bandwidth effects).

The interferometric observation should meet two requirements:

1. The baselines of the array should be large enough to reliably resolve the phenomenon of interest with a signal of sufficient quality.
2. The source must be constrained in the number of degrees of freedom to **much less** than the number of independent measurements of the visibility function.

The number of independent measurements of the visibility function is substantially increased by Earth rotational synthesis with a ground based interferometer. However, one should bear in mind that not every aspect angle constitutes an independent measurement of the source visibility function since the source Fourier transform has a certain correlation length.

7.1. INVERSION AND DECONVOLUTION

The mapping of a source involves the following steps:

1. The starting point is a calibrated set of complex visibility data $\hat{I}(\overline{u}_i)$ (both Fourier amplitudes and phases need to be known for all baselines).
2. The visibility data are interpolated on a regular grid. Areas in the Fourier domain which are not covered by a baseline are set to zero. The resulting "gridded" visibility array is subject to a two-dimensional Fast Fourier transform. Alternatively, if the number of data points is small, a source map estimate is produced by adding the Fourier inverse of all elements in the list of visibility data.
3. The result of either procedure, the "dirty map" $\tilde{I}(\overline{\alpha})$, can be described by a Fourier integral:

$$\tilde{I}(\overline{\alpha}) = \int\int W(\overline{u})\hat{I}(\overline{u}) \exp -2\pi j\overline{u}\cdot\overline{\alpha}\, d\overline{u} , \tag{23}$$

where W is a weighting function which vanishes for all angular frequencies which are not covered by the array,

$$W(\overline{u}) \equiv \begin{cases} 1, \text{ if } \overline{u} \in \{\overline{u}_{ik}\} \\ 0 \text{ otherwise} \end{cases} \tag{24}$$

The "dirty map" constitutes a representation of the source which fits the available data, but contains artifacts caused by the incomplete coverage of Fourier space by the array baselines.

4. The Fourier transform of the weighting function (eqn. 24) is computed:

$$A(\overline{\alpha}) = \int\int W(\overline{u}) \; \exp -2\pi j \overline{u} \cdot \overline{\alpha} \; d\overline{u} \; , \tag{25}$$

   The result $A(\overline{u})$ is called the "dirty beam". This is the estimated response of the synthesized array to a point source, whose visibility function would be unity for all baselines.
5. A deconvolution step remove as much as possible the artifacts of the dirty beam from the dirty map.

The generation of the dirty map and the dirty beam is referred to as "inversion" of the visibility data. The artifacts in the dirty maps are due to the data missing in the gaps in the Fourier domain which are not covered by baselines. The purpose of the deconvolution step is in simple terms to fill in those gaps with reasonable data. Generally, the gaps could be filled with any set of random numbers as long as the baseline data remain untouched. This freedom would result in a manifold of possible inversions, all of which are consistent with the data, and which would be indistinguishable. Fortunately, there are also other constraints. For instance, there may be information about the support of the source, i. e., the set of coordinates for which the source intensity does not vanish. Only a very small subset of the manyfold would be consistent with the support. Another important constraint is that the image should be consistent with a physical source, i. e., it cannot be negative. In other terms, its spectrum must have a reduced spectrum (eqn. 3). The latter turns out to be a very powerful constraint, but can be deceiving if the data contains noise, so it should be used as a strong constraint mainly for high quality data.

The deconvolution process in synthesis imaging is different from deconvolution applied to data which are taken with filled aperture telescopes. In the latter case, there usually is complete information present up to the diffraction limit of the telescope, possibly distorted in amplitude and phase. Cases where Fourier information is missing at frequencies below the diffraction limit are already considered ill-posed. Simple deconvolution techniques apply Wiener filtering, i. e., a re-weighting of the signal according to a known modulation transfer function while considering the signal-to-noise ratio. This method cannot be applied to synthesis imaging.

7.2. A SIMPLE DECONVOLUTION TECHNIQUE: HÖGBOM'S CLEAN ALGORITHM

The CLEAN algorithm was invented by J. Högbom in 1974 and has since then improved considerably. It is conceptually simple, easy to implement, and quite intuitive. Therefore it serves here as a key example to illuminate the deconvolution process.

It is important to realise that this algorithm implicitely introduces a model for the source to be composed of an arbitrary number of point sources. It is this assumption which helps the algorithm to fill in missing information into the gaps in the Fourier domain, because point sources have a very well-behaved Fourier transform. The dirty map is understood as the convolution of the point sources with the dirty beam and can be represented as the sum of dirty beams for each point source, at its position and with the appropriate strength. The goal therefore is to replace each dirty beam in the map by a point source. The algorithm is iterative and operates in image space ("map domain"). It starts with the original dirty map and involves the following steps:

1. Find the strength and the position of the current maximum in the dirty map.
2. Scale the dirty beam so it matches the strength of the maximum. Multiply scaled beam by a "loop gain" factor which is usually considerably less than 1. Subtract the result at the position of the maximum from the dirty map.
3. Make an entry of the strength times loop gain and the position coordinates in a list of point sources.
4. Examine the remaining dirty map and decide whether the remaining peaks still exceed the noise in the map. If this is the case, return to step 1. Otherwise, proceed with step 5.
5. Generate a two-dimensional representation of the point sources in the list and convolve the result with a function which represents the resolution of the array (e. g., a Gaussian with the width of the core of the dirty beam).
6. In order to judge the fidelity of the outcome of the steps above, it is advisable to add the residuals from the iteration (steps 1 ... 4) to the final result. This allows one to judge the significance of weaker peaks.

The algorithm works well on a collection of point sources and an array which provides many baselines. Because of its nature, it is not very suitable for sources which contain diffuse, extended structure. Cleaned maps of such objects tend to have a clumpy appearance. It helps to use a small loop gain and many iterations under these circumstances. However, the main problem may not be the algorithm *per se*, but rather a common shortcoming of synthesis arrays. The Fourier domain coverage with short baselines (just above the cutoff of a single telescope) is usually not there, because telescopes cannot look through each other's domes. This is the domain where extended sources contribute their signal, so lack of coverage hurts most. There are modifications to CLEAN ("multi-resolution CLEAN") which endeavour to improve the performance on extended sources.

Figs. 13 and 14 show some examples of mapping with the VLT Interferometer, using CLEAN for the deconvolution.

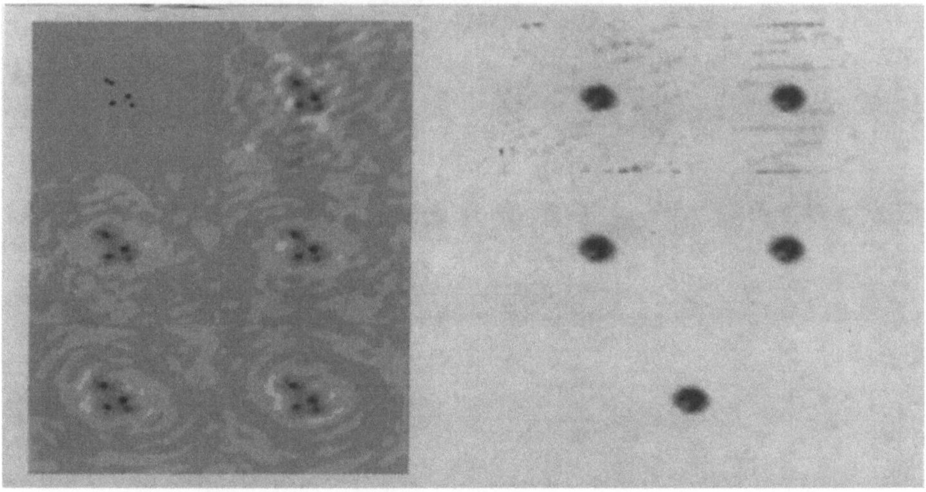

Figure 13. Left: model of a small stellar cluster (top left) and CLEAN maps of high SNR observations with the VLTI (four 8m and three 1.8m telescopes) at source declinations of +15 (top right), 0 (middle left), -15 (middle right), -35 (bottom left) and -65 degrees (bottom right), to show the influence of UV coverage on the mapping of a simple object. Right: CLEAN maps of high SNR observations of a maculated stellar disk with the VLTI at source declinations of +15 (top left), 0 (top right), -15 (middle left), -35 (middle right) and -65 degrees (bottom), to illustrate the influence of UV coverage on the mapping of an extended object.See fig. 14 for the model of the star (rotated by 90 degrees).

7.3. MODEL FITTING

In many cases, and most certainly with optical synthesis arrays, the number of simultaneous basclines will be small. Rather than making an attempt to invert the data in order to obtain maps of the source, fitting a model of the source to the data in the least squares sense may be the better approach. It is here where the concept of a limited number of degrees of freedom in an image becomes most apparent.

When fitting models, sources are described by a collection of simple shapes with few parameters. For example, a delta-function like point source is characterised by two coordinate parameters and the value of its integral. An extended "point" source may have a circularly Gaussian shape, adding a single width parameter to the previous three. More extended sources include asymmetries, like an elliptical Gaussian where in addition, the ratio of minor to major axes and the orientation must be specified. Simple shapes like these serve to generate models of more complicated sources. Any prior knowledge about the source helps in constraining the outcome of the fitting

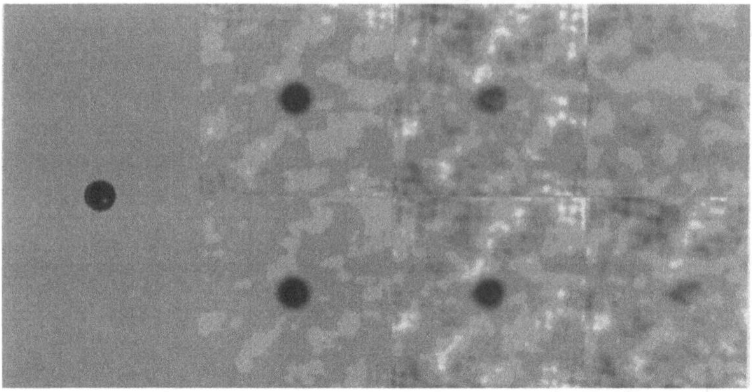

Figure 14. Model (far left) and CLEAN maps of observations of a stellar disk with the VLTI and various levels of *rms* phase noise in degrees: 10 (bottom near left), 30 (top near left), 60 (bottom near right), 90 (top near right), 120 (bottom rat right) and 180 (top far right), to illustrate the influence of phase noise on the mapping of an extended object.

process. For example, the number of components in a multiple system or the presence of an extended disk may already be known before the analysis of the observations. Inspection of the direct visibility data or a cleaned map may also help. A good starting model is critical for the success of the fitting process.

Once a list of components and their starting parameters is established, one can define the parameters which are to be fitted. Some parameters may be well known, e. g. the total flux of the source or flux or relative positions of some of the components, it is a good idea to leave them fixed during the first stages of iteration. The others can then be subject to any of a number of linear or non-linear least-squares fitting algorithms. An estimate of the Fourier transform of the source is computed from the model for the baselines which have been observed. Visibility moduli and closure phases are calculated from the Fourier data, and are compared with the measured visibilities and closure phases. The parameters are then varied to improve the fit. The goal of the fit is to minimize the discrepancies between modelled and observed visibility data to a point where they are consistent with the noise. If such a fit cannot be obtained, the starting model may be wrong, in that components of the source are missing. Then it helps to gradually modify the model by adding more components.

It is clear that the model fitting process relies on prior knowledge, otherwise the results would depend on guesswork. In most cases, however, one has a clear picture about the nature of the source which helps in judging the fidelity of a model. Care should also be taken that the number of parameters remains limited. It is always possible to fit a model with many parameters to any set of data, but one should have no illusions about the quality of the final result.

Figures 15 and 16 show examples of model fitting. The simulated source consists of a point-like object and an off-center, extended structure which could represent a central star and a surrounding oblique disk (fig. 15). The ratio of "disk" to "star" fluxes are 10, 1, and 0.1, from left to right. The top row shows a representation of the models while the bottom row shows the corresponding CLEANed maps. It is seen that recovering extended structure like the disk poses considerable difficulties to the CLEAN algorithm.

These sources can be represented with a simple model. The degrees of freedom are the width of the central starlike source, the position, ellipticity and orientation of the disk, and the ratio of the disk and star fluxes. Introducing an initial model (fig.16 middle) into the iterative fitting process may result in a fit (fig.16 bottom) which closely resembles the "truth" object (fig.16 top) even for the difficult case where the disk flux is only one tenth of that of the central source. In this case, inversion and deconvolution is unable to recover the disk.

8. Conclusions

This lecture introduces the basic concepts for generating maps of astronomical sources with an interferometer operating at optical and infrared wavelengths. The imaging algorithms described here were inherited from radio interferometry. The similarity between radio and optical interferometry cover all areas which relate to distributed arrays, including incomplete information in the UV plane and the resulting imaging and deconvolution techniques. As far as imaging is concerned, the major difference to Radio interferometry is the sensitivity of the detection to an extended field of sources, not all of which may participate in the interference process.

Imaging with an optical array is, at the time of this writing, in its infancy. Several high resolution images of simply structured sources have

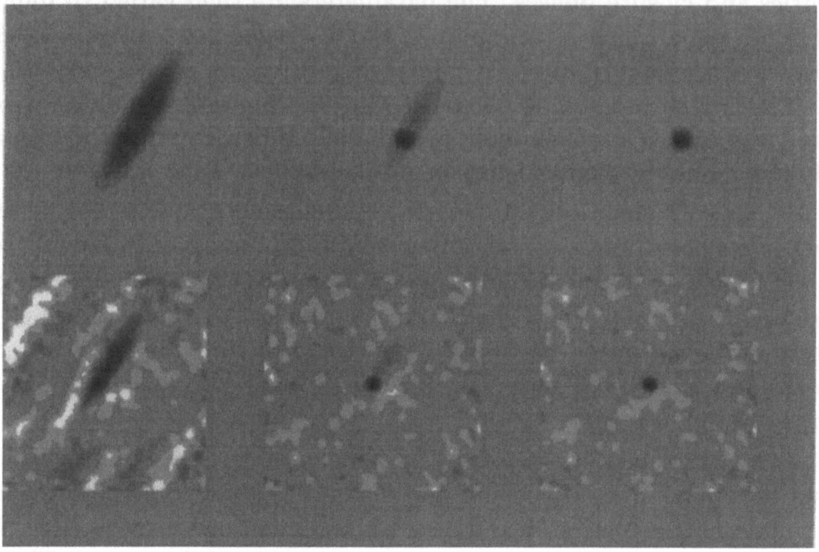

Figure 15. Models (top) and CLEAN maps of a star and a disk with the VLTI. Flux ratios of disk to star are 10 (left), 1 (middle) and 0.1 (right). While the high flux disk is recovered by the CLEAN deconvolution, it is hardly seen at equal fluxes and not at all seen for the low disk flux.

been successfully obtained from telescopes fitted with masks and using the mapping techniques described here already in the early 1990's. This demonstrated that imaging with limited UV data is possible at visible wavelengths. Recently, phase closure imaging of double stars has been demonstrated by two groups who use distributed arrays of separate telescopes.

With the advent of more short wavelength array interferometers and more simultaneous elements for existing interferometers, imaging will become a more important issue. Due to the different nature of problems at shorter wavelengths, it is fair to say that the current state of the art is comparable to Radio interferometry imaging in the early Sixties. Today, radio interferometers deliver maps with a dynamic range as good as 10^5. Problems of fundamental nature, related to photon noise and the lack of loss-free

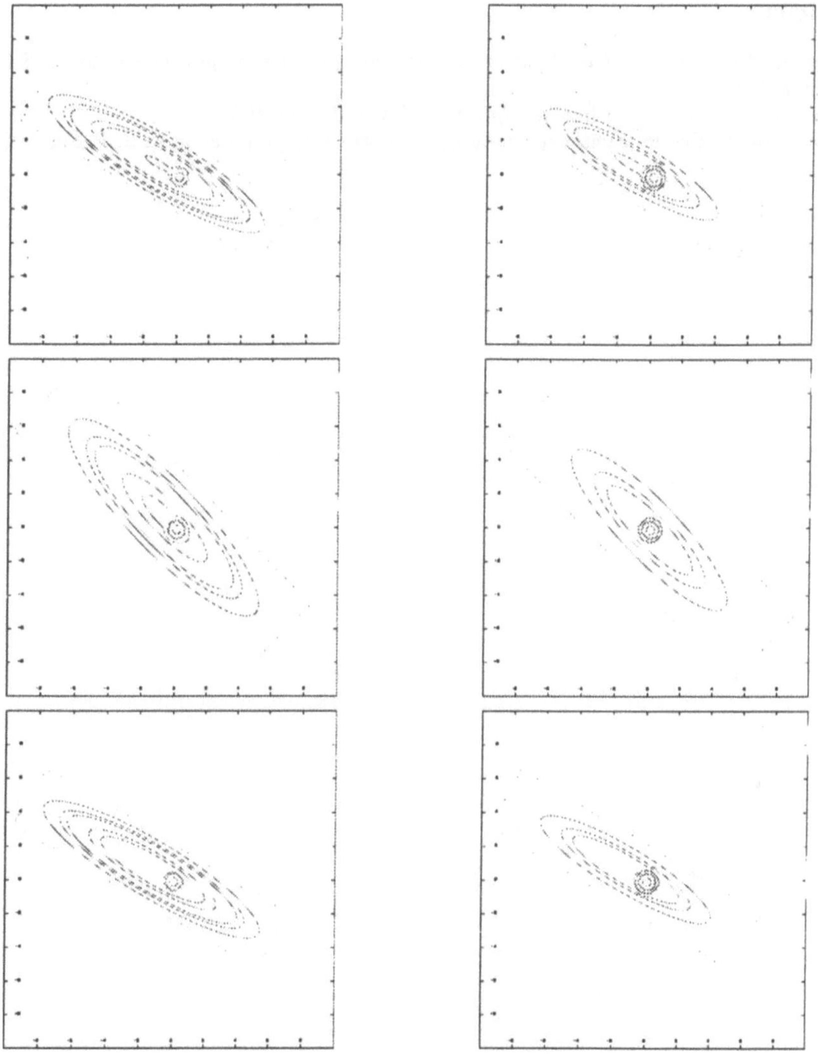

Figure 16. Source (top), input model with some parameters deliberately misadjusted (middle), and iterated model (bottom) of observations of a star and surrounding disk with the VLTI. Disk to star flux ratio: left 1, right 0.1.

coherent optical amplifiers, may prevent a similarly impressive performance of optical arrays. However, observing in the short wavelength domain which is rich of physical information of spectral nature, different mapping techniques which are better adapted may offset the disadvantages in the optical regime and may ultimately be equally successful.

References

1. Bracewell, Ronald N., *The Fourier Transform and Its Applications,* 2nd. Ed., McGraw-Hill 1992.
2. Goodman, Joseph W., *Statistical Optics,* Wiley and Sons 1985.
3. Perley, Schwab, Bridle, *Synthesis Imaging in Radio Interferometry,* A. S. P. Series 6, 1989.

ASTROMETRY WITH O/IR INTERFEROMETERS

M. SHAO
Jet Propulsion Laboratory

Abstract. This paper is a tutorial on astrometry with long baseline ground based interferometers. There are two types of astrometric measurements with ground based interferometers, global astrometry and narrow angle astrometry. The ultimate limits for both types of measurements are set by the atmosphere. However the instrumental considerations for the two types of measurements are quite different. This paper gives an overview of the instrumental aspects of these two different types of astrometric instruments.

1. Introduction

Astrometry with an optical interferometer bears a much closer resemblence to astrometry with a radio interferometer than astrometry with a single telescope. Because of the significant differences, global astrometry and narrow angle astrometry will be treated separately.

An interferometer combines the light from two telescopes coherently by directing the light towards a central beam combiner. The light in one or both arms is usually delayed by an optical delay so that the light from the star via the two arms are equal to roughly a wavelength. Astrometry is performed by measuring the position of the delay line where stellar fringes are formed.

The delay line position is related to the position of the star and the baseline vector, by the well known relation from radio astronomy:

$$X = B * S + c, \tag{1}$$

where B is the baseline vector (x,y,z), S is a unit vector to the star, and C is the delay offset. The delay offset represents the zero point of the delay

A-M. Lagrange et al. (eds.), High Angular Resolution in Astrophysics, 109–114.
© 1997 *Kluwer Academic Publishers.*

line. If the zero point of the delay line is delay line position when fringes are observed for a star directly overhead, c=0.

2. Global Astrometry

Global astrometry at optical and IR wavelengths is almost identical to VLBI astrometry at radio wavenlengths. Earth rotation can be used to change the baseline orientation of the interferometer. The baseline vector B in equation 1 is fixed to the Earth and rotates with a 24 hr period. The delay X also varies as a 24 hr sin wave. That sine wave has three measureable parameters, amplitude, phase and DC offset.

$$X = amp * sin(2 * pi * t/24hrs + phi) \tag{2}$$

If four stars are observed over 24 hrs, that data set has 8 unknown variables, the baseline vector and delay offset represents 4 more. The result is 12 measurements (4 stars*3/star) with 12 unknowns. Absolute R. A. can't be measured by this technique.

Data reduction for optical astrometry is virtually identical to its radio counterpart. A least squares software technique is used to solve for the positions of the stars and the baseline using as raw input the delay vs time data of each star. The least squares fit is actually to the delta positions (deviation from the catalog position) rather than the positions themselves.

More precisely, the least squares solution is the solution to the matrix equation

$$M(i) = A(i,j) * P(j), \tag{3}$$

where M is a vector of the difference in the delay where white light fringes are detected and the predicted delay for the nominal catalog position of the stars and the nominal baseline vector, P are the correction to the catalog positions of the stars and corrections to the nominal baseline, and A is a matrix of partial derivatives relating the stellar positions to delays.

2.1. INSTRUMENTAL ERRORS

A typical interferometer used for global astrometry on the ground would have a 40m baseline. One milliarcsec (mas) accuracy would imply that the optical delay has to be measured with 0.2 micron accuracy. That level of accuracy is challenging from both an instrumental and an atmospheric point of view.

From an instrumental point of view the most challenging aspect is baseline stability. Equation 1 and conventional radio interferometry analysis techniques make the assumption that the baseline of the interferometer is fixed to the solid Earth that rotates with a know angular velocity. For an interferometer with a 4000 km baseline, 1 mas accuracy implies the end points of the baseline are stable to 2cm. So as long as the radio telescope's pivot point is stable to 2cm astrometry at the 1 mas level is not a problem.

In the optical with a 40m baseline a number of effects can cause the baseline to shift by 0.2um. Thermal expansion of the telescope mount 1m above the surface of the Earth could shift the baseline by ten microns per deg C. The air temperature over the course of a night can easily change by 10 C. While the thermal problem is quite severe, other mechanical errors in the telescope mount can produce errors well in excess of 0.2um. Some examples are bearing imperfections. Ball bearings run in circular races. These races are not truely circular at the 0.2um level. Gravity causes the telescope structure to sag slightly. Telescopes are designed to point accurately and the gravity sag of a telescope is usually designed not to interfere with pointing. But of the 6 degress of freedom of the telescope optics, pointing only represents 2. Interferometry is sensitive to motion of 5 of the 6 degrees of freedom.

Because of the myraid sources of mechanical error in the telescopes, making a telescope mechanically perfect for 1mas global astrometry is a daunting task. The approach taken by groups doing global astrometry is not to make a perfect telescope but to build a metrology system to measure the imperfections of the telescope/mount at the 0.2um level. Then apply those corrections to the data analysis.

The basic approach is to place an optical fiducial, such as a retroreflector at the center of the telescope. Laser interferometers measure the position of that fiducial with respect to corner cubes burried 5 10 meters below the surface of the Earth.

Ultimately the accuracy of this is limited by the accuracy of the optical fiducial. If the fiducial is a open face corner cube with lambda/4 wavefront accuracy, the potential accuracy is 0.15um.

2.2. ATMOSPHERIC ERRORS

The 2nd major class of error in global astrometry is the atmosphere. Even for long baseline interferometers, the turbulent atmosphere will cause the position of a star to wander at 0.3 arcsec rms values for typical 1 arcsec seeing conditions.

While the instantaneous position can fluctuate a large amount, the positional uncertainty does decrease with longer integration time. Again for nominal seeing conditions at a good observatory, the atmospheric noise is typically 10 mas for a 1 hr integration.

Again there is a potential for improvement. The technique is two color observations. The principle of operation is that the index of air is slightly different at different wavelengths. The index of refraction of air is higher at shorter wavelengths. The position of the star is then:

$$Apparant - X = true - X + Wedge * (N - 1) \tag{4}$$

Apparant position is the position corrupted by the atmosphere and Wedge is the atmospheric wedge over the interferometer at the time of the measurement. N is the index of refraction of air. If we assume that we know the index of refraction of air at two wavelengths where the position of the star is measured, then we have two equations in two unknowns.

$$A-X-1 = T-X+Wedge*(N(l1)-1)A-X-2 = T-X+Wedge*(N(l2)-1) \tag{5}$$

The two color technique is not a complete solution. The index of air is not very different at different visible wavelengths. Because of the difficulty of doing interferometry in the blue, two color astrometry usually means using 500nm and 800nm as the two colors. The index of refraction at those two colors are 1.000294 and 1.000290. In equation 3, we solve for the true position using the measurements of the apparant positions at two colors. The noise in the measurement of the apparant position due to photon noise for example, is multiplied by a large factor (294-290)/(290) because the indices at the two wavelengths are not very different.

A second problem is that the index of air is not known exactly due to the presence of water vapor. In practice the gain in two color measurement accuracy has been limited to a factor of 3 5 better than one color measurements. Potentially accuracies of a few (2 5) mas is possible for a one hour integration time.

3. Narrow Angle Astrometry

Narrow angle astrometry, over fields of view of a few 10's of arcsec to a few 10's of arcminutes, is potentially much more accurate both from an instrumental and atmospheric point of view. The light from two stars near each other will traverse the same turbulent atmosphere before entering the interferometer. A differential measurement of the relative positions of the two stars should then not suffer most of the errors due to atmospheric turbulence. This is indeed the case when we measured the separation of a double star system in 1991. The rms error for a 1 sec integration for two stars 3 arcsec apart was 1.2mas (20 uas in 1 hr).

3.1. RADIO VS OPTICAL NARROW ANGLE ASTROMETRY

Narrow angle or relative astrometry with interferometers is fundamentally the same at radio and optical wavelengths. In the radio, the antenna is switched between two objects rapidly, faster than the coherence time of the atmosphere. The differential delay is then a measure of the angle between the two astrophysical objects, projected onto the baseline vector.

$$dX = X1 - X2 = B * (S1 - S2) \tag{6}$$

dX is the differential delay, B is the baseline vector and S1, S2 are unit vectors to the two stars. Note that the delay offset in equation 1 is not longer present when two delay measurements are differenced. In relative astrometry, we are only interested in measuring the vector S1-S2. If we make two measurements of dX with two different baseline vectors (preferably two roughly perpendicular baselines) that is sufficient to solve for S1-S2, if we know what the two baseline vectors are.

Phase reference interferometry is a powerful technique first developed for VLBI to increase the sensitivity of an interferometer. Observationally it is identical to relative astrometry, antennas are alternatively pointed at a bright object, then at a faint object. By using the bright object as a phase reference, one can extend the coherent integration time for the faint object significantly. The difference between the radio and optical comes from their different wavelengths.

If we take 1 cm and 1um as the nominal wavelengths of radio and optical techniques, we find that roughly speaking the isoplanatic angle, the angle in the sky where the atmospheric phase fluctuations can be considered identical is roughly 10,000 times larger for 1cm than 1um wavelength. At 1um, the isoplanatic angle is roughly 10 arcsec. At 1cm, the angle is roughly 30 degrees. The other difference is the switching time interval. The atmospheric coherence time at 1um is roughly 20 msec, the coherence time at 1cm is roughly 200 400 sec. The radio telescopes must swtich bewteen the faint star and the bright phase reference on a time scale faster than 400 sec. At 1um, the switching must be faster than 20 msec. Because it is not practical to re-point even a small 50cm telescope on a 10 msec time scale, optical phase reference interferometry has to use two separate optical systems to make simultaneous observations of the fringes of both stars.

The basic concept is the same. Both radio and optical interferometry make differential astrometric measurements using the phase reference technique to increase sensitivity to find "nearby" references. The difference is in two areas, narrow angle in the radio is a few degrees, narrow angle in O/IR is a few 10's of arcsec and fast switching in the radio is 100's of seconds and fast switching in O/IR is 10's msec.

3.2. BASELINE STABILITY

In global astrometry, we normally solve for the baseline vector along with the star positions because in general there is no way to determine the baseline vector with mas or sub-mas accuracy, other than by looking at the stars. But the requirements for differential astrometry is much less sever especially at optical wavelengths. Equations 2 is the reason why.

If we look at equation 2 more closely we can calculate the astrometric error that would result in having the wrong baseline. To first order, S1-S2=dX/B. A 1S1-S2 is 20 arcsec (for an IR astrometric interferometer), then 120 arcsec is 0.2 arcsec. If we are looking for 20 uas astrometry over a 20 arcsec field of view, the baseline vector must be known to one part in a million. The length of B (nominally 100m has to known to 1 part in a million (100 um) and the angle to 1 urad, or 0.2 arcsec.

Contrast this to 1 mas global astrometry where the baseline vector must be derived with 5 nano-radians, and 0.5 um accuracy for a 100m interferometer.

In differential astrometry, the baseline vector is not derived from the differential measurements, rather the interferometer does global astrometry over 30-40 degrees of known stars (eg from the Hipparcos catalog) to measure the baseline vector to at least 0.2 arcsec (and 100um). That baseline is then used in the data reduction for the differential measurement, at an accuracy to 20uas for the differential measurement.

The same relaxed baseline stability applies to radio differential astrometry, however because the angle between the target and phase reference in the radio is typically a few degrees apart, the tolerance is relaxed by a factor of 10 instead of 10,000 in the O/IR.

4. Summary

Ground based global astrometry will be limited to accuracies to a few milliarcsec by the atmosphere. However to achieve this level of accuracy, elaborate laser metrology is needed to monitor the baseline vector to ensure that it rotates with the Earth at the milliarcsec level.

Differential astrometry potentially can be as accurate as 10-20 uas using long baseline 100-200m IR interferometers. Phase reference techniques are needed to extend sensitivity to detect stars within 20 arcsec of a bright target for relative astrometry. This small field of view is dictated by atmospheric turbulence. But the small field results in much relaxed tolerance for baseline stability, several orders of magnitude less precise than what is needed for global astrometry.

IMAGE RECONSTRUCTION

From basic notions towards methodological strategies

A. LANNES
Director of Research at the CNRS
OMP, 14 Avenue E. Belin
31400 Toulouse
France

ABSTRACT. The aim of this tutorial is twofold: first, to provide basic tools for tackling the inverse problems of image reconstruction; and second, to present some advanced research in this field. All the general concepts presented in this paper are illustrated by considering the special case of the problems of image reconstruction in Fourier synthesis. In the corresponding theoretical framework, it is shown that CLEAN is a matching pursuit process which has two weak points: in situations of astrophysical interest (when dealing with extended objects and dilute apertures), the clean map is not stable in the corresponding object representation space, and the 'atoms' of this representation (translated versions of the clean beam) are not well suited for reconstructing the boundaries of the structuring entities of the object at the corresponding resolution level. As a result, CLEAN must be interrupted before the best possible fit is reached. WIPE is a Fourier synthesis method which overcomes these difficulties in a very efficient manner. How does WIPE wipe CLEAN clean? First, by introducing a global regularization principle based on the concept of resolution; and second, by conducting the matching pursuit process at the level of the 'elementary particles' (the high-resolution scaling functions) of the object space.

1. Introduction

WIPE is a Fourier synthesis method recently developed in radio imaging and optical interferometry [1, 2]. The name of WIPE is associated with that of CLEAN [3, 4], a heuristic deconvolution method intensively used in astronomy. (WIPE can also be used as a deconvolution technique.) As a matter of fact, WIPE was devised, quite independently, on the grounds of well known properties in harmonic analysis and band-limited interpolation [5]. There also exists a probabilistic interpretation of this method [6]. Nevertheless, to some extent, WIPE can equally well be regarded as an

A-M. Lagrange et al. (eds.), High Angular Resolution in Astrophysics, 115–143.
© 1997 *Kluwer Academic Publishers.*

updated version of CLEAN. This paper is aimed precisely at clarifying the relationship between these two methods. The theoretical tools presented in this context can also be used for the phase problem and its imbrication in aperture synthesis (see for example [7–9]).

In section 2, we first show that CLEAN lies in the family of 'matching pursuit' techniques [10]. In other words, the principle of the "iterative harmonic analysis" of CLEAN, as it is exhibited by Schwarz in [4], can be regarded as a special case of a more general approach. We then present the reconstruction methods that can be implemented for solving the related optimization problems, first, without any control of the propagation of errors (section 3), and afterwards, with a complete control of the stability parameters (section 4). The first weak point of CLEAN can thus be exhibited: in situations of astrophysical interest (for example, when observing extended complex objects with dilute apertures), the matching process of CLEAN is ill-conditioned. As shown in section 5, the regularization principle of WIPE, which can be applied to CLEAN as it is, remedies this lack of robustness. Unfortunately, CLEAN has another weak point: the basis functions used in the matching pursuit process, the 'clean beams,' are not well suited for reconstructing the boundaries of the structuring entities of the images at the selected resolution level. As a result, CLEAN must be interrupted before the best possible fit is reached. As indicated in section 5, the basic version of WIPE overcomes this difficulty in a simple and efficient manner. The multiresolution extension of WIPE, underlying the main aspects of this paper, reinforces the arguments justifying this methodological choice.

The guiding idea of our analysis is based on the theoretical framework presented in subsection 1.1. (The reader who is not familiar with the related basic notions is invited to consult appendices 1 to 4 in [5].) As CLEAN was first used as a Fourier synthesis technique, it was natural to compare the principles of CLEAN and WIPE in this context. Such a comparative study can only be started from a common statement of the problem. The corresponding formulation, which makes the theoretical framework more attractive, is presented in subsection 1.2. We then also specify the general conditions of the numerical simulations in support of our analysis.

1.1. THEORETICAL FRAMEWORK

In many inverse problems, the 'reconstructed object' is described in an 'object representation space' E generated by m vectors g_k selected among a family $\mathcal{E}_o$ of M vectors $g_1, g_2, \dots, g_M$. The latter may be regarded as the 'atoms' of the object representation under consideration (why not?). The linear space generated by all the vectors of $\mathcal{E}_o$, E_o, is a subspace of some Euclidean space: the 'object space' H_o. The 'data vector' ψ_d lies in another Euclidean space: the 'data space' K_d. (When the data are complex quantities, it is always possible to work in the real linear space underlying the complex linear space directly involved in the analysis.) To a first approximation, ψ_d is related to the object (or image) to be reconstructed via a linear operator A from H_o into K_d. The basic problem is to choose the vectors g_k in $\mathcal{E}_o$, and thereby to construct the object representation space E. The solutions are then obtained by minimizing on E the quadratic functional:

$$q : H_o \to \mathbb{R}, \qquad q(\phi) \triangleq \|\psi_d - A\phi\|_d^2. \tag{1}$$

Clearly $\|\cdot\|_{\rm d} = (\cdot \mid \cdot)_{\rm d}^{1/2}$ is the norm on $K_{\rm d}$; the scalar products on $K_{\rm o}$ and $K_{\rm d}$, as well as the norms, will be distinguished by the subscripts o and d, respectively.

Let F be the 'image' of E by A (the space of the $A\phi$'s, ϕ spanning E), A_E be the operator from E onto F induced by A, and ψ_F be the projection of $\psi_{\rm d}$ onto F (see Fig. 1). The vectors ϕ minimizing q on E, the solutions of the problem, are such that $A_E\phi = \psi_F$. They are identical up to a vector lying in the kernel of A_E . Note that $\ker A_E = E \cap \ker A$. (By definition, the kernel of A, $\ker A$, also referred to as the null space of A, is the set of vectors ϕ such that $A\phi = 0$.) The condition $\dim E \leq \dim K_{\rm d}$ is a necessary condition for $\ker A_E$ to be reduced to $\{0\}$.

As $\psi_{\rm d} - \psi_F$ is orthogonal to F, the solutions ϕ of the problem are characterized by the property:

$$\forall \varphi \in E, \qquad (A\varphi \mid \psi_{\rm d} - A\phi)_{\rm d} = 0 \qquad (\phi \in E).$$

On denoting by A^* the adjoint of A, this property can also be written in the form

$$\forall \varphi \in E, \quad (\varphi \mid r)_{\rm o} = 0 \qquad r \triangleq A^*(\psi_{\rm d} - A\phi) \quad (\phi \in E),$$

where r is regarded as a residue. The solutions ϕ are therefore characterized by the fact that the corresponding residue is orthogonal to E, *i.e.*,

$$\forall g_k \in \mathcal{E}, \quad (g_k \mid r)_{\rm o} = 0. \tag{2}$$

This condition is of course equivalent to $P_E r = 0$, where P_E is the projection (operator) of $H_{\rm o}$ onto E. The solutions of the problem are therefore the solutions of the equation on E:

$$P_E A^* A_E \phi = P_E A^* \psi_{\rm d} . \tag{3}$$

For any $\varphi \in E$ and any $\psi \in K_{\rm d}$, we have

$$(A\varphi \mid \psi)_{\rm d} = (\varphi \mid A^*\psi)_{\rm o} = (\varphi \mid P_E A^*\psi)_{\rm o} ,$$

hence $A_E^* = P_E A^*$. This explicitly shows that equation (3) is the 'normal equation:'

$$A_E^* A_E \phi = A_E^* \psi_{\rm d} . \tag{4}$$

When the problem is well posed, A_E is a one-to-one map ($\ker A_E = \{0\}$); the solution is then unique: there exists only one vector $\phi \in E$ such that $A_E\phi = \psi_F$. This vector, ϕ_E , is said to be the least-squares solution of the equation $A_E\phi$ '=' $\psi_{\rm d}$.

In this case, let $\delta\psi_F$ be a variation of ψ_F in F, and $\delta\phi_E$ be the corresponding variation of ϕ_E (see Fig. 1). As shown in appendix 1, the robustness of the reconstruction process is then governed by the inequality

$$\frac{\|\delta\phi_E\|_{\rm o}}{\|\phi_E\|_{\rm o}} \leq \kappa_E \frac{\|\delta\psi_F\|_{\rm d}}{\|\psi_F\|_{\rm d}} \tag{5}$$

where κ_E is the condition number of A_E :

$$\kappa_E \triangleq \frac{\sqrt{\mu'}}{\sqrt{\mu}} ; \tag{6}$$

μ and μ' respectively denote the greatest lower bound and the least upper bound of $\|A_E\phi\|_d^2$ for the ϕ's with norm unity in E:

$$\mu \triangleq \inf_{\|\phi\|_o=1} \|A_E\phi\|_d^2 \qquad\qquad \mu' \triangleq \sup_{\|\phi\|_o=1} \|A_E\phi\|_d^2 . \tag{7}$$

The closer to 1 is the condition number, the easier and the more robust is the reconstruction process. As

$$\|A_E\phi\|_d^2 = (\phi \mid A_E^* A_E\, \phi)_o ,$$

μ and μ' are the smallest and largest eigenvalues of $A_E^* A_E$, respectively.

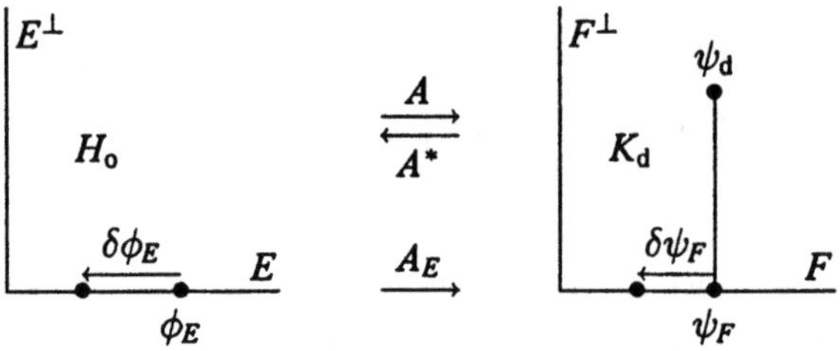

FIGURE 1. Uniqueness of the solution and robustness of the reconstruction process. Operator A is an operator from the object space H_o into the data space K_d. The object representation space E is a particular subspace of H_o (see text). The image of E by A, the range of A_E, is denoted by F. In this Euclidean representation, ψ_F is the projection of the data vector ψ_d onto F. The inverse problem must be stated so that A_E is a one-to-one map from E onto F, the condition number κ_E having a reasonable value (see Eq. (5)).

In the analysis of the different methods that can be devised and implemented for solving the problem, three main aspects must be examined:

1) the precise definition of ψ_d, A and E;

2) the selected minimization technique;

3) the robustness of the reconstruction process.

The first point is essential for the interpretation of the results. It is related to the statement of the problem and, if need be, to its 'regularization;' ψ_d is not necessarily the vector of the experimental data. In such a case, the definition of A must of course take into account the corresponding manipulations. As regards E, which may also be involved in the regularization of the problem, it is important to note that this space may be constructed, in a global manner or step by step, interactively or automatically. As suggested by the statement of point 2, many different techniques can be used for minimizing q on E; some of these are certainly more efficient than others, but in this case, the choice is not crucial. The last point concerns the study of the propagation of errors. The part played by inequality (5) in the development of the corresponding error analysis shows that a good reconstruction procedure must also provide, in particular, the condition number κ_E.

1.2. FORMULATION OF THE PROBLEMS OF FOURIER SYNTHESIS IN ASTRONOMY

In the problems of Fourier synthesis encountered in astronomy, the 'object function' of interest, ϕ_o, is a real-valued function of an angular position variable $\boldsymbol{\xi} = (\xi_1, \xi_2)$. The corresponding object functional variable ϕ lies in some object space H_o in which the functions are defined at a very high level of resolution. In this sense, one may say that H_o emulates the Hilbert space of square integrable real-valued functions $L^2_{\mathbb{R}}(\mathbb{R}^2)$. More precisely, H_o is characterized by two key parameters: the extension $\Delta\xi$ of its field, and its resolution scale $\delta\xi$; $\delta\xi$ is therefore much smaller than the resolution limit that can be reasonably expected at the end of the reconstruction process: $\delta\xi \triangleq \Delta\xi/N$, where N is some power of 2. The larger is N, the more oversampled is the object domain. The object space is the space of functions taking their values on the object domain. To define this space more explicitly, we first introduce the finite grid (see Fig. 2):

$$\mathbb{G} \triangleq \mathbb{L} \times \mathbb{L}, \qquad \mathbb{L} \triangleq \left\{ p \in \mathbb{Z} : -\frac{N}{2} \leq p \leq \frac{N}{2} - 1 \right\}.$$

On each pixel $\boldsymbol{p}\,\delta\xi$ ($\boldsymbol{p} \in \mathbb{G}$), we then center a scaling function of the form

$$e_{\boldsymbol{p}}(\boldsymbol{\xi}) \triangleq e_0(\boldsymbol{\xi} - \boldsymbol{p}\,\delta\xi) \tag{8}$$

where

$$e_0(\boldsymbol{\xi}) \triangleq \operatorname{sinc}\left(\frac{\xi_1}{\delta\xi}\right)\operatorname{sinc}\left(\frac{\xi_2}{\delta\xi}\right). \tag{9}$$

It is easy to verify that these functions form an orthogonal set with respect to the scalar product on the space of square integrable functions. Note that

$$\|e_{\boldsymbol{p}}\|_o^2 = (\delta\xi)^2 \qquad (\forall \boldsymbol{p} \in \mathbb{G}). \tag{10}$$

In our comparative analysis of CLEAN and WIPE, H_o is the Euclidean space generated by the basis vectors $e_{\boldsymbol{p}}$, $\boldsymbol{p}$ spanning $\mathbb{G}$ (see Fig. 2). The functions ϕ lying in this space can therefore be expanded in the form:

$$\phi = \sum_{\boldsymbol{p} \in \mathbf{G}} x_{\boldsymbol{p}} e_{\boldsymbol{p}} \qquad (x_{\boldsymbol{p}} \in \mathbb{R}). \tag{11}$$

In the general context of this paper, the functions $e_{\boldsymbol{p}}$, which play the role of interpolation or scaling functions, can be regarded as the 'elementary particles' of the object representation. Evidently, other orthogonal sets of scaling functions can be used (cf. subsection 4.3 in [1]). Let us finally note that H_o is a subspace of $L^2_{\mathbb{R}}(\mathbb{R}^2)$; its inner product can be explicitly expressed in the form (cf. Eqs. (8), (9), (10) and (11)):

$$(\phi_1 \mid \phi_2)_o = (\delta\xi)^2 \sum_{\boldsymbol{p} \in \mathbf{G}} x_{1,\boldsymbol{p}}\, x_{2,\boldsymbol{p}}\,.$$

Thus,

$$\|\phi\|_o^2 = (\delta\xi)^2 \sum_{\boldsymbol{p} \in \mathbf{G}} x_{\boldsymbol{p}}^2\,.$$

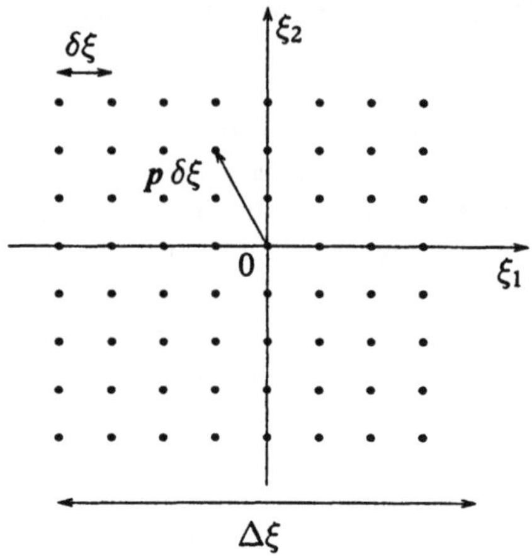

FIGURE 2. Object grid $\mathbb{G}\,\delta\xi$ for $N = 8$.

The Fourier transform of ϕ is defined by the relationship

$$\widehat{\phi}(\boldsymbol{u}) \triangleq \int \phi(\boldsymbol{\xi}) \exp(-2\mathrm{i}\pi \boldsymbol{u} \cdot \boldsymbol{\xi})\, \mathrm{d}\boldsymbol{\xi},$$

where $\boldsymbol{u}$ is a two-dimensional angular spatial frequency: $\boldsymbol{u} = (u_1, u_2)$. According to Eqs. (11), (8) and (9), we therefore have

$$\widehat{\phi} = \sum_{p \in \mathbb{G}} x_p \widehat{e}_p$$

where

$$\widehat{e}_p(\boldsymbol{u}) = \widehat{e}_0(\boldsymbol{u}) \exp\left(-2\mathrm{i}\pi \boldsymbol{p} \cdot \frac{\boldsymbol{u}}{\Delta u}\right)$$

with $\Delta u \triangleq 1/\delta\xi$ and

$$\widehat{e}_0(\boldsymbol{u}) = \frac{1}{(\Delta u)^2} \mathrm{rect}\left(\frac{u_1}{\Delta u}\right) \mathrm{rect}\left(\frac{u_2}{\Delta u}\right).$$

The 'experimental data' $\psi_e(\boldsymbol{u})$ are blurred values of $\widehat{\phi}_o(\boldsymbol{u})$ on a finite list of frequencies in the Fourier domain: $\mathcal{L}_e \triangleq \{\boldsymbol{u}_e(1), \boldsymbol{u}_e(2), \ldots, \boldsymbol{u}_e(N_e)\}$. As ϕ_o is a real-valued function, it is natural to define $\psi_e(-\boldsymbol{u})$ as the complex conjugate of $\psi_e(\boldsymbol{u})$. The 'experimental frequency list' $\mathcal{L}_e$ is defined consequently: if $\boldsymbol{u} \in \mathcal{L}_e$, then $-\boldsymbol{u} \in \mathcal{L}_e$ (except for $\boldsymbol{u} = \boldsymbol{0}$: in the convention adopted in the sequel, either the null frequency does not lie in $\mathcal{L}_e$, or there exists only one occurrence of this point.) The experimental frequency coverage generated by $\mathcal{L}_e$ is therefore centrosymmetric (see Fig. 3a).

The experimental data vector ψ_e lies in the 'experimental data space' K_e. By definition, K_e is the real Euclidean space underlying the space of complex-valued functions on $\mathcal{L}_e$ such that $\psi(-\boldsymbol{u}) = \overline{\psi}(\boldsymbol{u})$. This space, whose dimension is equal to N_e, is endowed with the scalar product

$$(\psi_1 \mid \psi_2)_e \triangleq \sum_{\boldsymbol{u} \in \mathcal{L}_e} \overline{\psi}_1(\boldsymbol{u}) \psi_2(\boldsymbol{u})\, W(\boldsymbol{u})\, (\delta u)^2, \tag{12}$$

where W is a given weighting function, and $\delta u \triangleq \Delta u/N$. Note that this scalar product is real valued and therefore symmetric: $(\psi_1 \mid \psi_2)_e = (\psi_2 \mid \psi_1)_e$. In CLEAN, the data space K_d reduces to K_e, whereas in WIPE, K_d is a natural extension of K_e (cf. section 5).

Let $\mathcal{H}$ be the Fourier domain: $\mathcal{H} \triangleq (-\Delta u/2,\ \Delta u/2)^2$. In aperture synthesis, the 'frequency coverage to be synthesized' is a centrosymmetric region $\mathcal{H}_s$ contained in $\mathcal{H}$. In Fig. 3a, this region is the disc centred on the origin. In this context, CLEAN and WIPE share a common objective, that of the image to be reconstructed. This image, ϕ_s, is defined so that its Fourier transform is quadratically negligible outside $\mathcal{H}_s$. More explicitly, ϕ_s is defined by the convolution relation of the form:

$$\phi_s \triangleq s \star \phi_o . \tag{13}$$

The 'synthetic beam' s is a function resulting from the choice of $\mathcal{H}_s$: the 'clean beam' in CLEAN, the 'neat beam' in WIPE (see Fig. 3b). The Fourier data corresponding to ϕ_s are then defined by the relationship:

$$\psi_s(u) \triangleq \widehat{s}(u)\psi_e(u) \qquad (\text{on } \mathcal{L}_e). \tag{14}$$

Clearly, $\psi_s(u)$ lies in $K_e \subseteq K_d$.

The neat beam is a sort of optimal clean beam: the optimal apodized point spread function that can be designed within the limits of the Heisenberg principle — Apodization theory is concerned with the control of the distribution of light over the exit pupil of an optical system in order to achieve a suppression of side lobes of the diffraction pattern [11].

More precisely, the neat beam s is a centrosymmetric function lying in the object space H_o, and satisfying the following properties:

(i) The 'energy' of $\widehat{s}$ is concentrated in $\mathcal{H}_s$. In other words, $\widehat{s}$ is small outside $\mathcal{H}_s$ in the mean-square sense. Given χ^2 less than 1, but close to 1 (say $\chi^2 = 0.97$), s satisfies a condition of the form :

$$\frac{1}{\| s \|^2} \int_{\mathcal{H}_s} |\widehat{s}(u)|^2 \, du = \chi^2 \tag{15}$$

where $\| s \|^2 = \| \widehat{s} \|^2 = \int_{\mathbb{R}^2} |\widehat{s}(u)|^2 \, du$.

(ii) The effective support $\mathcal{D}_s$ of s is as small as possible with respect to the choices of $\mathcal{H}_s$ and χ. The idea is of course to have the best possible resolution.

(iii) The normalization condition $\widehat{s}(0) = 1$, so that :

$$\int_{\mathbb{R}^2} \phi_s(\xi)\, d\xi = \int_{\mathbb{R}^2} \phi_o(\xi)\, d\xi . \tag{16}$$

In the terminology adopted in this paper, an atom such as s is a finite linear combination of elementary particles e_p. The integers p involved in this linear combination lie in a subset $\mathbb{D}_s$ of $\mathbb{G}$ (see Figs. 2 and 3b). As explicitly shown in [2] (section 2), the computation of the corresponding coefficients does not raise any particular difficulty.

In a first approach to Fourier synthesis, equation (13) suggests that the basis functions of the object representation space E should be translated versions of s: a finite number of

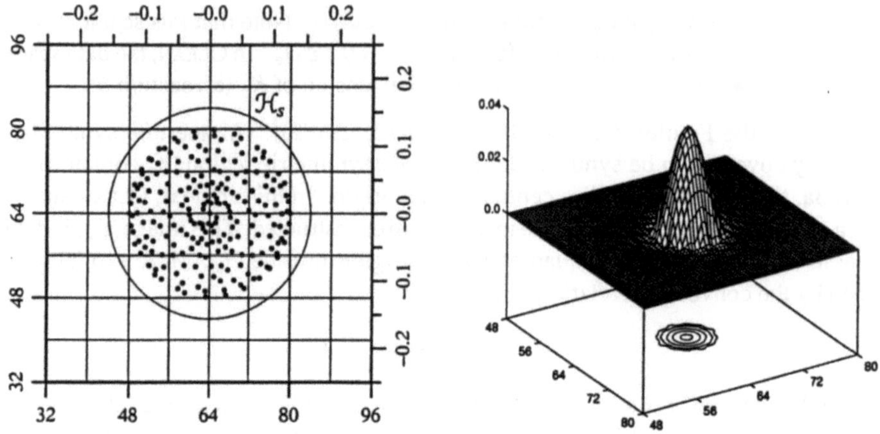

(a) (b)

FIGURE 3. General conditions of the simulation; a: experimental frequency coverage and frequency coverage to be synthesized; b: representation of the corresponding neat beam (for $\chi^2 = 0.97$). The experimental frequency list corresponding to the experimental frequency coverage shown in (a) includes 211 frequency points ($\dim K_e = 211$). The Fourier grid is of the form $\mathbb{G}\,\delta u$ where $\delta u = \Delta u/N$ with $N = 128$. The neat beam s represented in (b) corresponds to the frequency coverage to be synthesized $\mathcal{H}_s$. It is centred in the object grid $\mathbb{G}\,\delta\xi$ where $\delta\xi = 1/\Delta u$. The resolution limit of the reconstruction process (the full width of s at half maximum) is of the order of $1.5/\Delta u_s$ where Δu_s is the diameter of $\mathcal{H}_s$.

atoms centred on the nodes of grid $\mathbb{G}$. This very natural idea, which is exploited as it is, in the matching pursuit principle of CLEAN (cf. subsection 2.2), may be completely relaxed in WIPE. More precisely, the matching pursuit process may be performed at the level of the elementary particles. The resolution limit of the reconstruction process is then kept under control thanks to an appropriate regularization principle (cf. subsection 5.1).

The simulations presented in this paper correspond to the general conditions defined in Fig. 3. The 'object function' ϕ_o was defined as a set of 'δ functions' centred on the nodes of grid $\mathbb{G}\,\delta\xi$ with $N = 128$. The image to be reconstructed ($\phi_s = s \star \phi_o$) then lies in H_o. Figure 4 gives an idea of the complexity of this image.

The Fourier data were blurred by adding a Gaussian noise. More precisely, for all the $\boldsymbol{u} \in \mathcal{L}_e$ (see Fig. 3a), the standard deviation of $[\psi_e(\boldsymbol{u}) - \widehat{\phi}_o(\boldsymbol{u})]$ was set equal to 5% of the total flux of the object: $\sigma_e(\boldsymbol{u}) = \widehat{\phi}_o(\mathbf{0})/20$. The weighting function W introduced in Eq. (12) was explicitly defined by the formula

$$W(\boldsymbol{u}) \triangleq \frac{w(\boldsymbol{u})}{\displaystyle\sum_{\boldsymbol{u}' \in \mathcal{L}_e} \operatorname{sinc}^2\Big(\frac{u_1 - u_1'}{\delta u}\Big) \operatorname{sinc}^2\Big(\frac{u_2 - u_2'}{\delta u}\Big) w(\boldsymbol{u}')} \tag{17}$$

where

$$w(\boldsymbol{u}) \triangleq 1/\sigma_s(\boldsymbol{u}), \qquad \sigma_s(\boldsymbol{u}) \triangleq \widehat{s}(\boldsymbol{u})\sigma_e(\boldsymbol{u}). \tag{18}$$

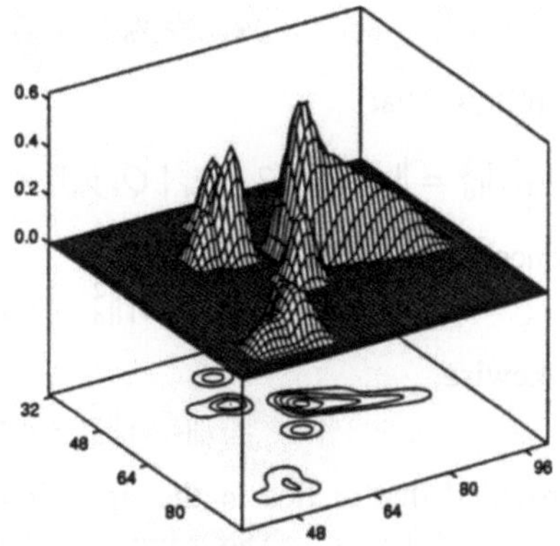

FIGURE 4. Image to be reconstructed, ϕ_s, at the resolution level defined in Fig. 3. Note that the portion of the field represented here is twice as large as that defined in Fig. 3b.

2. Reconstruction via matching pursuit methods

Among the various iterative methods that can be implemented for finding an approximation to the image (or the object) to be reconstructed, there exists a very slow algorithm which is based on a matching pursuit strategy. As will be clarified in this section, this algorithm is nothing but an aborted version of a particular algorithm minimizing q on E_o (see the introduction of subsection 1.1). The corresponding iterative process must never be used in practice for solving the problem. Its slow convergence may however be of interest for initializing the choice of the object representation space E. It is therefore important to analyse its principle (subsection 2.1), and in particular, to show that CLEAN is an algorithm of this type (subection 2.2).

2.1. RECONSTRUCTION PRINCIPLE

Let $h_j \triangleq Ag_j$ be the virtual data vector corresponding to the object atom g_j (cf. subsection 1.1), and Q_j be the projection of K_d onto the space generated by h_j:

$$Q_j\psi = \eta_j^2\,(h_j \mid \psi)_d\, h_j \qquad (\psi \in K_d) \tag{19}$$

where

$$\eta_j \triangleq \| h_j\|_d^{-1} = (g_j \mid A^*Ag_j)_o^{-1/2}\,. \tag{20}$$

The guiding idea is to determine the projection of ψ_d onto $F_d \triangleq AE_o$ via the elementary projections Q_j.

Let us consider the iteration in F_d:

$$\psi_0 = 0, \qquad \psi_{n+1} = \psi_n + \omega Q_k(\psi_d - \psi_n), \quad (\omega > 0); \tag{21}$$

ω is a relaxation parameter to be defined. At each iteration, $k \equiv k_n$ is chosen so that

$$\|Q_k(\psi_d - \psi_n)\|_d = \max_{1 \le j \le M} \|Q_j(\psi_d - \psi_n)\|_d\,. \tag{22}$$

If $Q_k(\psi_d - \psi_n) = 0$, then $\psi_n = \psi_{F_d}$ (the projection of ψ_d onto F_d) and the problem is solved.

Let us set $y_n \triangleq \psi_d - \psi_n$ and $z_n \triangleq \psi_{F_d} - \psi_n$. As $Q_k y_n = Q_k z_n$, we have from Eq. (21):

$$y_{n+1} = y_n - \omega Q_k y_n \qquad z_{n+1} = z_n - \omega Q_k z_n .$$

It follows that

$$\|y_{n+1}\|_d^2 = \|y_n\|_d^2 - 2\omega(y_n \mid Q_k y_n)_d + \omega^2\|Q_k y_n\|_d^2 = \|y_n\|_d^2 - 2\omega\|Q_k y_n\|_d^2 + \omega^2\|Q_k y_n\|_d^2 ,$$

hence

$$\|y_{n+1}\|_d^2 = \|y_n\|_d^2 - \omega(2-\omega)\|Q_k y_n\|_d^2 . \tag{23}$$

Likewise,

$$\|z_{n+1}\|_d^2 = \|z_n\|_d^2 - \omega(2-\omega)\|Q_k z_n\|_d^2 .$$

Provided that ω lies in the open interval $(0, 2)$, $\omega(2-\omega)$ is strictly positive. Then, $\|z_{n+1}\|_d \leq \|z_n\|_d$. The sequence $\{\beta_n\}_{n=0}^{\infty}$, where $\beta_n \triangleq \|z_n\|_d^2$, therefore converges towards some nonnegative number β. As shown in appendix 2, β proves to be equal to 0. As a result, $z_n \to 0$ as $n \to \infty$. The iterates (21) then converge towards ψ_{F_d}.

The maximal value of $\omega(2-\omega)$ is attained for $\omega = 1$. To increase the convergence speed of the projection operation, ω may be set equal to this optimal value. The corresponding algorithm, $\psi_{n+1} = \psi_n + Q_k(\psi_d - \psi_n)$, is nothing but a traditional matching pursuit process (a first introduction to this type of algorithm is to be found in [10]). The principle of this basic algorithm can be sketched as indicated in Fig. 5. The corresponding geometrical illustration corresponds to the special case where F_d is of dimension 2. On starting from the origin, the first step is defined by the projection of ψ_d onto the closest direction h_j, here h_1. Note that this projection is the same as that of ψ_{F_d}. The first iterate ψ_1 is then determined. The next step is defined similarly by considering the projections of the residual vector $\psi_d - \psi_1$, and so forth, whence this iterative scheme. As illustrated in Fig. 6, the projection onto F_d can be performed more rapidly with the aid of the conjugate-gradients method. The related property is that ψ_{F_d} is then attained in number of iterates less than or equal to the dimension of F_d (cf. [12]).

As $y_n = \psi_d - \psi_n$, we have from Eq. (19),

$$Q_k y_n = \eta_k^2 (Ag_k \mid \psi_d - \psi_n)_d Ag_k ,$$

hence

$$Q_k y_n = \eta_k^2 (g_k \mid r_n)_o Ag_k \tag{24}$$

where $r_n \triangleq A^*(\psi_d - \psi_n)$. The relaxed matching pursuit iteration (21) can therefore be written in the form

$$\psi_0 = 0, \qquad \psi_{n+1} = \psi_n + \omega\eta_k^2 (g_k \mid r_n)_o Ag_k .$$

Clearly, this sequence is the image by A of the object sequence (in E):

$$\phi_0 = 0, \qquad \phi_{n+1} = \phi_n + \omega\eta_k^2 \rho_{n,k} g_k , \tag{25}$$

where

$$\rho_{n,k} \triangleq (g_k \mid r_n)_o \qquad r_n = A^*\psi_d - A^*A\phi_n . \tag{26}$$

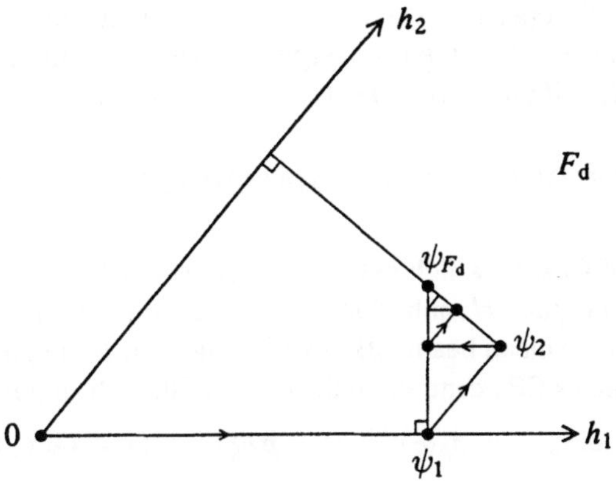

FIGURE 5. Projection via the matching pursuit method ($\omega = 1$)

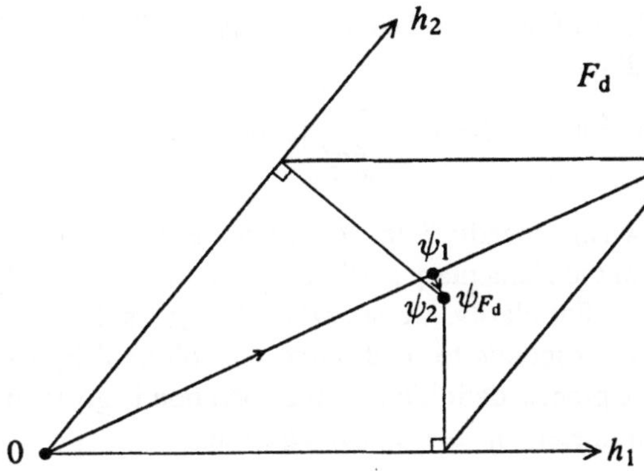

FIGURE 6. Projection via the conjugate-gradients method

According to its definition, the residue r_n is obtained via the iteration:

$$r_0 = A^* \psi_d , \qquad r_{n+1} = r_n - \omega \eta_k^2 \rho_{n,k} A^* A g_k . \tag{27}$$

As, from Eqs. (24) and (26), $Q_k y_n = \eta_k^2 \rho_{n,k} A g_k$, we have (cf. Eq. (20)):

$$\|Q_k y_n\|_d^2 = \eta_k^4 \rho_{n,k}^2 (g_k \mid A^* A g_k)_o = \eta_k^2 \rho_{n,k}^2 .$$

On setting (cf. Eq. (1))

$$q_n \triangleq q(\phi_n) = \|\psi_d - A\phi_n\|_d^2 = \|\psi_d - \psi_n\|_d^2 = \|y_n\|_d^2 ,$$

it follows from Eq. (23) that q_n is obtained through the iteration:

$$q_0 = \|\psi_d\|_d^2 , \qquad q_{n+1} = q_n - \omega(2 - \omega) \eta_k^2 \rho_{n,k}^2 . \tag{28}$$

Provided that ω lies in the open interval (0, 2), the iterates q_n converge towards the minimal value of q on E_0. Sequence (25) then converges towards a solution ϕ of the problem; ϕ is the unique solution ϕ_{E_0}, if and only if A_{E_0} is a one-to-one map.

2.2. PRESENTATION OF CLEAN AS A MATCHING PURSUIT ALGORITHM

In our formulation of CLEAN, which essentially follows that of Högbom [3], the object space is the Euclidean space H_o introduced in subsection 1.2. The vectors g_j are then translated versions of the clean beam CB $\equiv s$ (see Fig. 3b). More precisely, the elements of $\mathcal{E}_o$ are the clean beams CB_p centred on the nodes of the 'clean box' $\mathbb{G}_c\,\delta\xi$:

$$\text{CB}_p \equiv s_p\,, \qquad s_p(\xi) \stackrel{\Delta}{=} s(\xi - p\,\delta\xi) \qquad (p \in \mathbb{G}_c \subset \mathbb{G}).$$

The data space K_d coincides with the experimental data space K_e, and A with the experimental Fourier sampling operator: $(A_e\phi)(u) \stackrel{\Delta}{=} \widehat{\phi}(u)$ on $\mathcal{L}_e$. As the image to be reconstructed is defined as the convolution of the original object by the clean beam (Eq. (13)), the data vector ψ_d must be defined as the experimental data vector ψ_e damped by the Fourier transform of the clean beam: $\psi_d = \psi_s$ (cf. Eq. (14)). We then have $q \equiv q_e$ with (cf. Eqs. (1), (12) and (17)):

$$q_e(\phi) \stackrel{\Delta}{=} \|\psi_s - A_e\phi\|_e^2 = \sum_{u\in\mathcal{L}_e} |\psi_s(u) - \widehat{\phi}(u)|^2\, W(u)\,(\delta u)^2. \tag{29}$$

As explicitly shown in appendix 3, the 'dirty map' is the map of the scalar components of $A_e^*\psi_e$ in the basis of the elementary particles e_p. In this context, $A_e^*\psi_s$ may be referred to as the 'dusty map.' For clarity, we set $\text{DM}_e \equiv A_e^*\psi_e$ and $\text{DM}_s \equiv A_e^*\psi_s$. Likewise, the action of $A_e^*A_e$ corresponds to a 'discrete convolution' by the 'dirty beam' DB: $A_e^*A_e\phi = \text{DB} \,\underline{\star}\, \phi$ (the precise definition of this operation is given in appendix 3). Thus, from Eq. (20), the parameters $\eta_p \equiv \eta_j$ are all equal to:

$$\eta = (\text{CB} \mid \text{DB} \,\underline{\star}\, \text{CB})_o^{-1/2}. \tag{30}$$

The relaxed matching pursuit iteration (25) can then be written in the form

$$\phi_0 = 0, \qquad \phi_{n+1} = \phi_n + \omega\eta^2 \rho_{n,p}\,\text{CB}_p\,, \tag{31}$$

where (from Eq. (26))

$$\rho_{n,p} = (\text{CB}_p \mid r_n)_o \qquad r_n = \text{DM}_s - \text{DB} \,\underline{\star}\, \phi_n\,. \tag{32}$$

Clearly, ρ_n (the map of the $\rho_{n,p}$) is nothing but the 'discrete intercorrelation' of r_n with CB.

The residue r_n and the quadratic errors q_n are respectively obtained via the iterations (27) and (28):

$$r_0 = \text{DM}_s\,, \qquad r_{n+1} = r_n - \omega\eta^2\rho_{n,p}(\text{DB} \,\underline{\star}\, \text{CB}_p) \tag{33}$$

and

$$q_{e,0} = \|\psi_s\|_e^2 = \sum_{u\in\mathcal{L}_e} |\psi_s(u)|^2\, W(u)\,(\delta u)^2, \qquad q_{e,n+1} = q_{e,n} - \omega(2-\omega)\eta^2\rho_{n,p}^2\,. \tag{34}$$

Note that $\text{DB} \,\underline{\star}\, \text{CB}_p = \text{DB}_p \,\underline{\star}\, \text{CB} = (\text{DB} \,\underline{\star}\, \text{CB})_p$.

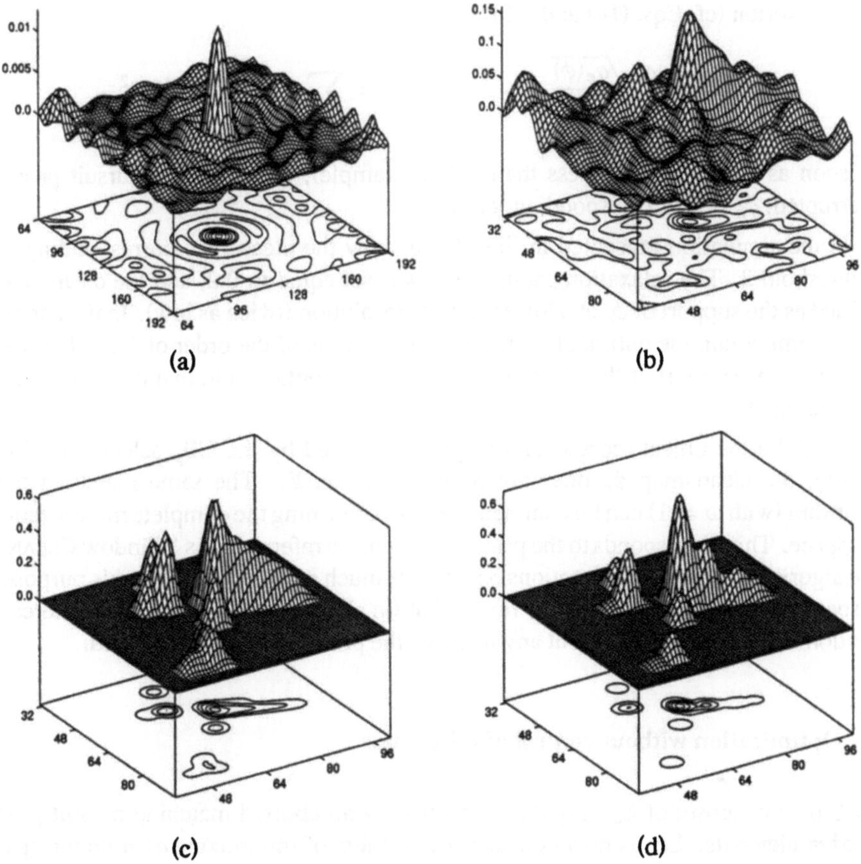

FIGURE 7. Image reconstruction via CLEAN with $\omega = 0.2$; a: dirty beam; b: dusty map; c: image to be reconstructed (Fig. 4); d: clean map for $C_e = 1.99$ (the fit criterion C_e is defined in Eq. (35)). In the conditions of this simulation (see Fig. 3), the optimal fit threshold of CLEAN is of the order of 1.75. For a lower threshold, the support of the clean map is no longer contained in that of the image to be reconstructed. In the framework of the analysis presented in this tutorial, the residual maps r_n or ρ_n are not be added to the clean map.

In the classical presentation of CLEAN, the convolution by the clean beam is performed *a posteriori*, whence some small differences in these iterations (cf. appendix 4). In particular, in the version of CLEAN presented here, p is chosen (at each iteration) so that $|\rho_{n,p}| = \max_{p' \in G_c} |\rho_{n,p'}|$.

The process is interrupted as soon as $q_{e,n}$ is less than a threshold value related to the level of the noise in the Fourier domain. In our implementation of CLEAN, we introduce the fit criterion (cf. Eqs. (18) and (29)):

$$C_e(\phi) \triangleq \frac{\sqrt{q_e(\phi)}}{\|\sigma_s\|_e} \qquad \|\sigma_s\|_e^2 = \sum_{u\in\mathcal{L}_e} \sigma_s^2(u)\, W(u)\, (\delta u)^2. \tag{35}$$

As soon as $C_e \triangleq C_e(\phi_n)$ is less than 2 (for example), the matching pursuit process is interrupted; ϕ_n is the corresponding 'clean map.'

In the simulation presented in Fig. 7, we show the clean map corresponding to the fit threshold 2. The relaxation parameter ω was set equal to 0.2, and the clean box was defined as the support of ϕ_s at a lower level of resolution (twice as low). In the conditions of this simulation, the optimal fit threshold of CLEAN is of the order of 1.75. For a lower threshold, the support of the clean map is no longer contained in that of the image to be reconstructed.

Let E be the object representation space generated by the CB_p selected by CLEAN. Clearly, the clean map ϕ_n does not minimize q_e on E. The same matching pursuit algorithm (with $\omega = 1$) can be confined to E for performing the complete minimization on this space. This corresponds to the principle of what is referred to as 'Window CLEAN' [4]. The algorithms presented in sections 3 and 4 are much more efficient for this purpose, but as specified in section 5, they only reveal that (in situations of astrophysical interest) the solution thus obtained is without any interest: the problem is ill-conditioned.

3. Optimization without control of robustness

Let $\mathcal{E}$ be any subset of $\mathcal{E}_o$, say that generated by an aborted matching pursuit process; $\mathcal{E}$ has m elements. Let us now consider the problem of minimizing $q(\phi)$ on the space E generated by the g_k, k spanning

$$\mathbb{J} \triangleq \{k : g_k \in \mathcal{E}\}.$$

By definition, E is the range of the operator:

$$S: \mathbb{R}^m \to H_o, \qquad S\alpha \triangleq \sum_{k\in\mathbb{J}} \alpha_k g_k\,. \tag{36}$$

In the case where $\mathbb{R}^m$ is equipped with its standard scalar product, the adjoint of S is explicitly defined by the relationship:

$$(S^*\phi)_k = (g_k \mid \phi)_o \qquad \text{(for all } k \in \mathbb{J}).$$

Indeed, for any $\phi \in H_o$, we have from Eq. (36):

$$(S\alpha \mid \phi)_o = \sum\nolimits_{k\in\mathbb{J}} \alpha_k\, (g_k \mid \phi)_o \equiv \sum\nolimits_{k\in\mathbb{J}} \alpha_k\, (S^*\phi)_k\,.$$

In what follows, S is not necessarily a one-to-one map from $\mathbb{R}^m$ onto E: the vectors g_k lying in $\mathcal{E}$ are not necessarily linearly independent.

Let α now be a vector minimizing on $\mathbb{R}^m$ the quantity $\|\psi_d - AS\alpha\|_d^2$. Then, the vector $\phi = S\alpha$ minimizes q on E. According to Eq. (2), the vectors α in question are such that

$$\forall k \in \mathbb{J}, \qquad \big(g_k \mid A^*(\psi_d - AS\alpha)\big)_o = 0. \tag{37}$$

These vectors are therefore the solutions of the normal equation

$$S^*A^*(\psi_d - AS\alpha) = 0 \qquad (\alpha \in \mathbb{R}^m), \tag{38}$$

i.e., the least-squares solutions of the equation $AS\alpha$ '=' ψ_d.

In most cases encountered in image reconstruction, the conjugate-gradients method is the best suited technique for solving Eq. (38). The version of this method presented below provides $\phi = S\alpha$.

ALGORITHM 1:

Step 0: Set $\epsilon = 10^{-6}$ (for example) and $n = 0$;
choose a natural starting point ϕ_0 in E;
compute $r_0 = A^*\psi_d - A^*A\phi_0$, $q_0 = \|\psi_d - A\phi_0\|_d^2$;
$\rho_{0,k} = (g_k \mid r_0)_o$ (for all $k \in \mathbb{J}$);
set $\delta_{0,k} = \rho_{0,k}$ (for all $k \in \mathbb{J}$).

Step 1: Compute $d_n = \sum_{k\in\mathbb{J}} \delta_{n,k} g_k$,

$$z_n = A^*Ad_n,$$
$$\zeta_{n,k} = (g_k \mid z_n)_o \quad (\text{for all } k \in \mathbb{J}),$$
$$\omega_n = \sum_{k\in\mathbb{J}} |\rho_{n,k}|^2 / \sum_{k\in\mathbb{J}} \delta_{n,k}\zeta_{n,k},$$
$$q_{n+1} = q_n - (2r_n - \omega_n z_n \mid \omega_n d_n)_o,$$
$$\phi_{n+1} = \phi_n + \omega_n d_n,$$
$$r_{n+1} = r_n - \omega_n z_n,$$
$$\rho_{n+1,k} = \rho_{n,k} - \omega_n \zeta_{n,k} \quad (\text{for all } k \in \mathbb{J});$$

if $\max_{k\in\mathbb{J}}\{|\rho_{n+1,k}|/\|g_k\|_o\} < \epsilon\,\|r_{n+1}\|_o$, termination.

Compute $\gamma_n = \sum_{k\in\mathbb{J}} |\rho_{n+1,k}|^2 / \sum_{k\in\mathbb{J}} |\rho_{n,k}|^2$,

$$\delta_{n+1,k} = \rho_{n+1,k} + \gamma_n \delta_{n,k} \quad (\text{for all } k \in \mathbb{J});$$

increment n and loop to step 1. □

Throughout this algorithm, r_n is the residue of the equation $A^*\psi_d - A^*A\phi = 0$ for $\phi = \phi_n$. Likewise, q_n is the value of $q(\phi)$ at the same iterate:

$$r_n = A^*\psi_d - A^*A\phi_n, \qquad q_n = \|\psi_d - A\phi_n\|_d^2.$$

The iteration in q_n results from the identity:

$$\begin{aligned} q(\phi + \delta\phi) &= q(\phi) - 2(\psi_d - A\phi \mid A\,\delta\phi)_o + \|A\,\delta\phi\|_d^2 \\ &= q(\phi) - 2(A^*\psi_d - A^*A\phi \mid \delta\phi)_o + (A^*A\,\delta\phi \mid \delta\phi)_o. \end{aligned}$$

The sequence of vectors ϕ_n converges towards a solution of the problem with all the remarkable properties of the conjugate-gradients method (see for example [12]). In practice, E is chosen so that A_E is a one-to-one map. The uniqueness of the solution can easily be verified by modifying the starting point of the algorithm. The stopping criterion is based on the fact that the final residue must be practically orthogonal to all the g_k's (Eq. (37)); the cosine of the angle between the vectors r_n and g_k is equal to $\rho_{n,k}/(\|g_k\|_o \|r_n\|_o)$. Here, as $\mathbb{R}^m$ is endowed with its standard scalar product, this algorithm cannot provide the condition number of A_E. (The transposition of what is presented in subsection 4.2 would give the 'generalized condition number' of AS.) We therefore recommend to use algorithm 1 only when κ_E is approximately known.

REMARK 3. — Let us consider the special case where A is the identity operator on K_d (which then coincides with H_o). The problem is then to find $P_E\phi_d$, the projection of $\phi_d = \psi_d$ onto E. Note that κ_E is then equal to unity. In this case, algorithm 1 collapses to

ALGORITHM 2:

Step 0: Set $\epsilon = 10^{-6}$ (for example) and $n = 0$;
set $\phi_0 = 0$ and $r_0 = \phi_d$;
compute $q_0 = \|\phi_d\|_o^2$,
$\rho_{0,k} = (g_k \mid r_0)_o$ (for all $k \in \mathbb{J}$);
set $\delta_{0,k} = \rho_{0,k}$ (for all $k \in \mathbb{J}$).

Step 1: Compute $z_n = \sum_{k\in\mathbb{J}} \delta_{n,k} g_k$,

$$\zeta_{n,k} = (g_k \mid z_n)_o \quad \text{(for all } k \in \mathbb{J}),$$

$$\omega_n = \sum_{k\in\mathbb{J}} |\rho_{n,k}|^2 / \sum_{k\in\mathbb{J}} \delta_{n,k}\zeta_{n,k},$$

$$q_{n+1} = q_n - (2r_n - \omega_n z_n \mid \omega_n z_n)_o,$$

$$\phi_{n+1} = \phi_n + \omega_n z_n,$$

$$r_{n+1} = r_n - \omega_n z_n,$$

$$\rho_{n+1,k} = \rho_{n,k} - \omega_n \zeta_{n,k} \quad \text{(for all } k \in \mathbb{J});$$

if $\max_{k\in\mathbb{J}}\{|\rho_{n+1,k}|/\|g_k\|_o\} < \epsilon \|r_{n+1}\|_o$, termination.

Compute $\gamma_n = \sum_{k\in\mathbb{J}} |\rho_{n+1,k}|^2 / \sum_{k\in\mathbb{J}} |\rho_{n,k}|^2$,

$$\delta_{n+1,k} = \rho_{n+1,k} + \gamma_n \delta_{n,k} \quad \text{(for all } k \in \mathbb{J});$$

increment n and loop to step 1. □

This algorithm converges towards the projection of ϕ_d onto E with all the properties of the conjugate-gradients method.

In principle, the projection operation can also be performed by using the matching pursuit iteration (25). In this case, on setting ω equal to its optimal value, this iteration collapses to

$$\phi_0 = 0, \qquad \phi_{n+1} = \phi_n + \eta_k^2 \rho_{n,k} g_k,$$

where

$$\eta_k = \|g_k\|_o^{-1} \qquad \rho_{n,k} = (g_k \mid r_n)_o \qquad r_n = \phi_d - \phi_n .$$

The residues r_n are then obtained via iteration (27):

$$r_0 = \phi_d , \qquad r_{n+1} = r_n - \eta_k^2 \rho_{n,k} g_k ;$$

and likewise for q_n (cf. Eq. (28)):

$$q_0 = \|\phi_d\|_o^2 , \qquad q_{n+1} = q_n - \eta_k^2 \rho_{n,k}^2 .$$

At each iteration, it is then natural to choose k so that $\eta_k|\rho_{n,k}| = \max_{j\in\mathbb{J}} \eta_j|\rho_{n,j}|$. In the general case where the g_k's ($k \in \mathbb{J}$) do not form an orthogonal set, the conjugate-gradients algorithm is of course preferable.

4. Optimization with control of robustness

For clarity, let us now assume that A_E is a one-to-one map. The method presented in section 3 then yields a solution α of Eq. (38), and thereby the solution of the problem: $\phi_E = S\alpha$. Unfortunately, as already mentioned, this method does not provide any information on the robustness of the reconstruction process. The most natural way of obtaining this information is to find ϕ_E , directly, as the solution of the normal equation (4):

$$B\phi = \phi_d , \tag{39}$$

where

$$B \triangleq A_E^* A_E = P_E A^* A_E \qquad \phi_d \triangleq A_E^* \psi_d = P_E A^* \psi_d . \tag{40}$$

In this section, we present the corresponding developments. To conduct our analysis, the eigenvalues of B are ordered so as to form a nondecreasing sequence (cf. Eq. (7)):

$$\mu \triangleq \mu_1 \leq \mu_2 \leq \cdots \leq \mu' \triangleq \mu_m . \tag{41}$$

As A_E is assumed to be a one-to-one map, μ is strictly positive. In the general case where the g_k generating E do not form an orthogonal set, the reader must keep in mind the fact that the action of P_E can be performed with the aid of algorithm 2.

4.1. RECONSTRUCTION ALGORITHM

The problem is solved by using the conjugate-gradients method (cf. section 2.3 in [12]). Starting from any ϕ_0 in E, the iterates ϕ_n converge to ϕ_E in at most m iterations, ϕ_n getting closer to ϕ_E at each iteration. In this algorithm, d_n is the 'direction of research' in iteration $n+1$, whereas ω_n is the corresponding 'parameter of exact line search;' r_n is the residue of the normal equation (39) for $\phi = \phi_n$:

$$r_n \triangleq \phi_d - B\phi_n .$$

As $B\phi_E = \phi_d$, we have $r_n = B(\phi_E - \phi_n)$, hence:

$$\|\phi_E - \phi_n\|_o \leq \frac{1}{\mu}\|r_n\|_o .$$

Denoting by μ_e an estimate of μ, we therefore have:

$$\frac{\|\phi_E - \phi_n\|_o}{\|\phi_n\|_o} \lesssim \epsilon_n, \qquad \epsilon_n \triangleq \frac{\|r_n\|_o}{\mu_e\|\phi_n\|_o}.$$

Let us introduce an acceptable error threshold ϵ_ϕ for $\|\phi_E - \phi_n\|_o/\|\phi_n\|_o$. Clearly, the iterative process can be interrupted as soon as ϵ_n is less than ϵ_ϕ ; ϵ_n therefore plays the role of a convergence estimator. The estimate of μ is refined throughout the iterative process as indicated in subsection 4.2. The corresponding algorithm can then be summarized as follows:

ALGORITHM 3:

Step 0: Set $\epsilon_\phi = 10^{-4}$ (for example) and $n = 0$;
choose a natural starting point ϕ_0 in E;
compute $r_0 = \phi_d - B\phi_0$;
set $d_0 = r_0$.

Step 1: Compute $z_n = Bd_n$,

$$\omega_n = \|r_n\|_o^2/(d_n \mid z_n)_o,$$

$$\phi_{n+1} = \phi_n + \omega_n d_n,$$

$$r_{n+1} = r_n - \omega_n z_n .$$

$$\epsilon_{n+1} = \|r_{n+1}\|_o/(\mu_e\|\phi_{n+1}\|_o);$$

if $\epsilon_n \leq \epsilon_\phi$, termination.

Compute

$$\gamma_n = \|r_{n+1}\|_o^2/\|r_n\|_o^2,$$

$$d_{n+1} = r_{n+1} + \gamma_n d_n.$$

Increment n and loop to step 1. □

4.2. EFFECTIVE OBJECT REPRESENTATION SPACE

In the conjugate-gradients method, the n-dimensional subspace of E generated by the conjugate directions $d_0, \ldots, d_{n-1}$,

$$E_n \triangleq \text{span}\{d_\ell\}_{\ell=0}^{n-1},$$

is referred to as the Krylov space of order n. According to a well known property (cf. Properties 2 and 3.1 in [12]), ϕ_n minimizes $q(\phi)$ on E_n.

Provided that n is sufficiently large, the least-squares solutions in E and E_n are very close to one another. At the end of the reconstruction process, E_n is therefore the effective object representation space. The dimension of this space, as well as the robustness of the reconstruction process, depends upon the localization of the eigenvalues of B, and more precisely, on the relative weight of the projections of r_0 onto the corresponding eigenspaces. We are thus led to consider the operator

$$B_n : E_n \to E_n\,, \quad B_n\phi \triangleq P_n A^* A\phi,$$

where P_n is the projection (operator) onto E_n.

The residues $r_0, \ldots, r_{n-1}$ form an orthogonal basis for E_n (see appendix 4 in [12]). As established in [2] (appendix 2), the matrix of B_n expressed in this basis is tridiagonal (this matrix is of course symmetric). Its diagonal and upper-diagonal elements are respectively given by the relationships

$$b_{n:\ell,\ell} = \begin{cases} \dfrac{1}{\omega_\ell} & (\ell = 0) \\[2ex] \dfrac{1}{\omega_\ell} + \dfrac{\gamma_{\ell-1}}{\omega_{\ell-1}} & (1 \le \ell \le n-1) \end{cases}$$

and

$$b_{n:\ell-1,\ell} = -\frac{\sqrt{\gamma_{\ell-1}}}{\omega_{\ell-1}} \qquad (1 \le \ell \le n-1).$$

The eigenvalues of B_n can therefore be calculated very easily with the aid of the QR algorithm (cf. section 11.3 in [13]). Let us order these eigenvalues as those of B (see Eq. (41)):

$$\mu_{n,1} \le \mu_{n,2} \le \cdots \le \mu_{n,n}\,.$$

By referring to the eigenvalue analysis based on the notion of 'minmax numbers' (cf. appendix 5 in [5]), it is easy to show that

$$\mu \le \mu_{n+1,1} \le \mu_{n,1}\,, \qquad \mu_{n,n} \le \mu_{n+1,n+1} \le \mu'.$$

Provided that the projections of $\phi_{\rm d}$ onto the eigenspaces corresponding to μ and μ' are different from zero, a condition which is always numerically satisfied in practice, $\mu_{n,1}$ and $\mu_{n,n}$ respectively tend to μ and μ' as n tends to m (see Fig. 3 in [2]).

In our reconstruction processes, the eigenvalues of B_n are computed at each iteration. (The cost for this is negligible compared to that of the action of B.) As soon as $(\mu_{n,1} - \mu_{n+1,1})/\mu_{n,1}$ is less than say 10^{-3},

$$\mu_e = \mu_{n,1}\,, \qquad \mu'_e = \mu_{n,n}\,,$$

are very good approximations to μ and μ', respectively. In most cases, the termination test of the basic algorithm is then satisfied (see Fig. 3 in [2]).

5. How WIPE wipes CLEAN clean

We now have all the tools for analysing the weak points of CLEAN as well as the tricks of WIPE allowing the corresponding difficulties to be overcome.

In situations of astrophysical interest, CLEAN is implemented with a value of the relaxation parameter ω much less than 1 (say 0.2). The basis vectors s_p selected in the matching pursuit process then define an acceptable object representation space E. Unfortunately, the problem is often ill conditioned; A_E is a one-to-one map, but its condition number is very large. For example, in the simulation presented in Fig. 7d, κ_E is equal to 45.08. As a result, ϕ_E is a very perturbed version of the clean map. This is unsatisfactory. Indeed, in this situation, the clean map can only be regarded as an image for which the fit criterion C_e (introduced in Eq. (35)) is of the order of the threshold value (say 2). In other words, the clean map must be accepted as it is, without any reference to a stable optimization process. The interpretation of the results may then be doubtful. As indicated in subsection 5.1, the regularization principle of WIPE remedies this lack of robustness, but the regularized version of CLEAN thus obtained is still different from WIPE. Indeed, CLEAN has another weak point: the boundaries of the structuring entities of the image may not be correctly represented in the clean map. In such situations, the matching pursuit strategy of CLEAN is not well suited for solving the problem. For example, in the particular case of the Fourier data of our simulation, the best possible fit corresponds to a value of C_e of about 0.9 (see Fig. 9 further on). Even with a good support constraint (a reasonable 'clean box'), CLEAN cannot reach this fit threshold with a satisfactory representation of the image support. In the same situation, as shown in subsection 5.2, WIPE reaches this threshold without any difficulty.

5.1. HOW WIPE REGULARIZES CLEAN

In the basic version of CLEAN presented in subsection 2.2, the functions $\phi \in E$ are linear combinations of atoms s_p. In the sense defined in subsection 1.2, the 'energy' of the Fourier transform of each atom s_p is concentrated in $\mathcal{H}_s$ (cf. Eq. (15)). The intrinsic instability of CLEAN is related to the fact that this property does not necessarily hold for any linear combination of such atoms. This difficulty arises any time the distances between these atoms are much smaller than the corresponding resolution limit. This is precisely the case when dealing with extended objects and a small relaxation parameter ω.

To overcome this difficulty, WIPE suggests that CLEAN should define the reconstructed image as the function minimizing on E the functional

$$q(\phi) \triangleq q_e(\phi) + q_r(\phi) \tag{42}$$

where (see Eq. (29))

$$q_e(\phi) \triangleq \sum_{u \in \mathcal{L}_e} |\psi_s(u) - \widehat{\phi}(u)|^2 W(u)\,(\delta u)^2$$

and

$$q_r(\phi) \triangleq \sum_{u \in \mathcal{L}_r} |\widehat{\phi}(u)|^2 \, (\delta u)^2. \tag{43}$$

The experimental criterion q_e constrains ϕ (the model) to be consistent with the damped Fourier data, while the regularization criterion q_r penalizes the high-frequency components of ϕ. The elements of the regularization frequency list $\mathcal{L}_r$ are located outside $\mathcal{H}_s$, at the nodes of grid $\mathbb{H}_r \, \delta u$ where

$$\mathbb{H}_r \triangleq \{r \in \mathbb{Z}^2 : r\,\delta u \in \mathcal{H},\; r\,\delta u \notin \mathcal{H}_s\}.$$

In the traditional version of CLEAN, q reduces to q_e. The minimization of q_e on E then reveals the ill-conditioned character of the problem.

The regularized version of CLEAN corresponding to criterion (42) can be formulated in the theoretical framework presented in subsection 1.1. To clarify this point, let us introduce the data vector:

$$\psi_d(u) \triangleq \begin{cases} \psi_s(u) & \text{on } \mathcal{L}_e; \\ 0 & \text{on } \mathcal{L}_r. \end{cases}$$

This vector lies in a data space K_d endowed with the scalar product:

$$(\psi_1 \mid \psi_2)_d \triangleq \sum_{u \in \mathcal{L}_e} \overline{\psi}_1(u)\psi_2(u)\, W(u)\, (\delta u)^2 + \sum_{u \in \mathcal{L}_r} \overline{\psi}_1(u)\psi_2(u)\, (\delta u)^2.$$

The Fourier sampling operator A is then the operator:

$$A : H_o \to K_d, \qquad (A\phi)(u) \triangleq \begin{cases} \widehat{\phi}(u) & \text{on } \mathcal{L}_e; \\ \widehat{\phi}(u) & \text{on } \mathcal{L}_r. \end{cases}$$

According to Eqs. (42), (29) and (43), $q(\phi)$ can then be effectively written in the form $\|\psi_d - A\phi\|_d^2$ (see Eq. (1)). The problem is then stated in terms of Fourier interpolation (cf. [1] and [5]). This is why q is of the form $q_e + \alpha q_r$ with $\alpha = 1$. In this context, it is important to note that in the definition of q_e (Eq. (29)), where ψ_s is defined in Eq. (14), the weighting function $W(u)$ takes into account the local redundancy of u up to δu (Eq. (17)).

The algorithms described in sections 3 and 4 can be used for minimizing q on E. The action of A^*A is that of a convolutor. The corresponding point spread function, the 'dusty beam,' has two components: the dirty beam and the 'regularization beam.' The latter is induced by the regularization list $\mathcal{L}_r$. With regard to the dusty map, note that $A^*\psi_d \equiv A_e^*\psi_s$ (cf. subsection 2.2).

In the simulation presented in Fig. 8, we compare the clean map of Fig. 7d (for which $C_e = 1.99$) with the image provided by this optimization process ($C_e = 1.15$). The condition number is now reasonable: $\kappa_E = 2.62$. As shown in the next subsection, the construction of the object representation space E can, however, be refined.

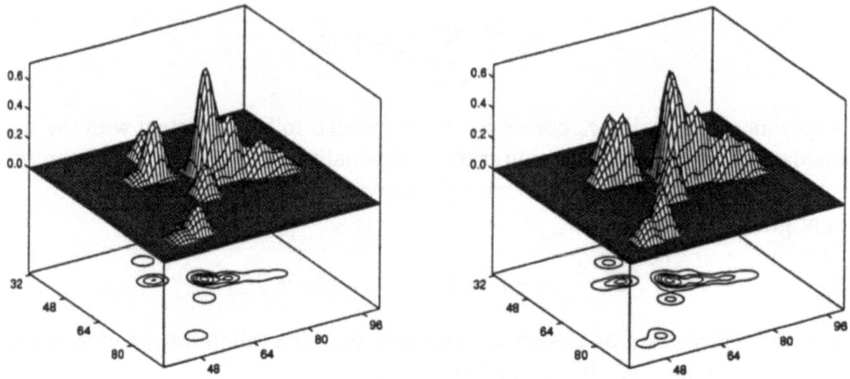

(a) (b)

FIGURE 8. Image reconstruction via the regularized version of CLEAN; a: traditional clean map for $C_e = 1.99$; b: improved clean map ϕ_E provided by the regularized version of CLEAN ($C_e = 1.15$). These images have to be compared with the image to be reconstructed (Fig. 7c). As shown in Fig. 9b, the matching pursuit process of WIPE can still refine this image.

5.2. HOW WIPE RELAXES THE MATCHING PURSUIT PROCESS OF CLEAN

In the regularized version of CLEAN described in subsection 5.1, the calculation of the condition number κ_E requires the action of the projection P_E at each iteration of algorithm 3. This projection is performed at the cost of the conjugate-gradients iterations of algorithm 2. This first remark suggests that E should be redefined as the linear space generated by the elementary particles $e_{p'}$ of all the atoms s_p selected in the matching pursuit process of CLEAN. The projection onto E is then trivial since these elementary particles form an orthogonal set. As the resolution limit of the reconstruction process is then controlled by the regularizer q_r (cf. Eq. (43)), the choice of such an object representation space proves to be very natural. Moreover, the definition of E can then be refined by continuing (or even by conducting) the matching pursuit process at the level of the elementary particles. As specified below, this is what is precisely done in WIPE.

Let $\mathbb{D}$ then be the subset of $\mathbb{G}$ corresponding to the choice of $\mathcal{E}$. (The elementary particles generating E are centred on the nodes of $\mathbb{D}\,\delta\xi$.) We say that $\mathbb{D}$ is the 'discrete field (or support)' associated with the definition of E. Depending on the particular problems to be solved, this discrete field may be fixed from the outset (for example, in an interactive manner), or constructed step by step in a matching pursuit strategy.

In this last case, which corresponds to the basic version of WIPE, let us denote by $\mathbb{D}^{(i)}$ the discrete field obtained at the end of the ith step of the construction of the object representation space. Let $\phi^{(i)}$ then be the solution of the problem in the corresponding object representation space $E^{(i)}$. In the basis of the elementary particles e_p (the interpolation basis of H_o), the scalar components of the residue $r^{(i)}$ are the quantities:

$$\rho_p^{(i)} \triangleq (e_p \mid r^{(i)})_o \qquad r^{(i)} \triangleq A^*\psi_s - A^*A\phi^{(i)} .$$

According to the definition of $\phi^{(i)}$, these coefficients vanish on $\mathbb{D}^{(i)}$ (see Eq. (2) with $g_k \equiv e_p, p \in \mathbb{D}^{(i)}$). One then has to decide whether the current field has to be extended. The current values of C_e and κ_E play an essential role in this decision. When the reconstruction procedure is not interrupted at this stage, WIPE uses algorithm 3 for computing the solution of the problem in the object representation space relative to the union of $\mathbb{D}^{(i)}$ with some set $\mathbb{D}' \subset \mathbb{G}$: $\mathbb{D}^{(i+1)} = \mathbb{D}^{(i)} \cup \mathbb{D}'$.

There exist many ways of selecting $\mathbb{D}'$. All are based on the examination of the distribution of the coefficients $\rho_p^{(i)}$ outside $\mathbb{D}^{(i)}$. For example, one may try to define $\mathbb{D}'$ as a connected region containing the 'pixel' $p_{\max}$ for which the maximum of these coefficients is attained. The simplest choice is then to define $\mathbb{D}'$ as the discrete field of the atom s centred on this pixel. With regard to the construction of the object representation space, the corresponding version of WIPE is then very similar to that of CLEAN.

In the matching pursuit steps where the field of the reconstructed image must be refined, it is natural to choose the nodes of $\mathbb{D}'$ along the boundaries of the structuring entities of the image. Let N_s be the number of particles involved in the linear combination defining the neat beam s (the number of nodes in $\mathbb{D}_s$). In the basic version of WIPE, the size of $\mathbb{D}'$, expressed in number of nodes, is defined as a fraction of N_s (say $N_s/2$), and the selected nodes are those for which the coefficients $\rho_p^{(i)}$ are the largest above some given threshold (half of the maximal value, for example). The field of the image (or object) to be reconstructed can thus be obtained in a natural manner.

The construction of the object representation space is interrupted as soon as the fit criterion $C_e(\phi^{(i)})$ is sufficiently small, for instance, less than or of the order of 0.85. The current field is then refined by proceeding to a morphological smoothing of its connected entities. In this classical operation of mathematical morphology, the discrete support of the neat beam, $\mathbb{D}_s$, is of course used as structuring element. The boundaries of the effective field of the 'neat map' (the reconstructed image) are thus defined at the appropriate resolution. In particular, the connected entities of size smaller than that of $\mathbb{D}_s$ are eliminated. As illustrated in Fig. 9, it is thus possible to reach the optimal value of C_e (0.88 in the simulation under consideration) with a satisfactory representation of the image field.

Let E be the object representation space at the end of the action of WIPE, and $\mathbb{D}$ be the corresponding discrete field. There exists a variant of WIPE, in which the object representation space is a particular subspace of E, that generated by all the atoms s_p whose discrete field is contained in $\mathbb{D}$. Note that these atoms are involved in the operation of morphological smoothing. In the conditions of the simulation presented in Fig. 9, the corresponding solution is very close to that provided by WIPE. As expected, the condition number is then slightly smaller (here, 3.36 instead of 3.83).

From the outset, the discrete field $\mathbb{D}$ may be taken equal to that of the clean box. One then uses the global version of WIPE in which the nonnegativity constraint is imposed (cf. subsection 4.3 in [2]). At the end of the corresponding reconstruction process, the fit

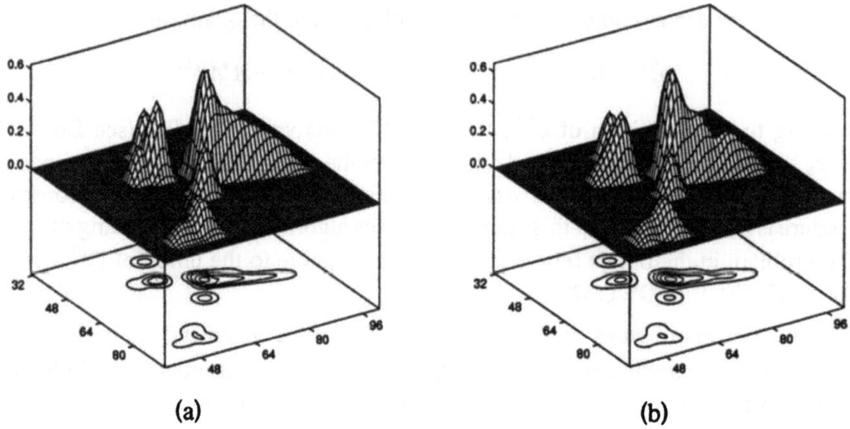

FIGURE 9. Image reconstruction through WIPE; a: image to be reconstructed; b: neat map (reconstructed image). The latter, for which $C_e = 0.88$ and $\kappa_E = 3.83$ (and which was obtained without any clean box), is to be compared to image (a) and to the maps presented in Fig. 8. The boundaries of the structuring entities of the image are now correctly restored, hence a better intensity distribution. The unreliable character of the oscillating perturbation along the main structuring entity of the reconstructed images is revealed by the image-eigenmode analysis provided by WIPE (see Fig. 6 in [2]).

criterion C_e is often smaller than its optimal value. As a result, the support of the image (or object) to be reconstructed is not well restored. A similar remark can be made for the Fourier synthesis methods in which the regularization principle is based on the concept of entropy. Moreover, the relative weights of the experimental and regularization criteria must then be carefully chosen (cf. [6] and [14]). The strategy adopted in the basic version of WIPE is therefore preferable; its implementation is simpler and more efficient.

The condition "C_e of the order of 1 with κ_E less than say 5, with a sufficiently small value of $\|\sigma_s\|_e/\|\psi_s\|_e$" often suffices to ensure a good solution to the problem, but strictly speaking, this is not a sufficient condition. The complete control must be based on a multiresolution strategy. The corresponding developments will be presented in a forthcoming paper.

Appendix 1. Notion of condition number

For any $\phi \neq 0$ in E, we have from the definitions of μ and μ' given in Eq. (7):

$$\mu \leq \frac{\|A_E\phi\|_d^2}{\|\phi\|_o^2} \leq \mu'.$$

For $\phi = \delta\phi_E$, the first inequality gives

$$\mu \leq \frac{\|\delta\psi_F\|_d^2}{\|\delta\phi_E\|_o^2},$$

whereas for $\phi = \phi_E$ the second yields

$$\frac{\|\psi_F\|_{\mathrm{d}}^2}{\|\phi_E\|_{\mathrm{o}}^2} \leq \mu'.$$

By combining these inequalities, it follows that

$$\mu \frac{\|\psi_F\|_{\mathrm{d}}^2}{\|\phi_E\|_{\mathrm{o}}^2} \leq \mu' \frac{\|\delta\psi_F\|_{\mathrm{d}}^2}{\|\delta\phi_E\|_{\mathrm{o}}^2},$$

hence:

$$\frac{\|\delta\phi_E\|_{\mathrm{o}}}{\|\phi_E\|_{\mathrm{o}}} \leq \frac{\sqrt{\mu'}}{\sqrt{\mu}} \frac{\|\delta\psi_F\|_{\mathrm{d}}}{\|\psi_F\|_{\mathrm{d}}}.$$

The ratio $\sqrt{\mu'/\mu}$ is referred to as the condition number of A_E.

Appendix 2. Convergence property

Using the notation introduced in subsection 2.1, we have

$$\|z_{n+1}\|_{\mathrm{d}}^2 = \|z_n\|_{\mathrm{d}}^2 - \omega(2-\omega)\|Q_k z_n\|_{\mathrm{d}}^2,$$

i.e., from Eq. (19):

$$\beta_{n+1} = \beta_n - \omega(2-\omega)\eta_k^2 (h_k \mid z_n)_{\mathrm{d}}^2.$$

Thus (cf. Eq. (22)),

$$\beta_{n+1} \leq \beta_n - C \max_{1\leq j\leq M} (h_j \mid z_n)_{\mathrm{d}}^2,$$

where $C \triangleq \omega(2-\omega)\min_{1\leq j\leq M}\eta_j^2$. As the relaxation parameter ω is supposed to lie in the open interval $(0,2)$, C is strictly positive. As shown in remark A2, there exists a positive constant C' such that for all z in F_{d}, we have:

$$\max_{1\leq j\leq M} (h_j \mid z)_{\mathrm{d}}^2 \geq C'\|z\|_{\mathrm{d}}^2.$$

As a result, $\beta_{n+1} \leq \beta_n - C''\beta_n$ with $C'' \triangleq CC' > 0$.

Let us now assume that β is different from 0. There then exists n such that

$$0 < \beta_n - \beta < C''\beta,$$

hence $\beta_{n+1} \leq \beta_n - C''\beta < \beta$. This is impossible, since β_{n+1} must be greater than β. Consequently, $\beta = 0$.

REMARK A2. The property in question can be established as follows. Consider the operator on F_{d}:

$$Rz \triangleq \sum_{j=1}^{M} (h_j \mid z)_{\mathrm{d}} h_j.$$

For any z and z' in F_d, we have:

$$(z \mid Rz')_\mathrm{d} = \sum_{j=1}^{M} (h_j \mid z')_\mathrm{d}\,(z \mid h_j)_\mathrm{d} = \sum_{j=1}^{M} (h_j \mid z)_\mathrm{d}\,(z' \mid h_j)_\mathrm{d} = (z' \mid Rz)_\mathrm{d}\,.$$

This identity shows that R is self-adjoint. Moreover, as $(z \mid Rz)_\mathrm{d} = \sum_{j=1}^{M} (h_j \mid z)_\mathrm{d}^2$, the condition $(z \mid Rz)_\mathrm{d} = 0$ implies $z = 0$; R is therefore positive definite. The fact that F_d is of finite dimension then implies that the smallest eigenvalue of R is strictly positive. Consequently, for any $z \in F_\mathrm{d}$,

$$\sum_{j=1}^{M} (h_j \mid z)_\mathrm{d}^2 \geq \lambda \|z\|_\mathrm{d}^2 \qquad (\lambda > 0).$$

It follows immediately that $\max_{1 \leq j \leq M} (h_j \mid z)_\mathrm{d}^2 \geq C' \|z\|_\mathrm{d}^2$ with $C' \triangleq \lambda / M$.

Appendix 3. Dirty map and dirty beam

We first show that the dirty map is the map of the scalar components of $A_\mathrm{e}^* \psi_\mathrm{e}$ in the basis of the elementary particles e_p. According to Eqs. (11) and (10), $A_\mathrm{e}^* \psi_\mathrm{e}$ can be expanded in the form $A_\mathrm{e}^* \psi_\mathrm{e} = \sum_{p \in G} x_{\mathrm{e},p}\, e_p$, where

$$x_{\mathrm{e},p} = \frac{1}{(\delta \xi)^2} (e_p \mid A_\mathrm{e}^* \psi_\mathrm{e})_\mathrm{o} = \frac{1}{(\delta \xi)^2} (A_\mathrm{e} e_p \mid \psi_\mathrm{e})_\mathrm{d}\,.$$

As

$$\widehat{e}_p(u) = \widehat{e}_0(u) \exp\left(-2i\pi p \cdot \frac{u}{\Delta u}\right)$$

with

$$\widehat{e}_0(u) = \frac{1}{(\Delta u)^2} \operatorname{rect}\left(\frac{u_1}{\Delta u}\right) \operatorname{rect}\left(\frac{u_2}{\Delta u}\right),$$

it then follows from Eq. (12) that

$$x_{\mathrm{e},p} = \frac{1}{(\delta \xi\, \Delta u)^2} \sum_{u \in \mathcal{L}_\mathrm{e}} W(u) \psi_\mathrm{e}(u) \exp\left(2i\pi p \cdot \frac{u}{\Delta u}\right) (\delta u)^2,$$

hence, since $\delta \xi\, \Delta u = 1$:

$$x_{\mathrm{e},p} = \sum_{u \in \mathcal{L}_\mathrm{e}} W(u) \psi_\mathrm{e}(u) \exp\left(2i\pi p \cdot \frac{u}{\Delta u}\right) (\delta u)^2.$$

This explicitly shows that $A_\mathrm{e}^* \psi_\mathrm{e}$ can be identified with the dirty map (see for example Fig. 7b).

The action of A_e^* corresponds to a 'back Fourier sampling operation.' The dirty map looks like the inverse Fourier transform of $W\psi_\mathrm{e}$, but from a mathematical point of view, it isn't. Indeed, $W\psi_e$ is a vector in the experimental data space K_e and not the distribution $\sum_{u\in\mathcal{L}_e} W(u)\psi_e(u)\delta_u$. When considering the basic versions of CLEAN and WIPE, this distinction may seem to be a 'mathematical stylishness,' but this is not the case, for example, in multifrequency Fourier synthesis (see the context of Eq. (68) in [2]).

Let us now consider the action of $A_e^*A_e$ on any $\phi \in H_o$. Setting $\Phi \triangleq A_e^*A_e\phi$, and expanding ϕ and Φ in the forms

$$\phi = \sum_{p\in\mathbb{G}} x_p e_p\,, \qquad \Phi = \sum_{p\in\mathbb{G}} X_p e_p\,,$$

we have:

$$X_p = \frac{1}{(\delta\xi)^2}(e_p \mid A_e^*A_e\phi)_o = \frac{1}{(\delta\xi)^2}\sum_{p'\in\mathbb{G}} x_{p'}(e_p \mid A_e^*A_e e_{p'})_o$$

$$= \frac{1}{(\delta\xi)^2}\sum_{p'\in\mathbb{G}} (A_e e_p \mid A_e e_{p'})_d x_{p'}\,.$$

By using the same arguments as above, it then follows that X_p can be written in the form

$$X_p = \sum_{p'\in\mathbb{G}} h_{e,p-p'} x_{p'}$$

where

$$h_{e,p} \triangleq \frac{1}{(\Delta u)^2}\sum_{u\in\mathcal{L}_e} W(u)\exp\left(2i\pi p\cdot\frac{u}{\Delta u}\right)(\delta u)^2.$$

Note that $h_{e,-p} = h_{e,p}$ and $(\delta u/\Delta u)^2 = 1/N^2$. Let $\mathbb{G}'$ be the grid twice as large as $\mathbb{G}$:

$$\mathbb{G}' \triangleq \mathbb{L}' \times \mathbb{L}', \qquad \mathbb{L}' \triangleq \{p \in \mathbb{Z} : -N \le p \le N-1\}\,.$$

The map of the coefficients $h_{e,p}$ on $\mathbb{G}'$ defines what is referred to as the dirty beam DB (see for example Fig. 7a). An expression such as

$$\mathrm{DB} \underline{\star} \phi$$

then denotes the vector (lying in H_o) whose scalar components are given by the discrete convolution:

$$\sum_{p'\in\mathbb{G}} h_{e,p-p'} x_{p'} \qquad (p \in \mathbb{G}).$$

As a result, in the general case where the nonzero components of ϕ are distributed all over grid $\mathbb{G}$, the operation $\mathrm{DB} \underline{\star} \phi$ is performed by implementing the FFT algorithm on grid $\mathbb{G}'$.

When N is large and the experimental frequency list very long, the direct calculation of the dirty map and the dirty beam may be very time-consuming. To save computer time, it is then preferable to use appropriate Fast Fourier Sampling techniques. The complete description of these FFS algorithms is given in [2] (section 3).

Appendix 4. On the traditional version of CLEAN

In the classical presentation of CLEAN [3, 4], the Fourier data are not damped by $\widehat{s}$, and the convolution by the clean beam is performed *a posteriori*. More precisely, the successive clean maps of the traditional version of CLEAN are given by the convolutions $\phi_n = \mathrm{CB} \underline{\star} \varphi_n$, where the iteration in φ is defined by the formula:

$$\varphi_0 = 0, \qquad \varphi_{n+1} = \varphi_n + \omega \frac{r_{\mathrm{e},n,p}}{h_{\mathrm{e},0}} e_p\,;$$

$r_{\mathrm{e},n,p}$ denotes the scalar component of the 'experimental residue' $r_{\mathrm{e},n} \triangleq \mathrm{DM_e} - \mathrm{DB} \underline{\star} \varphi_n$ at pixel p, and $h_{\mathrm{e},0}$ that of the dirty beam at the origin:

$$r_{\mathrm{e},n,p} = \frac{1}{(\delta\xi)^2}(e_p \mid r_{\mathrm{e},n})_\mathrm{o} \qquad h_{\mathrm{e},0} = \frac{1}{(\delta\xi)^2}(e_0 \mid A_\mathrm{e}^* A_\mathrm{e} e_0)_\mathrm{o}\,.$$

The relaxation parameter ω is referred to as the 'loop gain.' At each iteration, p is chosen so that $r_{\mathrm{e},n,p} = \max_{p' \in \mathrm{G_c}} |r_{\mathrm{e},n,p'}|$. The experimental residue $r_{\mathrm{e},n}$ is obtained iteratively according to the formula:

$$r_{\mathrm{e},0} = \mathrm{DM_e}\,, \qquad r_{\mathrm{e},n+1} = r_{\mathrm{e},n} - \omega \frac{r_{\mathrm{e},n,p}}{h_{\mathrm{e},0}} \mathrm{DB}_p\,.$$

Note that $\mathrm{DB}_p = \mathrm{DB} \underline{\star} e_p$.

The objective of the iteration in φ is to minimize the quadratic functional:

$$\|\psi_\mathrm{e} - A_\mathrm{e}\varphi\|_\mathrm{e}^2 = \sum_{u \in \mathcal{L}_\mathrm{e}} |\psi_\mathrm{e}(u) - \widehat{\varphi}(u)|^2\, W(u)\, (\delta u)^2.$$

This iteration is nothing but the matching pursuit iteration (25) in which the vectors g_k are basis vectors e_p, p lying in the clean box (cf. Eqs. (20) and (26); see also appendix 3). Let us note in passing that in the traditional version of CLEAN, "these basis vectors are more or less dealt with as δ functions" centred on the nodes of grid $\mathbb{G}$.

As far as the connection with our formulation of CLEAN is concerned, the related iterations in ϕ and r are of the form

$$\phi_0 = 0, \qquad \phi_{n+1} = \phi_n + \omega \frac{r_{\mathrm{e},n,p}}{h_{\mathrm{e},0}} \mathrm{CB}_p$$

and

$$r_0 = \mathrm{DM}_s\,, \qquad r_{n+1} = r_n - \omega \frac{r_{\mathrm{e},n,p}}{h_{\mathrm{e},0}} (\mathrm{DB} \underline{\star} \mathrm{CB}_p).$$

These iterations slightly differ from those introduced in our presentation of CLEAN (Eqs. (31) and (33)). The main difference is that the selected successive pixels p are not necessarily the same: in our version of CLEAN, p is chosen, at each iteration, so that $|\rho_{n,p}| = \max_{p' \in \mathrm{G_c}} |\rho_{n,p'}|$ (see Eq. (32)). According to the analysis presented in section 5, the weak points of CLEAN related to its intrinsic instability unfortunately remain the same. As the objective of CLEAN is to find an approximation to the object function convolved by the clean beam, it is therefore more natural to consider that, basically, CLEAN is the matching pursuit process described in subsection 2.2.

6. References

1. Lannes, A., Anterrieu, E. and Bouyoucef, K. (1994) Fourier interpolation and reconstruction via Shannon-type techniques. Part I: Regularization principle, *J. Mod. Opt.* **41**, 1537–1574.
2. Lannes, A., Anterrieu, E. and Bouyoucef, K. (1996) Fourier interpolation and reconstruction via Shannon-type techniques. Part II: Technical developments and applications, *J. Mod. Opt.* **43**, 105–138.
3. Högbom, J. A. (1974) Aperture synthesis with a non-regular distribution of interferometer baselines *Astron. Astrophys. Suppl.* **15**, 417-426.
4. Schwarz, U. J. (1978) Mathematical-statistical description of the iterative beam removing technique (method CLEAN) *Astron. Astrophys.* **65**, 345–356.
5. Lannes, A., Roques, S. and Casanove, M.-J. (1987) Stabilized reconstruction in signal and image processing. Part I: Partial deconvolution and spectral extrapolation with limited field *J. Mod. Optics* **34**, 161–226.
6. Maréchal, P. and Lannes A. (1996) Unification of some deterministic and probabilistic methodologies via the principle of maximum entropy on the mean, submitted to *Inverse Problems*.
7. Lannes, A. (1990) Remarkable algebraic structures of phase-closure imaging and their algorithmic implications in aperture synthesis *J. Opt. Soc. Am. A* **7**, 500–512.
8. Lannes, A. (1991) Phase and amplitude calibration in aperture synthesis. Algebraic structures *Inverse Problems* **7**, 261–298.
9. Anterrieu, E., Guilloteau, S. and Lannes, A. (1991) Spectral reduction of phasor closure data and stabilized image reconstruction via a new hybrid procedure, in F. Merkle (ed.), Proc. ESO conference on High Resolution Imaging by Interferometry II, Garching bei München, FRG, 889–897.
10. Mallat, S. and Zhang, Z. (1993) Non-linear adaptive time-frequency decomposition, in Y. Meyer and S. Roques (eds.), Progress on Wavelet Analysis and Applications, Éditions Frontières, Gif-sur-Yvette (France), pp. 155–164.
11. Slepian, D. (1965) Analytic solution of two apodization problems *J. Opt. Soc. Am.* **55**, 1110-1115.
12. Lannes, A., Casanove, M.-J. and Roques S. (1987) Stabilized reconstruction in signal and image processing. Part II: Iterative reconstruction with and without constraint. Interactive implementation *J. Mod. Optics* **34**, 321–370.
13. Press, W. H., Teukolsky, S. A., Vetterling, W. T. and Slannery B. P. (1992) Numerical recipies in C. The art of scientific computing, second edition, Cambridge, New York, New Rochelle, Melbourne and Sydney: Cambridge University Press.
14. Cornwell T. J. (1983) A method of stabilizing the clean algorithm *Astron. Astrophys.* **121**, 281–285.

RADIO INTERFEROMETRY

S.GUILLOTEAU
Institut de Radio Astronomie Millimétrique
300 Rue de la Piscine
F-38406 Saint Martin d'Hères

1. Introduction

Because of the relatively long wavelengths, the diffraction phenomenon is particularly critical at radio frequencies and very large antennas have to be built to obtain a significant angular resolution. For example, in the mm domain, a 15-m antenna has about the same angular resolution as the human naked eye in the visible. Yet, building even larger monolithic antennas is difficult because of the necessity to maintain the surface accuracy to a level consistent with the wavelengths. This becomes problematic because of two basic physical problems: gravity and stress over the antenna surface, which limit the size of fully steerable antennas, and thermal effects. The first problem is partly overcomed by *homologous* designs, in which the antenna deforms with elevation, but stays parabolic to some extent (though with a different focus), and/or by using newly invented material with higher stiffness-to-weight ratios, such as carbon fiber. Carbon fiber, as well as invar, have also been used to overcome the thermal limit, but the largest mm antennas such as the IRAM 30-m or the JCMT sub-mm telescope require thermal control. Yet, these telescopes are currently limited to about 10″ resolution, and building larger instruments become impractical.

Fortunately, the radio domain is well adapted to electronic devices. Amplifiers, mixers, fast samplers and digital logic, phase-locked local oscillators, and relatively cheap transport systems (through cables or more recently optical fibers) are readily available. Making heterodyne interferometry is possible, and the high angular resolution can be obtained by placing reasonable size elements at large distances from one another, and combining their outputs to produce interference fringes.

Radio interferometry has been described in many excellent textbooks

A-M. Lagrange et al. (eds.), High Angular Resolution in Astrophysics, 145–163.
© 1997 *Kluwer Academic Publishers.*

or workshop proceedings, and one cannot reasonably discuss all its aspects in a single lecture. Particularly recommended lectures are to be found in "Interferometry and Synthesis in Radio Astronomy", by A.R.Thompson, J.M.Moran and G.W.Swenson Jr., (Wiley-Interscience Editions), and "Synthesis Imaging", a proceeding from NRAO Summer School edited by R.A.Perley, F.R.Schwab and A.H.Bridle.

Rather than attempting to demonstrate or summarize all the properties of a radio interferometer, I have chosen to highlight very few essential properties with a detailed derivation. Other important properties are exposed but not demonstrated, and some particularities just briefly mentioned but not discussed. The goal is to offer potential users an overview of the main instrumental problems appearing specifically in data obtained with phase-linked radio interferometers (hence VLBI is excepted). The imaging process, which is treated independently in these series of lectures, is only briefly mentionned but should not be underestimated by the "astronomer", as it is an essential part of the analysis.

2. Heterodyne Detection

To understand the principle of radio interferometry, a minimum knowledge of its "basic block", i.e. the `antenna` is important. The antenna is a reflecting surface which focusses the incident electromagnetic waves. At its focus, a `receiver` system detects the incoming waves and transform them into useable (electronic) signals. The relation between the incident wave and the output of the receiver is essential.

2.1. BASIC PRINCIPLE

The `antenna` produces a Voltage proportional to the linear superposition of the incident electric field pattern. For a simple monochromatic case:

$$U(t) = E\cos(2\pi\nu t + \Phi) \tag{1}$$

In the `receiver`, a `mixer` superimposes the field generated by a `local oscillator` to the antenna output.

$$U_{LO}(t) = Q\cos(2\pi\nu_{LO}t + \Phi_{LO}) \tag{2}$$

The mixer is a non-linear element (diode) whose output is

$$I(t) = a_0 + a_1(V(t)+V_{LO}(t)) + a_2(V(t)+V_{LO}(t))^2 + a_3(V(t)+V_{LO}(t))^3 + ... \tag{3}$$

The second order (quadratic) term of Eq.3 can be expressed as

$$I(t) = \quad ...$$

$$
\begin{aligned}
&+\quad a_2E^2\cos^2(2\pi\nu t+\Phi)\\
&+\quad 2a_2EQ\cos(2\pi\nu t+\Phi)\cos(2\pi\nu_{LO}t+\Phi_{LO})\\
&+\quad a_2Q^2\cos^2(2\pi\nu_{LO}t+\Phi_{LO})\\
&+\quad \ldots
\end{aligned}
\tag{4}
$$

Developping the product of the two cosine functions, we obtain

$$
\begin{aligned}
I(t)= &\quad \ldots\\
&+\quad a_2EQ\cos(2\pi(\nu+\nu_{LO})t+\Phi+\Phi_{LO})\\
&+\quad a_2EQ\cos(2\pi(\nu-\nu_{LO})t+\Phi-\Phi_{LO})\\
&+\quad \ldots
\end{aligned}
\tag{5}
$$

There are obviously other terms in $2\nu_{LO}$, 2ν, $3\nu_{LO}\pm\nu$, etc... in the above equation, as well as terms at very different frequencies like ν, 3ν, etc... By inserting a filter at the output of the `mixer`, we can select only the term such that

$$\nu_{IF}-\Delta\nu/2\leq|\nu-\nu_{LO}|\leq\nu_{IF}+\Delta\nu/2 \tag{6}$$

where ν_{IF}, the so-called *Intermediate Frequency*, is a frequency which is significantly different from than the original signal frequency ν (which is often called the *Radio Frequency* ν_{RF}).

Hence, after mixing and filtering, the output of the `receiver` is

$$I(t)\quad\propto\quad EQ\cos(2\pi(\nu-\nu_{LO})t+\Phi-\Phi_{LO}) \tag{7}$$

or

$$I(t)\quad\propto\quad EQ\cos(2\pi(\nu_{LO}-\nu)t-\Phi+\Phi_{LO}) \tag{8}$$

i.e.

- changed in frequency: $\nu\rightarrow\nu-\nu_{LO}$ or $\nu\rightarrow\nu_{LO}-\nu$
- proportional to the original electric field of the incident wave: $\propto E$
- with a phase relation with this electric field: $\Phi\rightarrow\Phi-\Phi_{LO}$ or $\Phi\rightarrow\Phi_{LO}-\Phi$
- proportional to the `local oscillator` voltage: $\propto Q$

The frequency change, usually towards a lower frequency, allows to select ν_{IF} such that amplifiers and transport elements are easily available for further processing.

2.2. RECEIVER SIDEBANDS

The mixer described above accepts simultaneously frequencies which are (see Fig.1)

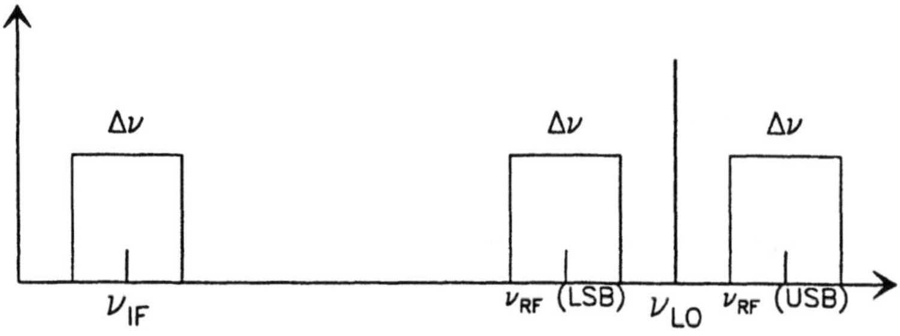

Figure 1. Relation between the IF, RF and local oscillator frequencies in an heterodyne system

- higher than the `local oscillator` frequency.
  This is called Upper Side Band (USB) reception
- lower than the `local oscillator` frequency.
  This is called Lower Side Band (LSB) reception

and cannot *a priori* distinguish between them. This is called Double Side Band (DSB) reception. Some receivers are actually insensitive to one of the frequency range, either because a filter has been placed at the receiver input, or because their response is very strongly frequency dependent. Such receivers are called Single Side Band (SSB) receivers.

An important property of the receiving system expressed by Eq.7-8 is that the sign of the phase is changed for LSB conversion. This property can be easily retrieved recognizing that the **Frequency** ν is the time derivative of the **Phase** Φ. Assume the phase $\Phi(t) = 2\pi\nu_0 t$ varies linearly with time, then

$$\nu_0 = \frac{1}{2\pi}\frac{d\Phi}{dt} \tag{9}$$

In this case, the signal

$$\begin{aligned} I(t) &\propto \cos(2\pi\nu t + \Phi(t)) \\ &\propto \cos(2\pi(\nu + \nu_0)t) \end{aligned} \tag{10}$$

is just another monochromatic signal with slightly shifted frequency. The above equations illustrate the time derivation relation between phase and frequency.

3. The Heterodyne Interferometer

Figure 2 is a schematic illustration of a 2-antenna heterodyne interferometer

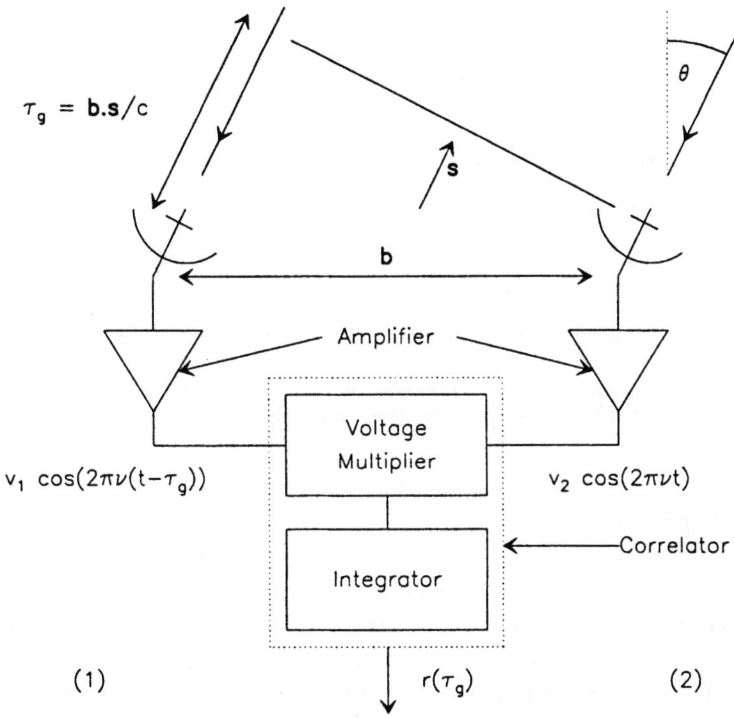

Figure 2. Schematic Diagram of a two-element interferometer

3.1. OVER SIMPLIFIED CASE

Let us forget the frequency conversion for some time, i.e. assume

$$\nu_{IF} = \nu_{RF}...$$

The input (amplified) signals from 2 elements of the interferometer are processed by a `correlator`, which is just a voltage multiplier followed by a time integrator. With one incident plane wave, the output $r(t)$ is

$$r(t) =< v_1 \cos(2\pi\nu(t - \tau_g(t)))v_2 \cos(2\pi\nu t) >= v_1 v_2 \cos(2\pi\nu\tau_g(t)) \quad (11)$$

where τ_g is obviously the geometrical delay,

$$\tau_g(t) = \mathbf{b}.\mathbf{s}/c \quad (12)$$

As τ_g varies slowly because of Earth rotation, $r(t)$ oscillates as a cosine function, and is thus called the *fringe pattern.* As we had shown before that v_1 and v_2 were proportional to the electric field of the wave, the correlator output (fringe pattern) is thus proportional to the power of the wave.

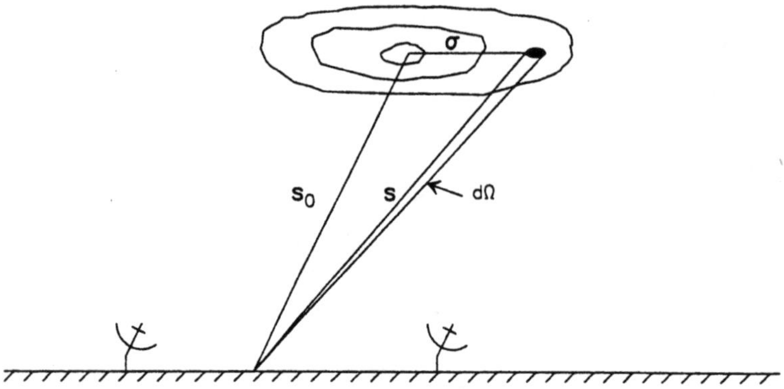

Figure 3. Position vectors used for the expression of the interferometer response to an extended source, schematically represented by the iso-contours of the sky brightness distribution.

3.2. SOURCE SIZE EFFECTS

The signal power received from a sky area $d\Omega$ in direction $\mathbf{s}$ is (see Fig.3 for notations) $A(\mathbf{s})I(\mathbf{s})d\Omega\Delta\nu$ over bandwidth $\Delta\nu$, where $A(\mathbf{s})$ is the antenna power pattern (assumed identical for both elements, more precisely $A(\mathbf{s}) = A_i(\mathbf{s})A_j(\mathbf{s})$ with A_i the voltage pattern of antenna i), and $I(\mathbf{s})$ is the sky brightness distribution

$$dr = A(\mathbf{s})I(\mathbf{s})d\Omega\Delta\nu\cos(2\pi\nu\tau_g) \quad (13)$$

$$r = \Delta\nu \int_{Sky} A(\mathbf{s})I(\mathbf{s})\cos(2\pi\nu\mathbf{b}.\mathbf{s}/c)d\Omega \quad (14)$$

Two implicit assumptions have been made in deriving Eq. 14. First, we assumed incident plane waves, which implies that the source must be in the far field of the interferometer. Second, we used a linear superposition of the incident waves, which implies that the source must be spatially incoherent. These assumptions are quite valid for most astronomical sources, but may be violated under special circumstances. For example VLBI observations of solar system objects would violate the first assumption, while observations of celestial masers could (in theory) violate the second one.

When the interferometer is tracking a source in direction $\mathbf{s_o}$, with $\mathbf{s} = \mathbf{s_o} + \sigma$

$$\begin{aligned} r = & \Delta\nu\cos(2\pi\nu\mathbf{b}.\mathbf{s_o}/c) \int_{Sky} A(\sigma)I(\sigma)\cos(2\pi\nu\mathbf{b}.\sigma/c)d\Omega \\ - & \Delta\nu\sin(2\pi\nu\mathbf{b}.\mathbf{s_o}/c) \int_{Sky} A(\sigma)I(\sigma)\sin(2\pi\nu\mathbf{b}.\sigma/c)d\Omega \end{aligned} \quad (15)$$

We define the *Complex Visibility*

$$V = |V|e^{i\Phi_V} = \int_{Sky} A'(\sigma)I(\sigma)e^{(-2i\pi\nu\mathbf{b}.\sigma/c)}d\Omega \tag{16}$$

which is the Fourier Transform of the modified brightness distribution $A'(\sigma)I(\sigma)$ where $A'(\sigma) = A(\sigma)/A_o$ is the normalised beam. Hence

$$\begin{aligned} r &= A_o\Delta\nu\cos(2\pi\nu\mathbf{b}.\mathbf{s_o}/c)|V|\cos(\Phi_V) + A_o\Delta\nu\sin(2\pi\nu\mathbf{b}.\mathbf{s_o}/c)|V|\sin(\Phi_V) \\ r &= A_o\Delta\nu|V|\cos(2\pi\nu\mathbf{b}.\mathbf{s_o}/c - \Phi_V) \end{aligned} \tag{17}$$

The amplitude and phase of the fringe pattern r thus contains the source structure information.

3.3. FINITE BANDWIDTH: DELAY TRACKING

If the radiation is non-monochromatic,

$$\begin{aligned} dr &= A_o|V|\cos(2\pi\nu\tau_g - \Phi_V)d\nu \\ r &= A_o|V|\int_{\nu_0-\Delta\nu/2}^{\nu_0+\Delta\nu/2}\cos(2\pi\nu\tau_g - \Phi_V)d\nu \\ r &= A_o|V|\frac{\sin(\pi\Delta\nu\tau_g)}{\pi\Delta\nu\tau_g}\cos(2\pi\nu\tau_g - \Phi_V)d\nu \end{aligned} \tag{18}$$

The fringe visibility is attenuated by a $\sin(x)/x$ envelope, called the bandwidth pattern, which falls off rapidly. For example, a 1% loss in visibility is obtained if $|\Delta\nu\tau_g| \simeq 0.078$, or with $\Delta\nu = 500$MHz and a baseline length $b = 100$m, when the zenith angle θ (defined in Fig.3) is 2 arcmin only. Thus, the ability to track a source for a significant hour angle coverage requires proper compensation of the geometrical delay when a finite bandwidth is desired.

3.4. DELAY TRACKING AND FREQUENCY CONVERSION

To correct for the geometrical delays, delay lines with moveable mirrors could in theory be inserted before the receiver system, as is done with optical interferometers. However, such delay lines would be huge, since the mirrors must be hundreds of wavelengths in diameters !... Fortunately, in an heterodyne interferometer, it is much easier to place the compensating delay after one (or several) frequency conversion(s), as illustrated in Fig 4. For USB conversion, the phase changes of the input signals before reaching the correlator are

$$\Phi_1 = 2\pi\nu\tau_g = 2\pi(\nu_{LO} + \nu_{IF})\tau_g \tag{19}$$

$$\Phi_2 = 2\pi\nu\tau_i + \Phi_{LO} \tag{20}$$

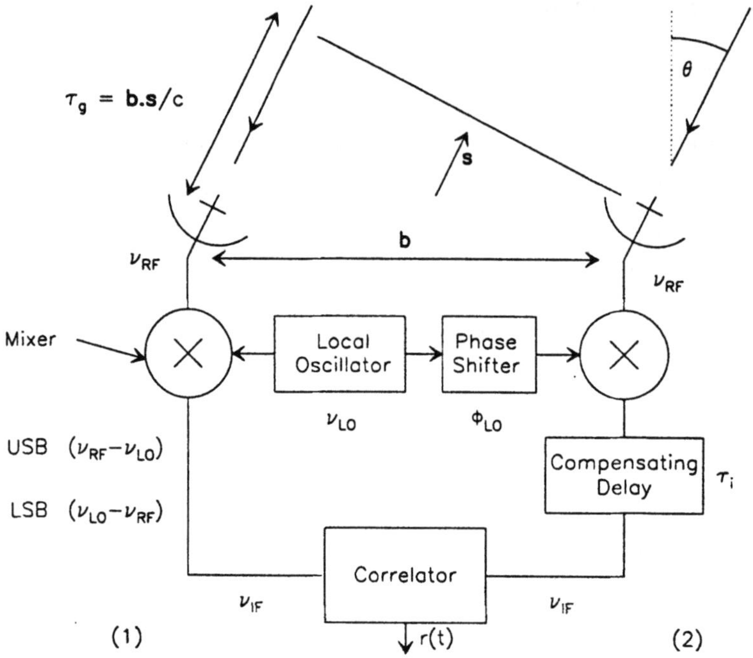

Figure 4. **2-element heterodyne interferometer with delay tracking after frequency conversion**

$\Delta\tau = \tau_g - \tau_i$ being the delay tracking error, the correlator output is

$$\begin{aligned} r &= A_o|V|\cos(\Phi_1 - \Phi_2 - \Phi_V) \\ USB \quad r &= A_o|V|\cos(2\pi(\nu_{LO}\tau_g + \nu_{IF}\Delta\tau) - \Phi_V - \Phi_{LO}) \\ LSB \quad r &= A_o|V|\cos(2\pi(\nu_{LO}\tau_g - \nu_{IF}\Delta\tau) - \Phi_V - \Phi_{LO}) \end{aligned} \tag{21}$$

When the two sidebands are superposed,

$$DSB \quad r = 2A_o|V|\cos(2\pi(\nu_{LO}\tau_g - \Phi_V - \Phi_{LO})cos(2\pi\nu_{IF}\Delta\tau) \tag{22}$$

the amplitude is modulated by the delay tracking error. Such a detection scheme would require very precise adjustments of the delay, and is often impractical. Fortunately, there are ways to **separate the sidebands** by periodic phase shifts of $\pm\pi/2$ applied to the `local oscillator` and demodulated in two different ways by the correlator. The demonstration of such a scheme is beyond the scope of this lecture.

However, the idea is to increment with time (by discrete steps of $\pi/2$) the phase of the signal in one way, and decrement it in the other way. Since a phase rate correspond to a frequency change (Eq.10), one can actually separate the two sidebands by shifting their original frequencies.

3.5. FRINGE STOPPING AND COMPLEX CORRELATOR

With the Earth rotation, the cosine term of Eq.17 modulates the correlator output quasi-sinusoidaly with a *natural fringe rate* of

$$\nu_{LO} d\tau_g/dt \simeq \Omega_{earth} \mathrm{b} \nu_{LO}/c \tag{23}$$

which is of order of 10 Hz for b= 300 m baselines and $\nu_{LO} = 100$ GHz, or $2''$ angular resolution (since the fringe rate only depends on the effective angular resolution). Such a rate is slightly too high for easy processing, e.g. digital sampling of the visibility. Note that in VLBI, where angular resolutions are of order 1mas, the fringe rate is even much higher.

Accordingly, the usual technique is to modulate the phase of the `local oscillator` Φ_{LO} such that $\Phi_{LO}(t) = 2\pi\nu_{LO}\tau_g(t)$ at any given time. Then

$$r_r = A_o|V|\cos(\pm 2\pi\nu_{IF}\Delta\tau - \Phi_V) \tag{24}$$

is a slowly varying output, which would be constant for a point source at the reference position (or delay tracking center). This process is called *Fringe Stopping.*

After fringe stopping, we can no longer measure the amplitude $|V|$ and the phase Φ_V separately. A second correlator, in which one of the signals has been phase shifted by $\pi/2$ becomes necessary. Its output is

$$r_i = A_o|V|\sin(\pm 2\pi\nu_{IF}\Delta\tau - \Phi_V) \tag{25}$$

With both correlators, we measure directly the real r_r and imaginary r_i parts of the complex visibility r. The device is thus called a "complex" correlator.

Note: A delay tracking error $\Delta\tau$ appears as a phase slope as a function of frequency, with

$$\Phi(\nu_{IF}) = \pm 2\pi\nu_{IF}\Delta\tau \tag{26}$$

with the + sign for USB conversion, and the – sign for LSB conversion.

4. Sensitivity

To be able to use an interferometer, it is also important to evaluate which objects can be detected. Hence, the sensivity of an array is a key parameter. Actually, two sensitivities have to be considered:

- the point source sensitivity, which is a well defined quantity;
- the surface brightness sensitivity, whose definition is less precise because the angular resolution of an interferometer is itself ill-defined.

For both historical and practical reasons, radio-astronomers use the "System Temperature" as the key parameter which helps to derive the point source sensitivity.

4.1. NOTION OF "SYSTEM TEMPERATURE"

By definition, the system temperature T_{sys} is such that the noise power of the antenna in a bandwidth $\Delta\nu$ is

$$P = kT_{sys}\Delta\nu \tag{27}$$

The output power of the receiver is proportional:

$$P_N = gkT_{sys}\Delta\nu \tag{28}$$

If, in addition to blank sky, the antennas "see" a source, the received power becomes $P_N + P_a$ where

$$P_a = gkT_a\Delta\nu \tag{29}$$

T_a is called the "antenna temperature" of the source.

This is not a purely conventional definition. An electric resistance at a physical temperature T actually delivers a noise power proportional to T. It can also be demonstrated that P_a is the power the antenna would deliver when observing a blackbody (filling its entire beam pattern) at the physical temperature T_a. Hence, the "antenna temperature" and "system temperatures" are very convenient quantities to estimate the sensitivity of radio telescopes.

As T_{sys} is actually the temperature of the "equivalent" blackbody seen by the antenna (in the Rayleigh Jeans approximation), it is easy to estimate:

$$\begin{aligned} T_{sys} &= T_{bg} && \text{cosmic background} \\ &+T_{sky} && \sim \eta_{fss}(1-e^{-\tau_{atm}})T_{atm}, \text{ sky noise} \\ &+T_{spill} && \sim (1-\eta_{fss}-\eta_{loss})T_{ground}, \text{ ground noise pickup} \\ &+T_{loss} && \sim \eta_{loss}T_{cabin}, \text{ optical losses in the receiver cabin} \\ &+T_{rec} && \text{receiver noise} \end{aligned} \tag{30}$$

It is often useful to have the system temperature corrected for atmospheric extinction (as if the antenna would be outside the atmosphere)

$$T^0_{sys} = e^{\tau_{atm}}T_{sys} = e^{\tau_{zenith}/sin(El)}T_{sys} \tag{31}$$

Of course, the antenna temperature (corrected for atmospheric extinction) is directly linked to the source flux density S_ν

$$T_a = \frac{\eta_a A}{2k}S_\nu \tag{32}$$

where A is the collecting area of the antenna, η_a is the aperture efficiency, k the Boltzman constant. The conversion factor S_ν/T_a only depends on the effective collecting area of the telescope; it is about 20 Jy/K for a 15-m telescope at 3mm, and slightly higher (because of the aperture efficiency) at shorter wavelengths.

Realistic numbers for the system temperature are 20-30 K at 10 GHz, 100 K at 100 GHz and 250 K at 230 GHz in the best conditions, or as a "rule of thumb" about 1K per GHz of observing frequency.

4.2. POINT SOURCE SENSITIVITY

In the weak source limit ($T_a << T_{sys}$), the sensitivity (to a point source) of a single baseline array is

$$\Delta S = \frac{\sqrt{2} k T_{sys}}{\eta A \sqrt{\Delta t \Delta \nu}} \tag{33}$$

assuming identical antennas, or more precisely $T_{sys} = \sqrt{T_{sys_1} T_{sys_2}}$ and $\eta A = \sqrt{\eta_1 A_1 \eta_2 A_2}$ if the antennas are different. Thus $\Delta S = \Delta S_0 / \sqrt{2}$, where

$$\Delta S_0 = \frac{2 k T_{sys}}{\eta A \sqrt{\Delta t \Delta \nu}} \tag{34}$$

is the sensitivity of one antenna of the baseline in total power mode for detection.

Note that ΔS is $\sqrt{2}$ times higher than the ΔS_0 corresponding to an antenna with the same **total** collecting area as those forming the baseline. This *loss* of sensitivity happens because the interferometer does not use the total power of each antenna, but just the correlated power between them.

When n identical antennas are connected, it is easy to demonstrate that the effective sensitivity becomes

$$\Delta S = \frac{\Delta S_0}{\sqrt{n(n-1)}} = \frac{2 k T_{sys}}{\eta A \sqrt{n(n-1) \Delta t \Delta \nu}} \tag{35}$$

ΔS tends toward the sensitivity of the antenna with the same *total* collecting area as the array when n is sufficiently large, because when the number of antennas becomes small relative to the number of baselines, neglecting the total power information becomes progressively less important.

4.3. EXTENDED SOURCE SENSITIVITY

To define an extended source sensitivity is more problematic, because it depends on the UV coverage and on the way the original data is actually combined. If the UV coverage is sufficiently uniform (i.e. the UV plane is homogeneously filled out to a certain radius), and if the visibilities are combined in a *natural* weighting mode (in which each original measurement is given a weight inversely proportional to the square of its noise), the surface brightness sensitivity of the array is well represented by a simple "beam dilution" formula:

$$\Delta T_b = \frac{(\theta_p)^2}{\theta_1\theta_2}\frac{\Delta T_s}{\sqrt{n(n-1)}} \tag{36}$$

where θ_p is the antenna primary beamwidth, $\theta_1\theta_2$ are the major and minor axis of the synthesized beam, n is the number of antenna, and δT_s is the brightness sensitivity of a single antenna used in total power mode:

$$\Delta T_s = 2\frac{T_{sys}}{\sqrt{\Delta\nu\Delta t}} \tag{37}$$

The brightness sensitivity can be related to the point source sensitivity with the approximate formula:

$$\delta T_b = \frac{\rho\lambda^2}{\theta_1\theta_2}\Delta S \tag{38}$$

where λ is the wavelength in mm, $\theta_1\theta_2$ are in arcseconds, and ΔS in the point source sensitivity in Jy, and ρ a factor of order 15. ρ actually depends on the details of the UV coverage.

A common error made when estimating the brightness sensitivity of an interferometer is to believe that smoothing the images (i.e. using a larger $\theta_1\theta_2$ in Eq.36-38) will improve the sensitivity. This is only partly true: since smoothing actually consists in giving less weight to the longest baselines, some the data is thrown out in this process, and the gain in sensitivity is much less than the geometrical gain expected from the synthesised beam sizes. Eq.36-38 are only valid when no smoothing (also called tapering) is applied. Moreover, smoothing **cannot recover** the structures which have been filtered out by the interferometer (i.e. structures which are larger than the angular resolution provided by the shortest interferometer baselines).

5. Using an Interferometer

5.1. UNDERSTANDING THE DATA PROPERTIES

From the schematic presentation of the interferometer principle, one should remember the following important results:

- **Sign of the Phase** depends on sideband conversion;
- **Frequency** is the time derivative of the **Phase**;
  This property helps considerably when trying to recover the proper phase signs in a system with several frequency conversions.

- **the Visibility** is the Fourier transform of the sky brightness *multiplied by the primary beam of the antennas*:
- **a Complex Correlator** directly measures the complex visibility;
- **the Correlator Output** is proportional
  - to the input power
  - to the product of the voltage pattern of the antennas
  - to the local oscillator power
  - to the receiver gain:
- **Finite Bandwidth** implies delay tracking;
- **Fringe Stopping** is necessary for high angular resolution;
- **a Delay Error** results in a phase slope as function of frequency.

Each of these properties plays a role in the data acquisition process, and is important to identify possible instrumental artefacts occuring during the acquisition process. The first and last properties are specially relevant.

Data obtained with an interferometer can be affected by several problems whose origin rests with one or several parts of the receiving equipment. Depending on the origin of the problem, it will affect the measurement of the visibility function in different ways. The most common effects include:

- Pointing Error;
  A pointing error will affect the visibility by a (real) scale factor for a point source, but the effect is more complex on an extended source (since in this case $A'(\sigma)I(\sigma)$ is not constant in time).
- Baseline Error;
  Since the geometrical delay τ_g is $\mathbf{b}.\mathbf{s}/c$, a poor knowledge of the baseline result in a delay tracking error.
- Focus Change;
  A focal length change on an antenna, due for example to thermal effects in the antenna structure, will also result in a delay tracking error (and not only in a phase change).
- Receiver Gain or LO power instabilities
  will introduce a real scale factor;
- Thermal expansion in the LO injection system
  introduce a variation in the phase Φ_{LO};
- Thermal expansion in the IF transport system
  introduce a variation in the delay τ_i at the IF frequency, and also in the phase Φ_{LO}, since the reference phase has to be carried along the IF.

5.2. CALIBRATION

Some of the above problems are avoided by proper design of the electronic systems. For example, the thermal effects in the IF are monitored and canceled out by measuring in real-time the effective electrical length of the IF transport cables. Pointing errors and focal change are minized by sufficiently frequent observations of sources of known positions. Other problems affecting the instrumental phase can (relatively easily) be calibrated by the observations of a continuum point source at regular time intervals. However, the structure and position of the reference source must be known to the relevant accuracy. With modern instruments capable of sub-arcsecond imaging, such as the VLA in the cm domain, or the IRAM interferometer in the mm range, the absolute uncertainty in the astrometric position of reference sources (usually quasars) can become a significant source of error ($0.05''$ or more).

In addition to the instrumental effects, the geometrical delay may vary because of changes in the refractive index of the atmosphere. A radio wavelengths, the **water vapor** is the most important time varying (dispersive) component of the atmosphere (the ionosphere plays a role at the longest wavelengths). These tropospheric effects are essentially random and can in general **not be calibrated out**. They are effectively contribute to a **seeing** limitation; the atmospheric seeing is almost wavelength independent.

5.3. BEATING THE SEEING EFFECTS

If the radio seeing is about $0.3''$ as in the optical domain, how is it possible for instruments such as the Very Large Array, or the MERLIN interferometer, to produce images with $0.1''$ resolution or even better? This question actually has 3 different answers, and a 4th solution also exists.

First, although in general the seeing is almost wavelength independent, the seeing can be perfect on some circumstances at the longest radio wavelengths, because the total pathlength fluctuations through the troposphere can be a small fraction of the wavelengths. On excellent sites, these circumstances can occur even for frequencies up to 200 GHz.

Second, the *timescales* involved in the atmospheric fluctuations resulting in seeing effects strongly depend on the frequency. In the cm domain, the most significant fluctuations occur on timescales of order of minutes or more. Sufficiently rapid calibration can correct for these fluctuations. In other words, the coherence time of the atmosphere is of order of minutes to hours, while it is a few milliseconds in the optical domain. This long coherence time (combined with stable clocks) actually allows VLBI networks to operate in a phase referencing mode much like phase-linked interferometers.

The third solution is a variant of the second one. Since the coherence time is long enough, there is potentially enough signal on the source itself to recover a good model of the source during one coherence time. This process is called "self-calibration" in radio astronomy, and is the exact equivalent of "adaptive optics" in the optical. Redundant arrays have sometimes been used to help the self-calibration process, but this method has now been recognized as less effective than using fully non-redundant arrays.

The fourth solution has only been tried on an experimental basis with the IRAM and BIMA mm arrays so far. It derives from the fact that most of the atmospheric fluctuations which are important at radio wavelengths are due to the water vapor. Water vapor is also the most important contributor to the (varying part of) the opacity in the mm domain. Hence, by monitoring the sky emissivity at for example 220 GHz, one can in principle derive the fluctuations of the water vapor content, and in turn **predict** the pathlengths fluctuations. This allows to correct for atmospheric phases *in real time* **without reference source**, i.e. to perform "blind adaptive optics". The technique breaks down when other atmospheric components play an important role either in the pathlength or in the sky emissivity, e.g. under cloud cover. The feasibility of this technique has been demonstrated with the IRAM interferometer, and it will most likely be generalized on all existing and planned mm interferometers.

5.4. INSTRUMENTAL EFFECTS IN THE IMAGING PROCESS

When the measured visibilities have been properly calibrated, and if possible corrected from seeing effects, it is customary to produce an image (although model fitting in the visibility plane is sometimes a better technique).

The imaging process consists in several steps:

- Gridding ;
  Since the visibilities are sampled along ellipses in the UV plane, it is necessary to resample them along a fixed grid in the UV plane to allow Fast Fourier Transform techniques to be applied. This resampling is done through a convolution kernel (in the UV plane). The process is called *gridding.*
- Weighting and Tapering:
  In this gridding process, one can apply arbitrary weights to the individual visibilities. Two weighting scheme are commonly used: *natural* weighting, where the weight is inversely proportional to the noise level of the ungridded visibility, and *uniform* weighting, where each gridded cell (unless it is empty) has the same weight. A combination, called *robust* weighting, is also possible: it consists (roughly speaking) to use uniform weighting, unless the original weight of the cell is lower than some threshold.
  Tapering is a similar process that *weighting*: it consists in multiplying the weights by a gaussian function in the UV plane, and hence is

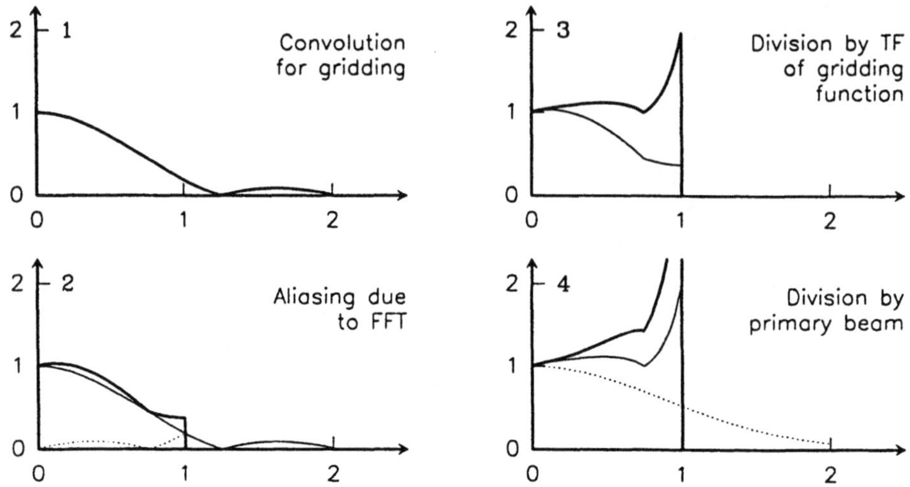

Figure 5. **Illustration of modification in the noise distribution introduced by the imaging process. The noise level is plotted against the distance from image center (in image half-width units).**

equivalent to a smoothing in the image plane.

- FFT and grid correction:
  The weights and weighted visibilities are Fourier transformed to the image plane, to produce respectively the synthesized (or dirty) beam and dirty image. The gridding is a convolution in the UV plane, hence a multiplication in the image plane: the dirty beam and dirty image are both corrected by the Fourier transform of the gridding function at this stage.
- Deconvolution:
  Since the dirty image is the Fourier transform of the product of the weights by the visibilities, it is the convolution of the sky brightness (*multiplied by the primary beam pattern of the antennas*) by the dirty beam. Deconvolution techniques must then be applied to the dirty image to recover the true sky brightness distribution. This step is **non-linear**, which means that photometry in interferometry is a difficult task. Possible deconvolution errors must be taken into account when analysing interferometric images.
- Division by Primary Beam.
  This last stage is often neglected, but should be applied, specially for relatively extended sources...

The above processing has some effects on the noise distribution in the image, which are illustrated in Fig.5. They result from the mathematical processing required by the use of the FFT, and produce a non-uniform noise, which in general strongly increases towards the edge of the image. Proper choices in the gridding function minimizes these effect without imposing to image unreasonably large areas. Note that the primary beam size is unrelated to the image size, but its effect should not be forgotten.

Finally, it should be recognized that most deconvolution techniques suffers from biases very similar to the "short spacing" problem. For example, CLEAN systematically underestimates the flux of weak extended sources. Although the origin of the effect is not due to loss of visibility, the resulting bias is very similar. In many images from millimeter arrays, this annoying property of CLEAN has been misinterpreted as "missing flux" due to lack of short spcaings in the interferometric images. Adding the short spacing information may indeed helped in such cases, but higher sensitivity images without short spacing can also recover completely the lost information. Accordingly, care should be excercised when determining flux densities from interferometric images.

5.5. PREPARING THE OBSERVING PROJECTS

Radio interferometry is applicable to various scientific purpose. One can consider basically two wavelengths range: the dm/cm range, and the millimeter range. They differ considerably in several ways:

- Scientific domains;
  Millimeter wave astronomy deals essentially with cold matter (molecules and dust, typical temperatures between 10 and 100 K), with low surface brightness. On the other hand, the centimeter range is the domain of synchrotron and maser emissions, with very high brightness; major exceptions are HI emission, and thermal lines from molecules like NH_3.
- Signal to Noise:
  The difference in brightness temperature makes a considerable difference in terms of achievable signal to noise. Millimeter interferometry is in general sensitivity limited, while centimeter interferometry is dominated by dynamic range effects: small residual calibration errors, source confusion, etc...
- Bandwidth:
  For continuum data, the effective bandwidth for cm arrays is typically 50 MHz, and that of mm arrays is 500 to 800 MHz. For spectral lines, the bandwidth is dictated by the doppler effect.
- Atmospheric properties;
  Phase fluctuations are to first order proportional to the frequency. The timescales involved are also important: the timescale for a given phase excursion is proportional to the wavelength. Hence, it is conceivable to

calibrate atmospheric phases with a reasonable calibration cycle (10-30 min) at dm wavelengths, but not in the mm range. Imaging qualities will thus be very different.

- Calibration accuracy;
  Strong calibrators are easily found in the centimeter domain, but not in the millimeter range. Amplitude calibration is difficult in the mm range.
- Number of antennas;
  Millimeter arrays have small numbers of antennas (4 to 9) and require complete synthesis to produce images. Cm arrays have very large number of antennas, and can provide "snapshot" images.
- Self-calibration:
  Self-calibration techniques require a minimum number of antennas (at least 6 for most algorithms), and also a minimum signal to noise. They are thus usually unapplicable to mm interferometry.
- Field of view.
  Cm arrays have large field of views. On the contrary, mm arrays have very limited field of view (often less than 1 arcmin), and may require the use of mosaicing techniques.

The choice of the wavelength range is dictated by the scientific goal. But the above properties should be considered when attempting to use a radio interferometer, and the observing strategy should be tailored to the goal. For the occasional user, a useful "checklist" is the following:

- *Will the result be Signal-to-noise limited ?*
  (Use Eq.35 and Eq.36-38, together with the "rule of thumb" $T_{sys} \simeq 1-2$K per GHz, to estimate the noise level). If sensitivity is not a limitation, the calibration accuracy will be the limit, and the best strategy could be to spend more time on calibration. A classical example of such a case is the measurement of velocity gradient in a strong molecular line. The main limitation is then the accuracy of the bandpass calibration (phase frequency response) since relative positions are governed by relative phases.
- *May dynamic range be the limitation ?*
  If yes, special care should be given to phase and amplitude calibration.
- *Is good UV coverage essential ?*
  For strong signals, the information may be obtained in a short integration time. The need for some UV coverage becomes then the driving factor for the duration of the observation.
- *Is the angular resolution appropriate ?*
  High angular resolution is one of the goal of interferometry, but the brightness sensitivity should not be overlooked.

- *Is the field of view appropriate ?*
  If it is insufficient, another instrument with smaller antennas may be preferable to the use of mosaicing techniques.
- *Can self-calibration techniques be used ?*
  If yes, the requirements on the basic instrumental calibration are somewhat relaxed.
- *Will the flux missed by short spacing impair the analysis ?*
  This limitation is often less severe than believed, for example when the missing flux corresponds to a brightness below the sensitivity threshold.

The above list is not exhaustive. All major arrays provide more detailed information for project preparation, and in case of doubt, the ultimate point in the checklist should be "Can somebody help me".

VERY LONG BASELINE INTERFEROMETRY

W. D. COTTON
NRAO
520 Edgemont Rd.
Charlottesville, VA 22903
USA

Abstract. The practice of Very Long Baseline Interferometry (VLBI) has undergone a great deal of development since the first demonstrations of the technique in 1967. High quality images including spectroscopy and polarimetry are now routinely produced for a variety of high brightness radio sources. The following gives some historical background as well as discussions of the observing technique and analysis of the data.

1. Introduction

The resolution of an imaging instrument is limited by the ratio of the linear size of its aperture to the observing wavelength. Radio wavelengths are relatively large, so enormous diameter apertures are required to obtain high resolution. The cost of an antenna increases as a large power of its diameter and economic considerations limit the maximum size of steerable antennas to about 100 meters. To obtain resolution finer than the several arcminutes produced by such an instrument, synthetic aperture techniques (i.e. interferometry) are required.

High resolution observations of the sky at radio frequencies can be made with the technique of Very Long Baseline Interferometry (VLBI). Since the initial demonstrations of the technique in 1967, VLBI has progressed from the simple detection of high brightness emission (brightness temperatures $> 10^{10}$ Kelvin) to detailed images with dynamic ranges of many thousands and resolutions of a few hundreds of microarcseconds.

The technique was first developed as one in which existing large radio antennas were used to form temporary interferometers. The signals were

A-M. Lagrange et al. (eds.), High Angular Resolution in Astrophysics, 165–191.
© 1997 *Kluwer Academic Publishers.*

time tagged using independent clocks and recorded on magnetic tape to be played back later for further processing. This technique allows using arbitrarily long baselines without the necessity of real time, high bandwidth communication links. Even the size of the earth does not limit the resolution possible as VLBI measurements using antennas in orbit start in the next year. Detailed discussions of the techniques of VLBI are given in [17] [40].

2. Historical Background

By the mid 1960s there was a clear need for high resolution radio telescopes. Quasar variability, interplanetary scintillations and opaque spectra suggested very high brightness temperatures in quasars and the discovery of interstellar OH masers indicated a need for high spatial resolution spectroscopy. Observations by the 127 km baseline Jodrell Bank (UK) radio–link interferometer indicated that there were many radio sources which were completely unresolved with a tenth arcsecond resolution [16] [30]. Higher resolution observations required baselines longer than were practical for connected element interferometers

As soon as the necessary technology was commercially available (1965-1966), several groups began working on independent clock with tape recorder based radio interferometry. The first successful observations were made in January 1967 by the University of Florida group [8] at a wavelength of 17 m of the decametric outbursts from Jupiter. Within a couple of weeks the Canadian group [7] successfully demonstrated their equipment on a short baseline at a wavelength of 67 cm. on quasars. This was quickly followed by successful observations by the NRAO-Cornell group at a wavelength of 49 cm [3], [9]. The NRAO-Cornell group in collaboration with the MIT/Haystack group soon began a series of observations at 18cm. of interstellar OH masers [29] and quasars using transcontinental baselines. A collaboration between Australian astronomers and Caltech/JPL began making observations at 13 cm wavelength in 1968 [19] on trans–Pacific baselines. In 1968 intercontinental baseline VLBI observations were being made using antennas in the US, Canada, Puerto Rico, England, Australia and Sweden. By the end of the 1960s VLBI was a rapidly developing technique involving antennas and research groups in a large number of countries. Detailed discussions of the drama of the early development of VLBI are given in [24], [23], [20], [6], [18].

For many years the technique was limited to cooperative experiments involving a number of groups and used existing large radio astronomical antennas. This loose organization became more formal over time to deal with the logistical problems associated with these observations and in the late 1970s and early 1980s consortia were formed to administer these observations. Individual observatories allocated substantial blocks of observing time to these observing networks. This practice continues to the present, although full time antennas and arrays have also been constructed.

The technique of VLBI has also proven of value to geodetic measurements as well as measuring UT1 and the orientation of the Earth's pole using extragalactic radio sources as a stable reference system. Subcentimeter accuracy surveying is possible using VLBI techniques over arbitrary baselines. Much of the technological development has been done by the Geodetic and Earth orientation communities and several VLBI networks are dedicated to these measurements.

2.1. CHRONOLOGY OF ASTRONOMICAL VLBI NETWORKS

- 1967 First demonstrations in Canada, USA;
- 1968 First intercontinental interferometers formed;
- Several years of Ad Hoc observations;
- 1976 USA Network Users Group:
  The NUG was the first of the organizations for coordinating VLBI observations using existing telescopes. This allowed observers to propose to a single organization rather than many and this organization was responsible for the detailed scheduling of the VLBI time allocated by the member observatories. Correlators were located at MIT (Haystack Observatory), Caltech, and NRAO. This organization was replaced by the VLBA.
- 1980 European VLBI Network:
  European radio observatories organized to form the EVN to facilitate VLBI in Europe. Observatories in Asia and South Africa have since joined the EVN. Joint or "global" sessions between the EVN and NUG (later the VLBA) have been scheduled several time a year. Correlators are located in Bonn, Bologna and a new facility (Joint Institute for VLBI in Europe, JIVE) is under construction in Dwingeloo, the Netherlands.
- 1986–1988 Demonstration Satellite VLBI;
  Experiments were performed using the NASA TDRSS communications satellite demonstrating the feasibility of VLBI observations from space using baselines in excess of twice the Earth's diameter.
- 1993 Australia Telescope:
  One of the components of the Australia Telescope is a part time VLBI network using a mixture of preexisting and new, dedicated antennas.
- 1995 Very Long Baseline Array:
  This array was built and is operated by the NRAO and is the first full time, dedicated VLBI array for astronomy.

– 1995 Coordinated Millimeter VLBI Array (CMVA).
  This network of existing antennas organizes observations at millimeter wavelengths.

2.2. RECORDING SYSTEMS

The heart of any VLBI system is its data recording technology and associated correlator. To date all astronomical observations have been made using magnetic tape. (Some special purpose atmospheric experiments using the ALSEP transponders on the moon used high speed paper tape.) The sensitivity of continuum VLBI observations is generally limited by the observing bandwidth which is in turn limited by the data recorder system. The requirements of VLBI observing far exceed the capacity of commercially available (or at least affordable) recorders so most VLBI recorders are custom built. The most important parameters of a record system are the recordable bandwidth which limits the sensitivity of the interferometer and the length of time it takes to fill a tape which sets logistical limits on the observations. The following describes some of the major recording systems used.

– Canadian analog recorder (1966 - 1980?):
  The early Canadian recorder system recorded analog signals from the antennas together with timing signals on Ampex video recorders.
– Mk I (1966 - 1976):
  The Mk I VLBI recorder was a high speed 7 track (later 9 track) standard computer tape drive. The recordable bandwidth was 360 kHz and a reel of tape was filled in 3 minutes. These tapes were originally correlated on a general purpose computer but a special purpose correlator was built at Haystack Observatory.
– Mk II (1972 - 1995):
  The Mk II [9] system went through several incarnations using modified video recorders. The early versions were 2 inch Ampex studio recorders which were replaced with 1 inch IVC recorders and the final versions were modified home VCR recorders. This system could record a 2 MHz bandwidth and tapes lasted from one to 4 hours. Mk II correlators were constructed at NRAO (3 station), MPI in Bonn (3 station), Caltech (5 station and 18 station) and Onsala Sweden (never used). One of the Caltech correlators (5 station) was moved to Bologna.
– Mk III (1979 - present):
  The Mk III recorders are based on Honeywell 1 inch instrumentation tape recorders using 28 record heads. This system was more flexible than earlier recorders but the maximum recordable bandwidth was 56 MHz and at this rate a tape was filled in 13 minutes. The MK

III system was developed at Haystack Observatory for NASA to be used in its geophysics program. Correlators are operated at Haystack Observatory, MPI in Bonn, Caltech (and Washington, DC for geodetic use). The Mk III has been upgraded using narrower record heads which allow many passes along the tape. The record time at peak recording rate was increased to 5 hours.

- VLBA (1993 - present):
  This record system also uses the Honeywell tape transport and is partially compatible with Mk III recorders but has 32 record heads. At a rate of 64 MHz a tape is filled in 5 hours. Higher recorder bandwidths are also possible. The VLBA correlator is at the NRAO's Array Operations Center in Socorro, NM, USA.
- S2 (1993? - present):
  The S2 system was designed and built by the Canadian VLBI group and used with the Australia Telescope. It is based on multiple VCR recorders and can record up to 64 MHz bandwidth. A set of tapes lasts 4 hours.
- K4 (1990 - present):
  The K4 system and its predecessor, the K2 system, is used by the Japanese VLBI community and is based on a commercial (Sony) recorder and will record 128 MHz of bandwidth.
- Mk IV (1997? -).
  The Mk IV system will replace the Mk III systems in geodetic networks and the EVN. This system is also based on the Honeywell transport and allows higher bandwidth and longer record times than the Mk III system.

3. What is VLBI?

Basically, VLBI is radio interferometry; that is, an array of devices to sample the wavefront at a few locations in an aperture. Special and general purpose computers are then used to apply the principles of physical optics to form an image [10]. Connected element radio interferometers require high bandwidth, stable communications links (cables, waveguide, radio links, fiber optics etc.) to transmit time and frequency information to each collecting element (antenna) and the received signal back to the correlator. Samples of the spatial coherence function of the source are formed by measuring the time lag cross–correlation of the signals from all combinations of antennas (baselines). The image of the source is related to the spatial coherence function by a Fourier transform–like relationship [36] [11].

The size of a connected element array, hence its resolution, is limited by the practical details of constructing a stable, high bandwidth communication network connecting the antennas with the correlator. VLBI is the technique of using independent time and frequency standards and recording the observed signal for playback and correlation at a later time. This technique removes the need for high bandwidth real time communications and thus allows baselines to be arbitrarily large. This technique does put very stringent requirements on the equipment. The time and frequency standard, or "clock", must be accurate to a few microseconds and have short term stability of a small fraction of the reciprocal of the observing frequency. The clock of choice is a hydrogen maser, but as these are quite expensive, rubidium clocks are sometimes used. The best hydrogen masers are more stable than the atmosphere and hence do not limit the coherence of the interferometer.

The sensitivity of an interferometer is determined in part by the bandwidth of the signal correlated, so the widest possible bandwidth is desired. The practice in VLBI is to sample the signal using either 2 or 4 levels which can be represented as 1 or 2 bits and then recording the digitized signal on tape. The sampling theorem requires sampling at twice the highest frequency in the sampled bandpass so a 1 bit sampled signal requires 2 Mbit per second of recorded data per MHz of observed bandwidth. Current high bandwidth VLBI systems can record in excess of 100 Mbit per second.

4. What VLBI can and cannot do

The resolution of an interferometer is a function of baseline length and the observing wavelength:

$$\theta \approx B/\lambda$$

where B is the length of the baseline and λ is the wavelength. In the radio region of the spectrum wavelengths are large (0.003 to $>$ 1 meter) but using VLBI the baselines can be very large (thousands of kilometers). Appropriate combinations of baseline length and wavelength can give resolutions of less than a milliarcsecond. Figure 1 gives the detailed relationships among resolution, baseline length and observing wavelength.

If the antennas measure the signals in two orthogonal polarizations the polarization of the source can be inferred from cross-correlations of the various received signals. The time lag correlation function determined by the correlator is related to the sky frequency by a Fourier transform relation, so imaging spectroscopy is a simple by-product of the technique.

It is useful to describe the brightness of a source in terms of the physical temperature of an equally bright black body of the same angular size; this is the brightness temperature. The higher the resolution, the more resolved a source at a given flux density and brightness temperature is. With baselines of thousands of kilometers, VLBI techniques resolve sources with brightness

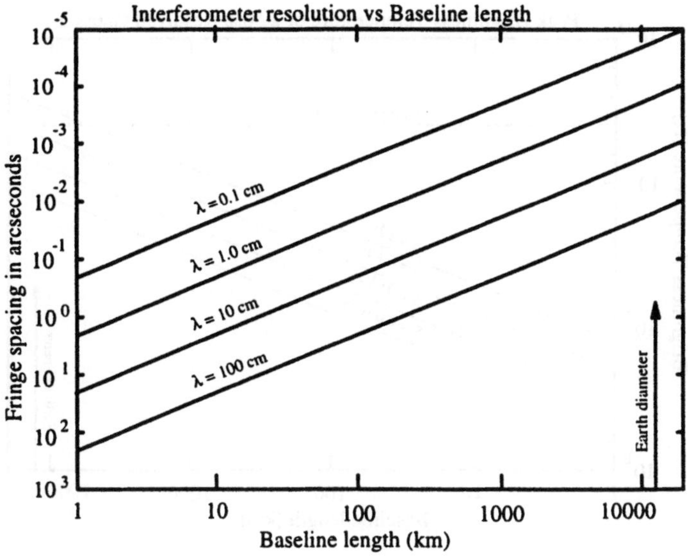

Figure 1. This figure shows the resolution of an interferometer as a function of baseline length and observing wavelength.

temperatures on the order of 10^9 to 10^{12} Kelvins. These sources clearly have non–thermal emission mechanisms. The one case of VLBI observations of thermal processes is measurements of absorption lines using a background non–thermal source.

What cannot be done with high resolution correlation interferometers is to image low brightness sources. Difficulties with observing low brightness sources with a high resolution interferometer come from two, related, effects. First, a correlation interferometer is a high pass spatial frequency filter; it simply does not respond to structure whose spatial scales are much larger that its fringe spacing (resolution). Figure 2 illustrates the brightness temperature of a moderately resolved source as a function of baseline length and source flux density. At significantly higher resolution, such a source will be completely resolved and not detectable. The second effect can be described in terms of an array whose size can be scaled to increase the resolution. The noise in an image in units of flux density per beam area remains constant as the size of an array is increased, but the source emission per beam area decreases as the source is spread over more beam areas. Above a certain resolution, the noise in each beam area dominates the source which is no longer apparent in the image.

Low brightness regions in a source imaged by a high resolution interferometer will not be visible in the image. If the source contains no high brightness regions then the source will not be detected at all. For example,

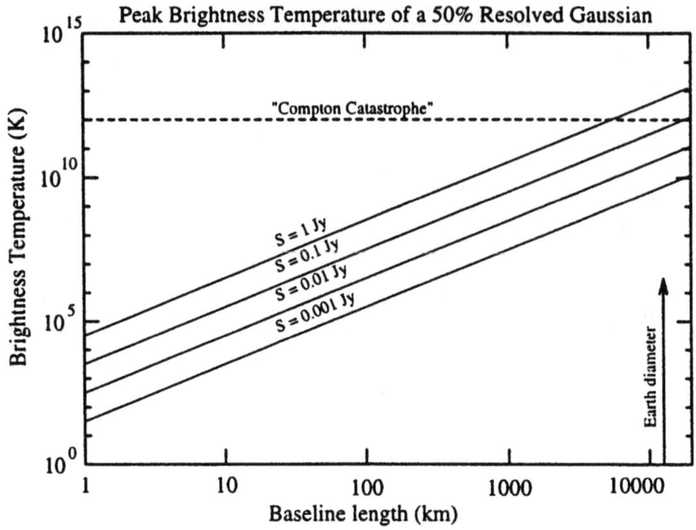

Figure 2. This figure gives the peak brightness temperature of a Gaussian source of a given flux density which is 50% resolved by an interferometer of a given length. The "Compton Catastrophe" brightness is approximately the brightness temperature at which the density of photons in a synchrotron source is high enough that Compton scattering off the relativistic electrons will quickly quench the source.

the normal photospheric emission from stars is not observable with VLBI baselines of thousands of kilometers.

5. How Radio Interferometry Differs from Optical/IR

Interferometry at radio wavelengths differs from that at higher frequencies principally in its relative ease and the maturity of the technology. The following list summarizes these differences:

- Large isoplanatic patch size:
  The angular size over which the atmospheric phase fluctuations are highly correlated is relatively large, from several to tens of degrees. This allows using calibration sources several degrees or more from the source of interest.
- Long atmospheric coherence time:
  The large coherence scale size in the atmosphere also means that atmospheric phase fluctuations are fairly slow. The time scales generally (but not always) are in the range of seconds to tens of minutes.
- Phase preserving amplifiers:
  Radio amplifiers can amplify signals without destroying the phase information in the signal and the phase can be measured with respect

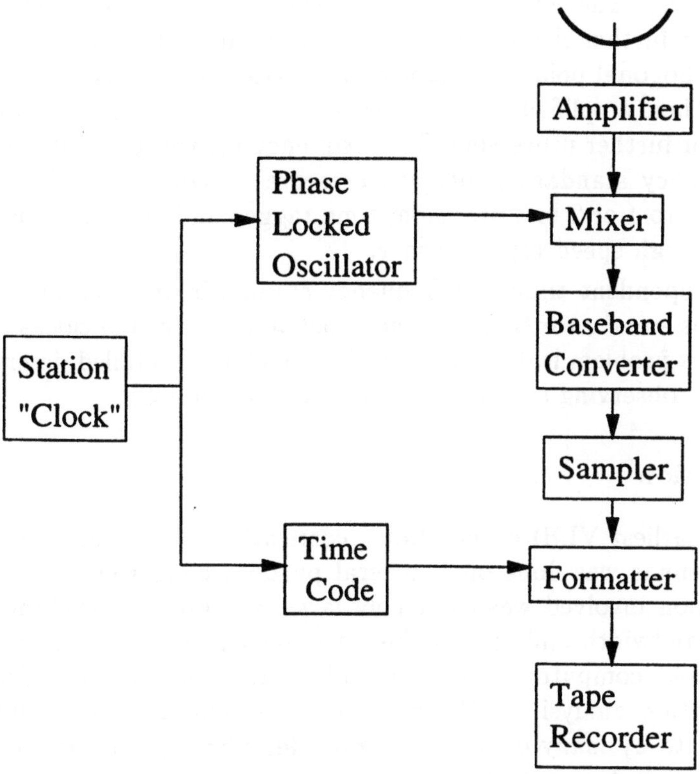

Figure 3. Block diagram of a typical VLBI observing configuration from the antenna at top to the tape recorder at the bottom.

to that of the frequency standard. This allows dividing the signal for many correlations with no loss of sensitivity.

- Mature technology.
  Radio technology at centimeter and longer wavelengths is quite well developed. State-of-the-art centimeter receivers contribute less to the total system noise than does the Big Bang. At longer wavelengths the disk emission from the Galaxy dominates.

6. Observing Instrumentation

The typical configuration of a VLBI observing system is shown in Figure 3. Each VLBI antenna collects the signal from the direction of the source and focuses it into the "feed" (named by engineers who think of antennas as transmitters). The feed will couple a single polarization mode of the signal into the further electronics. If polarization measurements are desired, then two orthogonal polarizations need be processed. The radio frequency signals are then amplified and converted in frequency (mixer, baseband converter) for further processing. The frequency conversion is phase locked to the frequency standard (clock) to preserve its coherence. The data is then sampled to 1 or 2 bits precision, time tagged and written to magnetic tape using a high speed tape recorder.

Since independent time and frequency standards are used at each antenna there is some uncertainty in the epoch and rate of the clocks at each antenna. This problem is dealt with in the correlator. Detailed descriptions of the VLBA observing systems are given in [35] and [32].

7. Correlation

In the very earliest VLBI observations, correlation of the signals from the pairs of antennas was done on a general purpose computer. The amount of computation involved was extremely large and grew dramatically with increasing bandwidth and the number of baselines correlated. In practice, special purpose computers called correlators are constructed to form the cross–correlation analysis of the data streams. This process reduces the volume of data by many orders of magnitude, allowing further processing to be accomplished on general purpose computers.

In VLBI the practice is to record the data on magnetic tape which allows it to be played back and correlated later, from days to years after the observations. The basic operation is to multiply the samples of a given wavefront from the two antennas and average them in time. Since the Earth rotates, the relative time delay between the arrival of a wavefront at one antenna and the other changes continually. Figure 4 illustrates the geometric delay between the arrival of a wavefront at the two antennas of an interferometer. In addition, the observed frequency of the observation at each antenna is modified by the Doppler shift due to the Earth's rotation. The correlator must align the two data streams to the closest sample and shift the frequency of one or both of the signals to compensate for the difference in observed frequency. For narrow band spectroscopic observations some of the frequency shifting is done in the antenna at observe time.

A correlator uses a geometric and perhaps atmospheric model to predict the difference in delay and frequency. Each antenna has an independent clock which introduces some uncertainty in the apparent delay and its time derivative. Additional uncertainties arise from errors in the assumed geometric and atmospheric models. The Earth is a rather unstable observing platform with many wiggles, wobbles and floating plates which are pre-

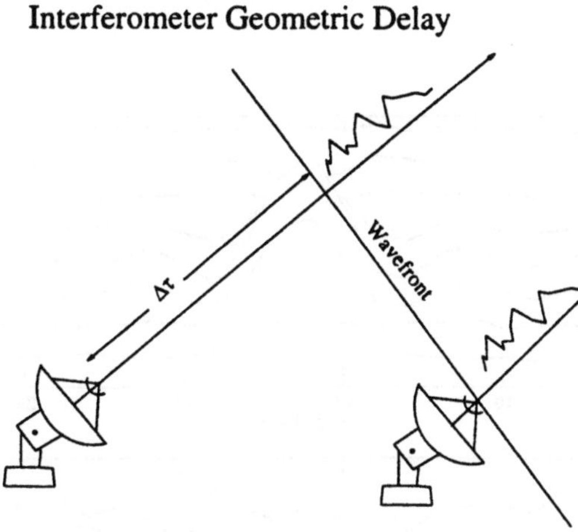

Figure 4. **This figure demonstrates the geometric delay ($\Delta\tau$) between the arrival of a wavefront at one antenna of an interferometer and the next. The fluctuations of the source signal are represented along the line from each antenna to the source.**

dictable only to a certain degree. In particular, the current orientation of the rotation axis and the phase of the earth's rotation are not predictable and must be measured. The correlator model will also be in error if the position of the target source is not known to a precision of the order of the resolution of the interferometer. For all these reasons the expected delay between the wavefront arrival times at the two antennas is not precisely known nor is its behavior in time. In the jargon of the radio astronomer, the data is said to contain residual delay and fringe rate errors. To compensate for these uncertainties, the cross–correlation of the signals is generally made at a range of lags and sampled sufficiently rapidly in time that the time variability of the delay error does not degrade the data. Since the data are digital samples, a range of lags may be introduced by shifting the two data streams with respect to each other by a number of integral samples. A detailed discussion of VLBI correlators can be found in [33].

7.1. SPECTROSCOPY AND CORRELATOR DESIGN

The time lag cross–correlation function is related to the sky frequency spectrum by a Fourier transform relation [11]. If sufficiently many lags were correlated, this relationship allows the observed visibility to be measured as a function of sky frequency with sufficient resolution for spectroscopic purposes. Equivalently, all VLBI data is spectroscopic in nature.

The time lag – sky frequency relationship can be exploited in the design of the correlator; that is, the order of the multiplication of the data streams and the Fourier transform may be reversed. A correlator in which

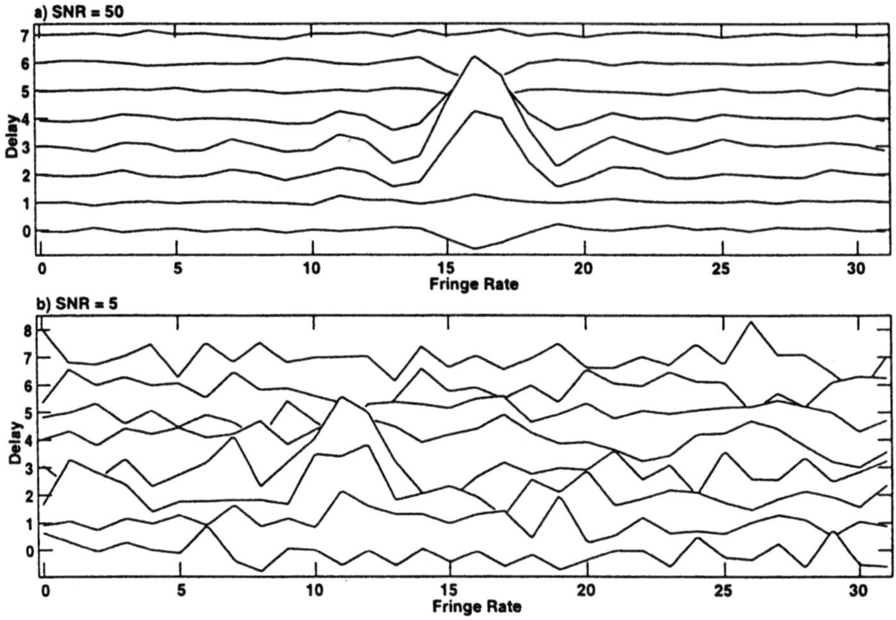

Figure 5. The amplitude of the interferometer response in residual delay – fringe rate space.
a) A relatively high signal-to-noise ratio case where the sin x/x response to the source is seen in both delay and fringe rate.
b) A low signal-to-noise ratio case where the source is just visible.

the multiplication is done before the Fourier transform is known as an XF (or lag) correlator and one in which the Fourier transform is done before the multiplication is known as an FX correlator. For large correlators, the FX design is more economical and both designs are currently in use. In either design, there are still residual delay and fringe rate errors which must be determined and removed.

7.2. FRINGE FITTING

The process of determining residual delay and fringe rate errors is known as "Fringe fitting" even though it has been some time since anyone actually saw a live "fringe." Due to the various Fourier transform relation involved with this data, the errors in the data can be thought of either as uncertainty of the location of the response to the source in delay–fringe rate space or as phase slopes in time and frequency. Figure 5 illustrates the response to a source in the delay – fringe rate plane.

The major contributions to the residual delay and fringe rate errors are contributions from the individual antennas. This allows for determining the delay and fringe rate errors on either a baseline or antenna basis. These methods are known as "baseline" [28] and "global" [34] fringe fitting with combinations possible [1].

7.3. POLARIZATION

The polarization state of the emission from a radio source can be determined if two orthogonal polarization modes of the signal are recorded. Generally right(R) and left (L) circular polarizations are used for this purpose. The visibilities of the source structure in the Stokes' parameters (I: total intensity, Q,U: linear polarization, V:Circular polarizations) are related to the correlations RR, LL, RL, LR:

$$V_I = 0.5 \times (RR + LL)$$

$$V_Q = 0.5 \times (RL + LR)$$

$$V_U = 0.5 \times i(LR - RL)$$

$$V_V = 0.5 \times (RR - LL)$$

8. Non-Imaging Uses of VLBI

Most of the discussion in this lecture is of the uses of VLBI in constructing images but the technique also has other applications.

- Model fitting source structure:
  Imaging requires fairly extensive observations on a large number of baselines. If the observations involve a small number of baselines or the source is marginally resolved, fitting models such as several elliptically shaped components with Gaussian profiles to the data may be the appropriate analysis technique.
- Astrometry:
  Absolute positions of sources to accuracies of a milliarcsecond can be obtained from analysis of the delay and fringe rate residuals from observations with a good distribution on the sky. Relative positions with accuracies of 10s of microseconds can be determined by interleaved observations of sources close together on the sky.
- Surveying:
  The delay and fringe rate residuals are functions of the locations of the antennas as well as the instantaneous orientation of the Earth. Positions of the antennas and the orientation of the earth can be obtained from an analysis of these residuals.

9. VLBI Imaging

VLBI, like any radio interferometer, uses antennas to sample the wavefront of the radiation from a source at selected points in the aperture of the array. The principles of physical optics are then used to form an image [10]. In simplified form ,the relationship is that the Fourier transform of the spatial auto–correlation function of the emission from the source is its image. This auto–correlation function of the radiation is known as the coherence function. The auto–correlation function is not measured directly but is assumed equivalent to the cross–correlation of the signal measured between pairs of antennas. Thus, each pair of antennas, or baseline, samples a single point of the coherence function in the spatial frequency domain. These measurements of the coherence function are called "visibilities" in the jargon of radio astronomers.

In order to Fourier transform the coherence function, it must be known at all points in the spatial frequency domain. However, a given baseline only samples the function at a single location determined by the separation and orientation of the two antennas as viewed from the source. Many baselines will sample the coherence at many locations and as the number of baselines for an N antenna is $N(N-1)/2$ the imaging ability of an array increases rapidly with increasing numbers of antennas. If the antennas track the source, then the observing geometry will change as the earth rotates, causing each baseline to trace an ellipse in the spatial frequency domain. Thus, image quality increases with increasing number of elements in the array and with the amount of time the source is observed.

It is not possible to completely sample the coherence function, so some approximations are in order. The usual approach is to define a sampled coherence function which is the measured value where it was measured and zero elsewhere. This is equivalent to multiplying the true coherence function (as corrupted by noise and other effects) with a sampling function which is unity where a measurement exists and zero elsewhere. The sampled coherence function is defined at all locations in the spatial frequency domain and can thus be Fourier transformed.

9.1. DECONVOLUTION

The image derived from the sampled coherence function is called the "dirty" image as it contains artifacts from the sampling process. According to the Convolution Theorem in Fourier theory, the Fourier transform of the product of two functions is the convolution of the Fourier transform of the two functions. The dirty image is therefore the convolution of the estimate of the sky with the "dirty beam" which is the transform of the sampling function.

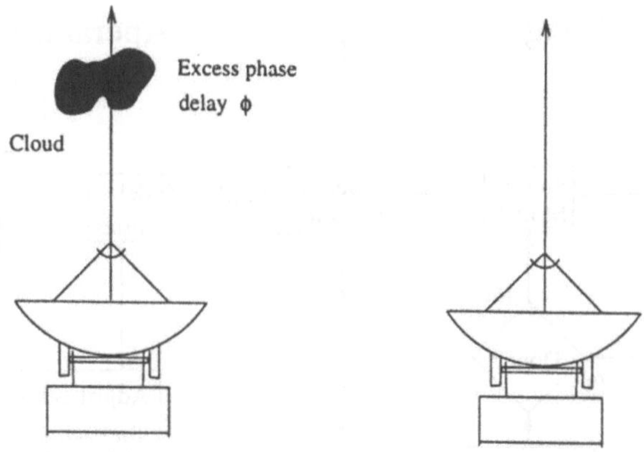

Figure 6. **This figure illustrates how a cloud can introduce a delay and phase offset in an individual antenna of an interferometer.**

A better estimate of the sky can be derived from the dirty image by means of a nonlinear deconvolution. The most popular deconvolution method in radio astronomy is the CLEAN algorithm and its variants which are discussed in [12]. The CLEAN algorithm is to iteratively decompose the image into a set of delta functions (CLEAN components) by searching for the maximum point in the residual image and subtracting a fraction of the brightness times the dirty beam centered at that location. When this process has converged, the delta functions are convolved with a Gaussian of the approximate size of the synthesized beam and added back to the final residual image. This restored image is known as the CLEAN image. Other algorithms are also sometimes used, one example is Maximum Entropy deconvolution [15] [12].

9.2. CLOSURE AND SELF-CALIBRATION

The preceeding discussion presumes that the measured complex visibilities are properly calibrated to be accurate measures of the coherence function. This is not initially the general case; in particular, the phase of VLBI visibilities are badly corrupted by the various errors in the correlation model, especially the atmospheric contribution. An illustration of atmospheric perturbation of the phase of the signal received at an antenna is shown in Figure 6. These errors are primarily introduced into the signal measured by the individual antennas and can be factored into antenna based components. This property has lead to the application of the surveying concept of "closure" in which the sum of the errors around a closed loop must be zero. Thus, the sum of the phases around the three baselines of a triangle will cancel the atmospheric and other errors and only be a function of the

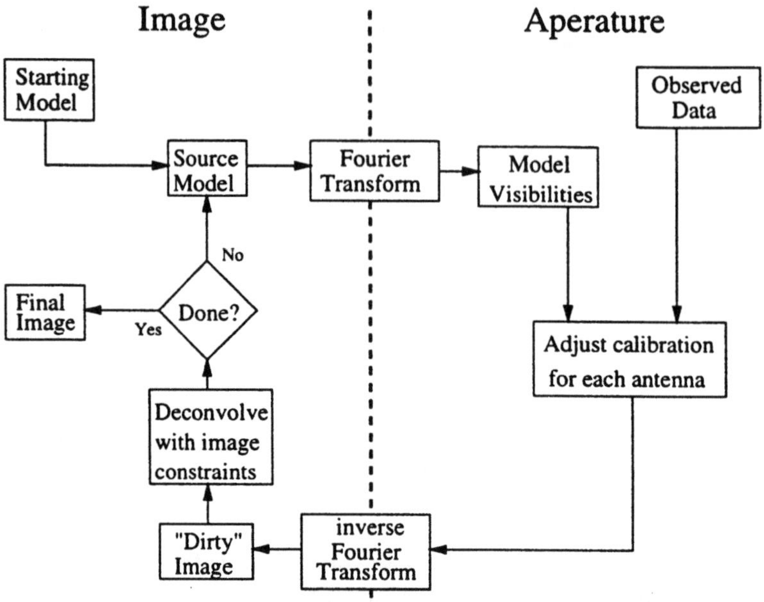

Figure 7. The flowchart of the self–calibration process from an initial source model and the observed data.

source structure. The application of this principle to radio astronomy was first noted in the 1950s [22] and was later applied to VLBI [31] [13].

This principle has been applied to an iterative imaging scheme for VLBI known as "hybrid mapping" or "self–calibration" but is more properly called adaptive optics. With a many element array there are many more baselines than antennas. If the source structure were known, then the system of equations relating the phases of the different baselines could be solved except for an unknown additive constant. The usual practice is to assume an initial model of the source and adopt a "reference" antenna whose phase is arbitrarily set to zero and the phases of other antennas determined with respect to the reference antenna. Figure 7 gives a sketch of the self–calibration technique. One drawback to this scheme is that position information cancels in the same way as atmospheric fluctuations and the absolute position information is lost. Another problem is that the antenna phase errors must be determined over a time period during which they do not change significantly and the source must be sufficiently strong for the antenna phase errors to be determined. The self–calibration plus deconvolution technique routinely produces images with a "dynamic range" (ratio of the peak brightness to the off source RMS noise) of several thousands and has produced dynamic ranges of several tens of thousands.

9.3. REFERENCE SOURCE CALIBRATION

The drawbacks in the self–calibration process, loss of position information and limited sensitivity, can be overcome if the atmospheric phase fluctuations can be estimated using observations of a “calibrator” source of known position and structure in the same isoplanatic patch as the target. Frequent switching between the sources allows measuring the atmospheric phase continously.

9.4. POLARIZATION

The polarized interferometer response to a signal is a combination of the (corrupted) source polarization and a spurious instrumental polarized response. The effects of the instrument must be measured and removed to recover the source polarization.

If all of the cross–polarized cross–correlations are measured on each baseline the Stokes visibilities can be derived and the source can be imaged and deconvolved in each of the Stokes parameters independently [14]. If only some of the cros–correlations are observed (say RL but not LR) it is possible to make and deconvolve an Q + iU image [15].

10. Sensitivity

The sensitivity of VLBI observations is a complex issue. The VLBI observations are limited by the sensitivity of the antennas and associated electronics, the recorded bandwidth and the duration of the observations as well as the calibration and imaging techniques. In the cases of fringe fitting and self–calibration there is also a minimum detectable source strength which is not simply related to the noise in the final image. Detailed discussions of sensitivity issues are given in [37] [39].

10.1. DETECTION THRESHOLDS

In both fringe fitting and self–calibration techniques calibration parameters are determined directly from observations of the source of interest. For this to be effective, the source must be detectable in a time over which the corrupting effects do not change by too large an amount; generally this is the atmospheric coherence time. This time interval depends on the details of the observations but is usually in the range of tens of seconds to tens of minutes. The noise in the observations on a single baseline (ij) is [37]:

$$\Delta S_{ij} = \frac{1}{\eta}\sqrt{\frac{T_{si}T_{sj}}{2\Delta t\Delta\nu}}$$

where η is the correlator efficiency (≈ 0.5), T_s is the system temperature in Jy (ranges from a few to several thousand), $\Delta\nu$ is the observing bandwidth and Δt is the solution interval time. Detections generally require a signal-to-noise ratio of five or greater.

10.2. IMAGE NOISE

The noise in the image is determined by the total amount of data and imaging techniques as well as the noise on individual baselines. The noise in an image determined from a homogenous N antenna array and certain assumptions about the imaging technique is given by [37]:

$$\Delta I = \frac{1}{\eta} \times \frac{T_s}{\sqrt{N(N-1)\Delta\nu t_{int}}}$$

where t_{int} is the total integration time. If self–calibration has been used then the noise in the data will have been increased by the process of determining the antenna phases and the noise in the image will be further increased by a factor of $\sqrt{(N-1)/(N-3)}$ [39].

10.3. SENSITIVITY EXAMPLES

As was discussed above, there are two levels of sensitivity; detection of the source in the coherence time of the interferometer for fringe fitting and self–calibration and the noise in the final image. The following gives illustrative examples of both of these. The baseline examples give the minimum detectable source strength and the image examples are the RMS of source fluctuations in an image whose dynamic range is not limited by residual calibration errors. All examples are for an observing wavelength of 6 cm and typical current wide bandwidth recordings.

Baseline:

VLBA baseline (5σ, 120 seconds, 6 cm, 64 MHz) $\approx$ 17 mJy
VLA – Arecibo (5σ, 120 seconds, 6 cm, 64 MHz) $\approx$ 1 mJy
VLA – Bonn (5σ, 120 seconds, 6 cm, 64 MHz) $\approx$ 1 mJy

Image:

VLBA (8 hours, 6 cm, 64 MHz, self-calibrate) $\approx 40\mu$Jy
EVN (8 hours, 6 cm, 56 MHz, self–calibrate) $\approx 12\mu$Jy

11. Science

It is beyond the scope of this lecture to discuss the scientific results obtained using VLBI techniques. Rather, it is the intent here to describe some of the areas of astronomical research which can usefully employ these techniques. The physical processes observed generally involve nonthermal emission as these are required to produce the very large brightness temperatures detectable by VLBI. These lines of scientific inquiries fall into two general categories; studying regions of high brightness temperature and using these regions as probes to measure something else.

11.1. INTERSTELLAR MASERS

Interstellar masers are relatively common [4] from a number of transitions and molecular species in HII regions, the envelopes of evolved stars and the nuclear regions of some galaxies. As the masing process requires rather special conditions, VLBI observations of the masing regions allows investigating the physical parameters in these environments with some detail. VLBI observations produce very accurate measurement of radial velocities as well as proper motions at the level of microseconds per year and the method of statistical parallaxes has been used to measure the distance to masing regions in our galaxy which is entirely independent of the optical distance scale [27]. Observations of masers in the nuclear regions of other galaxies can help establish very large mass densities in the nuclear region. The "megamaser" in NGC 4258 is currently among the best lines of evidence for massive black holes [26].

11.2. ABSORPTION OF CONTINUUM SOURCES

Many of the continuum sources observable with VLBI are in the relatively early universe (redshifts of 1 to 3) with correspondingly long lines of sight. Absorption of the background continuum emission in intervening galaxies and pregalactic clouds of gas allows detailed examination of the spatial and velocity structure in these galaxies and clouds [5].

11.3. GRAVITATIONAL LENSES

Galaxies and clusters of galaxies along the line of sight to quasars can form gravitational lenses resulting in multiple images of the background source. Detailed measurements of the distortions of the images produced by VLBI help to constrain the mass distribution in the lensing objects [38].

11.4. DISTANCES TO SUPERNOVAE

VLBI can be used to derive the distance to supernovae in other galaxies by comparing the apparent rate of expansion of the radio shell with the ejection velocity determined from the optical emission lines during the outburst. This method is independent of the optical distance scale and allows independent measurements of the Hubble constant. See [25] for a recent application of this technique.

11.5. ACTIVE GALACTIC NUCLEI

Synchrotron emission is very common in radio sources and can obtain brightness temperatures in the range of 10^9 to 10^{12} K observable by earth based VLBI. These bright sources are frequently observed in the nuclei of galaxies and in quasars. Above a brightness temperature of about 10^{12} K, rapid cooling of the source by inverse Compton production of gamma rays is expected. However, the gamma and X-ray emission is generally much less than expected by the simple models. The low X- and gamma ray luminosities as well as asymmetric radio structures and apparent motions in excess of the speed of light have lead to the conclusion that these radio source are highly relativistic plasma beams. These beams are assumed to be produced by a central massive black hole. For a recent review of AGNs see [2].

12. Future

The dimensions in which the technique of VLBI can be improved are: 1) higher resolution, 2) a wider range of frequencies, 3) increased sensitivity and 4) better image quality. Higher resolution can be achieved through either longer baselines or shorter wavelengths. Increasing the range of wavelengths requires improved electronics at the shorter wavelengths as well as improved techniques for removing propagation effects at both long and short wavelengths. Improved sensitivity can be obtained by upgrading some antennas to modern electronics, increasing the number of antennas used and improved calibration techniques. Some of these issues are discussed below.

12.1. ANTENNAS IN SPACE

Baselines of nearly earth diameter length are routinely used in VLBI so longer baselines require the use of antennas in space. The feasibility of this technique has been demonstrated using the NASA TDRSS communications satellite and the Japanese Institute of Space and Astronautical Science (ISAS) is preparing to launch the VSOP satellite in the Fall of 1996 to make VLBI observations with baselines up to several times the earth's diameter[21]. Other orbiting VLBI antennas are possible at a later date; the Russian RADIOASTRON satellite has been under development for a number of years.

12.2. SHORTER WAVELENGTHS

VLBI at longer wavelengths is complicated by propagation effects in the earth's ionosphere as well as in the interplanetary and interstellar medium. At shorter wavelengths (less that 1 cm) the sensitivity of the electronics is still less than the theoretical maximum and the observations are sensitivity limited. Better techniques for measuring atmospheric phase fluctuations are needed to allow observations in the atmospheric millimeter and submillimeter windows. On a longer term, VLBI entirely from outside the earth's atmosphere will help extend VLBI techniques to shorter wavelengths although there are no funded projects working on this at present.

12.3. INCREASED SENSITIVITY

At centimeter wavelengths antenna electronics are nearly as sensitive as is theoretically possible, and further improvements are most easily obtained through increasing the recorded bandwidth requiring development of tape recorder or communications technology. Not all antennas used for VLBI have state-of-the-art electronics and some improvements will be had by upgrading these antennas. Some reductions in image noise can be obtained from an increased number of antennas used and a reduction of the minimum detectable source can be obtained from better calibration techniques.

12.4. INCREASED IMAGE QUALITY

All imaging systems introduce some artifacts into the images produced; the lower the level of artifacts, the better the "quality" of the image. VLBI image quality can be improved through better sampling of the spatial frequency plane (more antennas, orbiting antennas), improved calibration techniques and improved imaging techniques.

13. Acknowledgments

The National Radio Astronomy Observatory is a facility of the National Science Foundation (USA). The author would like the thank K. I. Kellermann for discussions about the origins of the VLBI technique.

References

1. Alef, W. and Porcas, R. W. (1986) VLBI fringe-fitting with antenna-based residuals. *Astron. Astrophys.*, **168**, 356–368.
2. Antonucci, R. (1993) Unified models for active galactic nuclei and quasars. *Ann. Rev. Astron. Astrophys.*, **31**, 473–521.

3. Bare, C., Clark, B. G., Kellermann, K. I., Cohen, M. H., and Jauncey, D. L. (1967) Interferometer experiment with independent local oscillators. *Science*, **157**, 189–191.
4. Bowers, P. (1993) Sub–arcsecond observations of maser stars. In R. J. Davis and R. S. Booth (eds.), *Sub–arcsecond Radio Astronomy*, Cambridge University Press, Cambridge, pp. 52–58.
5. Briggs, F. H., Wolfe, A. M., Listz, H. S., Davis, M. M., and Turner, K. L. (1989) The spatial extent of the z=2.04 absorber in the spectrum of PKS0458-020. *Astrophys. J.*, **341**, 650–657.
6. Broten, N. W. (1988) Early days of Canadian long-baseline interferometry: reflections and reminiscences. *J. Roy. Astron. Soc. Can.*, **82**, 233–241.
7. Broten, N. W., Legg, T. H., Locke, J. L., McLeish, C. W. , Richards, R. S., Chisholm, R. M., Gush, H. P., Yen, J. L., and Gault, J. A. (1967) Long base line interferometry: a new technique. *Science*, **156**, 1592–1593.
8. Brown, G. W., Carr, T. D., and Block, W. F. (1968) Long–baseline interferometry of S-bursts from Jupiter. *Astrophys. Lett.*, **1**, 89–94.
9. Clark, B. G. (1973) The NRAO tape–recorder interferometer system. *IEEE Proc.*, **61**, 1242–1248.
10. Clark, B. G. (1989) Coherence in radio astronomy. In R. A. Perley, F. R. Schwab, and A. H. Bridle (eds.), *Synthesis Imaging in Radio Astronomy*, number 6 in ASP Conference Series, Astronomical Society of the Pacific, pp. 1–10.
11. Clark, B. G. (1995) Interferometers and coherence theory. In J. A. Zensus, P. J. Diamond, and P. J.Napier (eds.), *Very Long Baseline Interferometry and the VLBA*, number 82 in ASP Conference Series, Astronomical Society of the Pacific, pp. 73–92.
12. T. Cornwell and Braun R. Deconvolution. In R. A. Perley, F. R. Schwab, and A. H. Bridle (eds.), *Synthesis Imaging in Radio Astronomy*, number 6 in ASP Conference Series, Astronomical Society of the Pacific, pp. 167–183.
13. Cotton, W. D. (1979) A method of measuring compact structure in radio sources using VLBI observations. *Astron. J.*, **84**, 1122–1128.
14. Cotton, W. D. (1993) Calibration and imaging of polarization sensitive VLBI observations. *Astron. J.*, **106**, 1241–1248.
15. Cotton, W. D., Geldzahler, B. J., Marcaide, J. M., Shapiro, I. I., Sanroma, M., and Ruis, A. (1984) VLBI observations of the polarized radio emission from the quasar 3C454.3. *Astrophys. J.*, **286**, 503–508.
16. Davies, R. D., Rowson, B., Booth, R. S., Cooper, A. J., Gent, H., Adgie, R. L., and Crowther, J. H. (1967) Measurments of OH emission sources with an interferometer of high resolution. *Nature*, **213**, 1109–1110.
17. Felli, M. and Spencer, R. E. (eds). (1989) *Very Long Baseline Interferometry.* Number 283 in NATO ASI Series C. Kluwer Academic Publishers, Dordrecht.
18. Galt, J. (1988) Beginnings of long-baseline interferometry in Canada:, a perspective from Penticton. *J. Roy. Astron. Soc. Can.*, **82**, 242–247.
19. Gubbay, J., Legg A. J., Robertson, D. S., Moffet, A. T., Ekers, R. D., and Seidel, B. (1969) Trans–Pacific interferometer measurments at 2,300 MHz. *Nature*, **222**, 730–733.
20. Gush, H. P. (1988) Beginnings of VLBI in Canada. *J. Roy. Astron. Soc. Can.*, **82**, 221–232.
21. Hirabayashi, H., Inoue, M., and Kobayashi, H. (eds.) (1991) *Frontiers of VLBI.* University Academy Press, Inc., Toyko.
22. Jennison, R. C. (1958) A phase sensitive interoferometer technique for the measurment of the Fourier transforms of spatial brightness distributions of small angular extent. *Mon. Not. Roy, Astron Soc.*, **118**, 276–284.
23. Kellermann, K. I. (1992) Reflections on the Soviet–American VLBI program. In N. S. Kardashev (ed.)*Astrophysics on the Threshold of the 21st Century*, Gordon and Breach Science Publishers, Philadelphia, pp. 37–51.
24. Kellermann, K. I. and Cohen, M. H. (1988) The origin and evolution of the

N.R.A.O.–Cornell VLBI system. *J. Roy. Astron. Soc. Can.*, **82**, 248–265.

25. Marcaide, J. M., Alberdi, A., Ros, E., Diamond, P., Shapiro, I. I., Guirado, J. C., Jones, D. L., Kirchbaum, T. P., Mantovani, F., Preston, R. A., Rius, A., Schilizzi, R. T., Trigilio, C. , Whitney, A. R., Witzel, A., and Zensus, A. (1995) Expansion of SN 1993J. *Science*, **270**, 1475–1478.
26. Miyoshi, M., Moran, J., Herrnstein, J., Greenhill, L., Nakai, N., Diamond, P., and Inoue, M. (1995) Evidence for a black hole from the high rotation velocities in a sub–parsec region of NGC 4258. *Nature*, **373**, 127–129.
27. Moran, J. (1993) Distance measurments through observations of masers. In R. J. Davis and R. S. Booth, editors, *Sub-arcsecond Radio Astronomy*, Cambridge University Press, Cambridge, pp. 62–67.
28. Moran, J. M. (1976) Very long baseline interferometric observations and data reduction. Volume 12C of *Methods of Experimental Physics*, Academic Press, New York, pp. 235–237.
29. Moran, J. M., Crowther, P. P., Burke, B. F., Barrett, A. H., Rogers, A. E. E., Ball, J. A., Carter, J. C. , and Bare, C. C. (1967) Spectral line interferometry with independent time standards at stations separated by 845 kilometers. *Science*, **157**, 676–677.
30. Palmer, H. P., Rowson, B., Anderson, W., Donaldson, B., Miley, G. K., Gent, H., Adgie, R. L., Slee, O. B., and Crowther, J. H. (1967) Radio diameter measurments with interferometer baselines of one million and two million wavelengths. *Nature*, **213**, 789–790.
31. Readhead, A. C. S. and Wilkinson, P. N. (1978) The mapping of compact radio sources from VLBI data. *Astrophys. J.*, **223**, 25–36.
32. Rogers, A. E. E. (1995) VLBA: Formatter to tape. In J. A. Zensus, P. J. Diamond, and P. J.Napier (eds.), *Very Long Baseline Interferometry and the VLBA*, number 82 in ASP Conference Series, Astronomical Society of the Pacific, pp. 93–115.
33. Romney, J. D. (1995) Theory of corrrelation. In J. A. Zensus, P. J. Diamond, and P. J.Napier (eds.), *Very Long Baseline Interferometry and the VLBA*, number 82 in ASP Conference Series, Astronomical Society of the Pacific, pp. 17–37.
34. Schwab, F. R. and Cotton, W. D. (1983) Global fringe search techniques for VLBI. *Astron. J*, **88**, 688–694.
35. Thompson, A. R. (1995) VLBA: Antenna to data formatter. In J. A. Zensus, P. J. Diamond, and P. J.Napier (eds.), *Very Long Baseline Interferometry and the VLBA*, number 82 in ASP Conference Series, Astronomical Society of the Pacific, pp. 3–16.
36. Thompson, A. R., Moran, J. M., and Swenson, G. W. (1991, 1994) *Interferometry and Synthesis in Radio Astronomy.* Wiley–Interscience(first print.), Krieger (second print.), New York, Malabar (Florida).
37. Walker, R. C. (1989) Calibration methods. In M. Felli and R. E. Spencer, editors, *Very Long Baseline Interferometry*, number 283 in NATO ASI Series C, Kluwer Academic Publishers, pp. 141–162.
38. Walsh, D. . Observations of gravitational lenses. In R. J. Davis and R. S. Booth (eds.), *Sub-arcsecond Radio Astronomy*, Cambridge University Press, Cambridge, pp. 111-116.
39. Wrobel, J. M. (1995) VLBI observing strategies. In J. A. Zensus, P. J. Diamond, and P. J.Napier (eds.), *Very Long Baseline Interferometry and the VLBA*, number 82 in ASP Conference Series, Astronomical Society of the Pacific, pp. 411–427.
40. Zensus, J. A., Diamond, P. J., and Napier, P. J. (eds.) (1995) *Very Long Baseline Interferometry and the VLBA*. Number 82 in ASP Conference Series. Astronomical Society of the Pacific.

Discussion

F. Malbet: At much longer baselines, can you find unresolved objects with enough flux to calibrate the VLBI array?

Cotton: On the longest baselines there are not any completely unresolved sources and it's necessary to solve for a source model jointly with flux calibration using observations of the smallest sources.

F. Malbet: What is the most important quantity to measure to restore an image: visibility amplitude or phase? In the case of phase, is it mandatory to use chained measurements (of Steve Ridgway's talk), i.e. redundant arrays to get a good SNR on the phase measurement or are there other ways to get the phase?

Cotton: Phase is more important to imaging than amplitudes. Phase calibration can be done using redundant baselines but the uv coverage of a redundant array is much inferior to a non redundant array. Self–calibration can be used with a non redundant array.

F. Malbet: How do you define the FOV in VLBI? What are the limitations on the FOV?

Cotton: The most fundamental limits on the field of view are the power patterns of the individual antennas although this is rarely a problem with VLBI. Averaging in time and frequency can also distort objects far from the phase and delay tracking centers although both of these problems can be eliminated by correlating with higher frequency resolution and/or higher time resolution.

G. Daigne: At long wavelengths, is there not any situation where self–calibration cannot be used (interplanetary scintillations)?

Cotton: Self–calibration can be used whenever time variable phase fluctuations are the principle problem. If the image is broadened by "seeing" in which the signals have followed multiple paths with varying delays, then self–calibration will not help.

J. Eislöffel: You were telling us that with increased resolution one can observe only objects with higher brightness temperature, so effectively the sources that can be observed are a subsample of those that one can see at lower resolution. Then satellite VLBI will have access to only some of the sources that ground based VLBI can observe. What is the new science that can be done there that couldn't be done by ground based VLBI?

Cotton: VLBI using a satellite will allow exploration of brightness temperatures in excess of 10^{12} K where inverse Compton scattering of radio photons off the relativistic electrons in a static synchrotron source is expected to rapidly cool the source.

B. Junor: Are there limits to VLBI imaging performance imposed by the solar wind and the general interstellar medium?

Cotton: Both the interplanetary and interstellar media have seeing disks which fundamentally limit the resolution that can be obtained using VLBI.

B. Junor: Is the performance of a VLBI system limited by the local clocks? In other words, should we develop better standards than H masers?

Cotton: For astronomical observations where short term phase stability is very important, H masers are already more stable than the atmosphere, so better stability would not help much. For astrometric and geodetic measurements, better long term stability would help.

L. Roberts: In self–calibration, there was a decision if the calibration was finished or not, how is this decided?

Cotton: Self–calibration is usually considered finished when the result stops improving by some measure such as RMS noise in the image or by a χ^2 test of how well the model visibilities fit the data.

S. Ridgway: Will closure phases among a sufficiently large number of telescopes allow us to determine the individual telescope phases?

Cotton: In a non–redundant array, it is not possible to determine both the antenna and visibility (per baseline) phases using only phase closure. However, if the number of redundant baselines exceeds the number of antennas, this is possible but the uv coverage will be significantly inferior to that of a non–redundant array.

D. Mourand: In the self–calibration procedure, you loose the information on the position. How could you superimpose radio-maps (with 1 mas resolution) over optical images?

Cotton: The position of the radio source is generally known to an accuracy of much better than an arcsecond before VLBI observations are made. If self–calibration is used, positions are not known at the level of the size of the synthesized beam.

J. Hildebrand: With the new data cube technology being developed, do you believe that optical VLBI would be possible someday?

Cotton: Optical VLBI is quite unlikely in the foreseeable future. The stability requirements on the clocks greatly exceeds what is currently possible.

R. Costa: Is flux calibration possible for VLBI data? Can you do photometry with them?

Cotton: With some care the flux density calibration of VLBI data can be made accurate to a few percent. Photometry with an interferometer can be significantly in error if some of the source structure is completely resolved.

S. Loiseau: How do you synchronize the clocks?

Cotton: At the beginning of VLBI the clocks were synchronized by bringing them together. Today GPS systems are used. Fringe fitting calibrators is used for fine clock synchronization.

M. Chang: A minor point - you said that the multiplication process and the Fourier transform processes were commutable. Is this on a practical level or just a mathematical level?

Cotton: Both.

D. Le Mignant: Actually, what do you think would be the first thing to improve in VLBI? Resolution, uv plane coverage, or sensitivity?

Cotton: The resolution of VLBI should be improved by a factor of 2-3 this Fall (1996) with the launch of the VSOP satellite.

B. Koehler: Haven't you been caught by the virus of going to the moon, i.e. moon-moon or Earth-moon VLBI?

Cotton: In a few decades ,it may not be possible to do radio astronomy from the surface of the earth and the back side of the moon becomes interesting. VLBI limited to the surface of the moon is not very interesting at most radio wavelengths because the moon rotates too slowly. VLBI from the Earth to the moon probably has so much resolution that all sources would be completely resolved.

M. Feldt: How do you deal with baseline determination in a mixed array of ground–based and communications satellites? To what accuracy can these baselines be determined?

Cotton: Fringe fitting is generally useful for dealing with the large position uncertainties of satellites. If fringe fitting is used, then accurate satellite positions are not needed, since the satellites cannot change position often enough to use phase referencing techniques. The cost of fringe fitting satellite data with large position uncertainties is that the larger search windows in delay and rate require a higher SNR to reliably detect the source. This effect increases the minimum detectable source strength.

M. Le Louarn: Aren't there satellite tracking stations that could give you the satellite positions to determine the baselines?

Cotton: Satellite tracking stations do measure the location of the satellite, but JPL will only promise locations no worse than 80 meters which is very bad for use with VLBI.

LASER GUIDE STARS

R. FOY
Centre de Recherches Astronomiques de Lyon
Observatoire de Lyon
9 avenue Charles André, 69561 Saint-Genis-Laval, France

1. Introduction

The laser guide star (hereinafter referred as LGS) is aimed at providing a reference source bright enough for adaptive optics. This concept has recently begun to be widely discussed, but in fact it is not so new. The first theoretical study for the LGS concept was published in 1985 ([15]); in that paper most of the basic problems raised by the LGS were pointed out. The first publications of experimental work have been based on the first shoot at the sky from an astronomical telescope ([60]), at Mauna-Kea, and on the ATLAS experiment lead by ONERA in France ([17]; [52]). But this activity has been boosted by the LGS workshop held in Albuquerque ([22]), where a huge amount of work undertaken in US military laboratories has been partly declassified. Theoretical research and experiments, including shoots at the sky, had been funded within the framework of the Strategic Defense Initiative programme ("Star War"), starting in the early 80's. The US work was lead by the Lincoln Laboratory at MIT ([40]) and by the Starfire Optical Range of the Philips Laboratory at Albuquerque ([22]), in collaboration with several other laboratories and companies.

In this lecture, I will recall in the following section why the LGS is required for adaptive optics at modern large telescopes, and what are the ways currently investigated to implement it (section 3). Most of them rely on the excitation of sodium atoms in the mesosphere, the relevant properties of which will be summarized in section 4. Next I will recall the excitation process of sodium. At the telescope, the wavefront from the LGS differs from that observed for the astrophysical object; I will address in section 6 two major differences which raise two problems: the cone effect (also called focus isoplanatism in US labs) and the tilt indetermination. Section

A-M. Lagrange et al. (eds.), High Angular Resolution in Astrophysics, 193–229.
© 1997 *Kluwer Academic Publishers.*

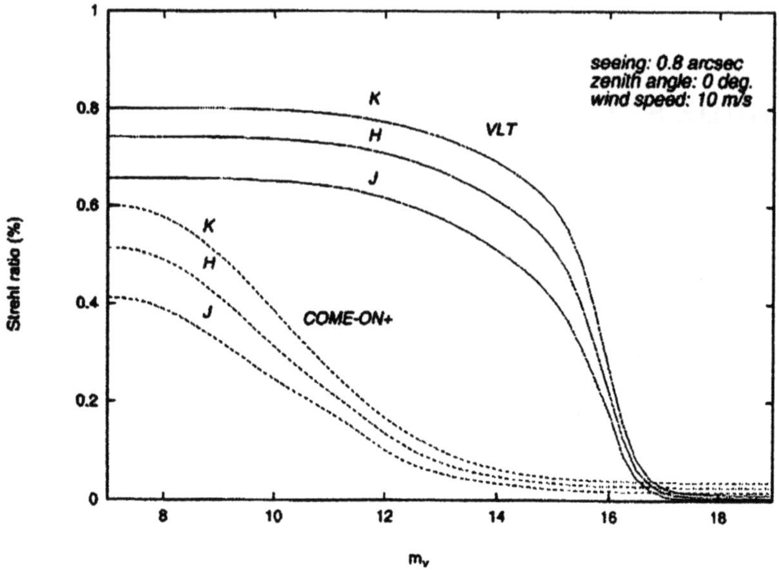

Figure 1. Strehl ratio versus visual magnitude of the source which feeds the wavefront sensor, at zenith, assuming that the seeing is 0.8"and the wind speed is 10 m/s. J, H and K refer to the photometric bands. Dotted lines: COME-ON PLUS adaptive optics at the ESO 3.6 m telescope. Solid line NAOS adaptive optics at the VLT. (From [59])

7 gives a description of current experiments on the sky. Operational issues are emphasized in the concluding section, with Research & Development programmes which are required for a full implementation of the LGS at large telescopes.

2. Why do we need artificial stars?

To feed the wavefront sensors, adaptive optics requires comparatively bright reference sources within the isoplanatic patch of the programme objects. This is the major limitation of adaptive optics devices currrently running at the foci of astronomical telescopes ([27], [14], [2], [49]). Figure 1 shows the Strehl ratio as a function of magnitude for the three photometric bands J ($1.25\mu m$), H ($1.65\mu m$) and K ($2.2\mu m$) computed for the COME-ON PLUS device at the ESO 3.60 m telescope and for the NAOS device under construction for the VLT.

Curves are plotted for $r_0 = 15cm$ and a wind speed $v = 10\,m/s$, assuming the telescope aims at zenith. Typically, the adaptive optics planned for the VLT should restore wavefronts with a residual rms phase error $\sigma_\phi = 1\,rad$ (that is to say at $\lambda/2\pi \approx \lambda/6$) in the near infrared with a ref-

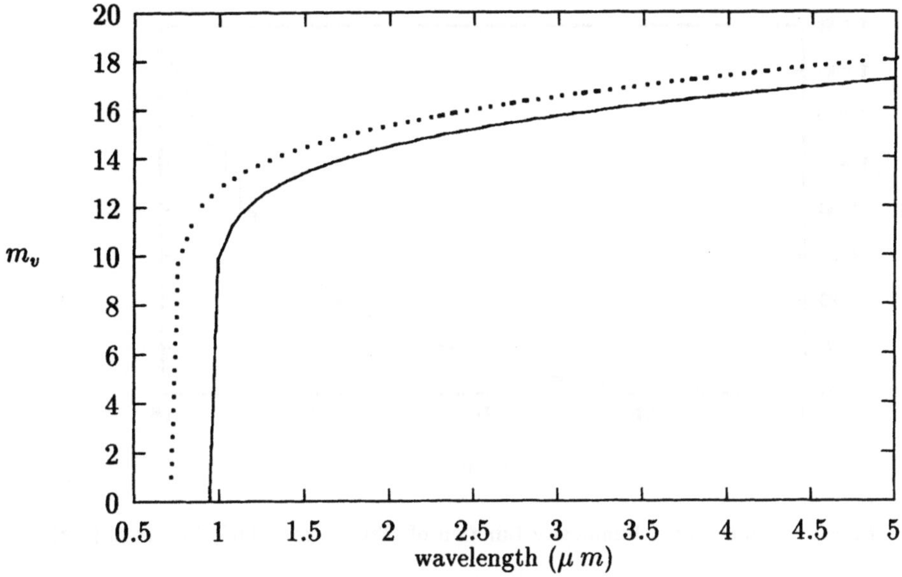

Figure 2. The expected limiting magnitude of the reference source to get a Strehl ratio of 0.4 (solid line) or 0.2 (dotted line) versus wavelength for the NAOS adaptive optics system at the ESO VLT ([33]).

erence source as faint as $m_v = 15$. Strehl ratio vanishes rapidly for fainter reference sources. Figure 2 shows the behaviour of the visual magnitude required for the reference source to get a Strehl ratio of 0.2 and of 0.4 as a function of the wavelength of observation, λ, of the programme object, as expected for the NAOS device. The necessity of bright reference sources is clear, particularly at short wavelengths. These curves have been computed using a standard model for Cerro Paranal ([33]): a seeing of 0.6", two turbulent layers located $3\,km$ and $8.5\,km$ above the summit with 53% of the integrated $C_N^2(h)$, and a weighted wind velocity of $9.8\,km/s$.

From the integrated luminosity function of QSO's ([59]), the number of QSO's brighter than $m_v = 15$ is roughly 20 (Fig. 3).

This is a typical example which requires the use of a reference source in the vicinity of the programme object. But an additional constraint for the reference source is the distance between the programme object and the reference; it has to be as small as possible, smaller than the isoplanatic patch of the programme object at the wavelength of the observation. The probability of finding a suitable reference source can be statistically estimated using models of the galactic stellar distribution ([4], [47]). Table 1 gives a sky coverage estimate assuming that i/ the seeing is 0.8″, ii/there is only photon noise, and iii/ the wavefront has infinite bandwith ([41]): these prob-

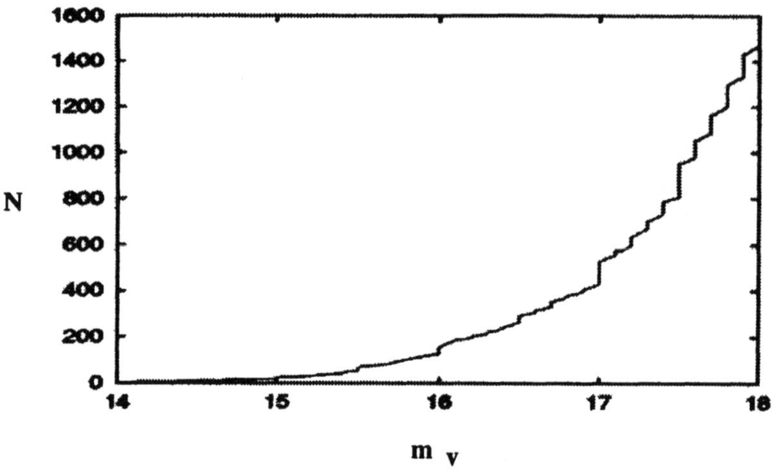

Figure 3. Integrated luminosity function of QSO's, from Théodore et al ([59]).

abilities are inacceptedly low as the telescope diameter increases. Indeed, for a given quality of the corrected wavefront (i.e.: $\sigma = 1\,rad$), the order of the highest Zernike polynomial which fits the wavefront increases with the diameter; then the correlation between the wavefronts from the programme object and from the reference for the highest order decreases: it results in an increased isoplanatism error. In other words, for a given isoplanatic angle r_0/h, the isoplanatism error increases with the telescope diameter since the required angular accuracy for the restored image increases (because the telescope angular diffraction limit decreases).

TABLE 1. Probability to get a reference source as a function of wavelength for a seeing of 0.8"at a 2 m and a 6 m telescopes, from [41].

wavelength(μm) D(m)	3	2	1	0.6
2	90%	20%	0.5%	0.04%
6	30%	5%	0.2%	0.01%

It is clear that to be able to take full benefitof adaptive optics over 100% of the sky, one needs to create an artificial reference source in the direction of the programme object to feed the wavefront sensor. The radiating process has to arise above the highest turbulent atmospheric layer which contributes significantly to the integrated $C_n^2(h)$, a height which ranges between 12 and 18 km (above sea level). Stratospheric turbulence reported in [6] is not confirmed by SCIDAR measurements ([62],[63]).

3. How to make artificial stars

Several processes have been proposed to create a bright spot in the sky which simulates a star. They all rely on the use of laser beams, and most of them on backscattering. Probably the simplest idea is to shoot at the telescope with a satellite borne laser ([24]). A milliwatt laser on board a satellite with a highly eccentric orbit shoots at the telescope, providing a reference source equivalent to a 5th magnitude star. This reference can be kept within the isoplanatic patch of the programme object for 5000 s or more, allowing an adaptively corrected 8 m telescope to reach magnitude 30. The drawback is the limited sky coverage, typically 1 square degree per year.

The LGS is produced by shooting at the sky in the direction of the programme object with a laser beam, either from the telescope itself (see Fig 4), or more frequently from an auxiliary telescope. The backscattered light looks like a star, but fuzzier either because the angular size of the diffraction limited laser beam is non-negligible with respect to the seeing disk, and/ør it is not diffraction limited, and/or it is degraded on its way to the mesosphere. Because of the thickness of the backscattering layer, the off-axis LGS image is elongated.

Backscattering processes are (see Fig 6) :

- Mie scattering, due to dust and aerosols; the density of these particles is very low above the inversion layer, so that this process is not efficient above $\approx 4\,km$ above sea level, which is much too low with respect to the altitude of turbulent layers. Sometimes, when violent volcano eruptions eject large amounts of ash into the stratosphere, strong Mie scattering can be observed for several months from layers $20 - 25\,km$ high, e.g.: the effect of the Pinatubo eruption in 1991 ([22]). Such eruptions are too scarce to allow us to rely on them for LGS.
- Rayleigh scattering due to molecules, mostly N_2. The backscattered flux decreases with air density, i.e.: with altitude. The limiting altitude ranges between 15 and 25 km, depending on the laser power and wavelength, and the acceptable wavefront phase error. The ATLAS experiment has used Rayleigh scattering with a doubled frequency NdYAG laser ($\lambda = 532\,nm$), at the Laser Lunar Ranging Station at the Observatoire de la Côte d'Azur (5); it has shooted at $10 - 15\,km$ altitudes ([32]). At the Mt Wilson 2.5 m telescope, a 50 W excimer laser will produce an array of three LGS 18 km above the site, at $\lambda = 351\,nm$

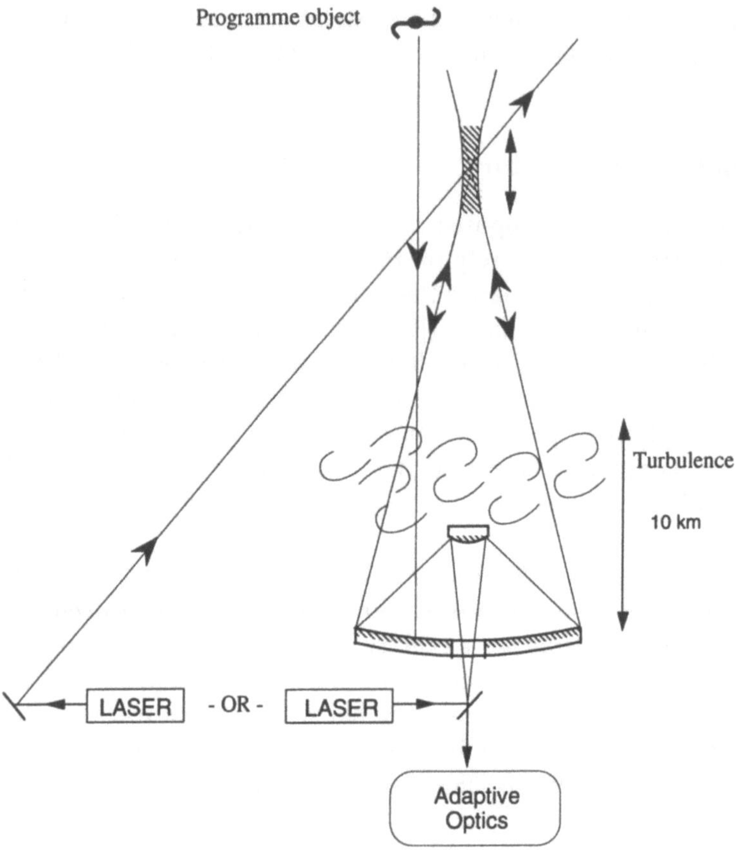

Figure 4. Principle of the laser guide star. The laser beam shoots at the sky above the turbulent layers, either from the large telescope or from an auxiliary telescope; backscattered light feeds the adaptive optics device with enough photons to allow us to get an image of the programme object corrected for turbulence degradation ([56]).

([61]). The Starfire Optical Range (SOR) has experimented two generations of LGS adaptive optics systems ([21]); in both cases, they used Copper vapor lasers, emitting at 510.6 and 578.2 nm, equal power, with a repetition rate of 5 kHz. Average powers were 75 W and 200 W respectively for the two generations.

- Fluorescence backscattering uses resonant excitation of atoms in the mesosphere in the range $90 - 100\,km$, far above the highest turbulent layers. Table 2 gives the cross section of three strong resonant lines of sodium, potassium and iron and the column density of these species ([38]).

  Because of the higher column density, sodium atoms return more flux than potassium atoms, and because of the higher cross section, they

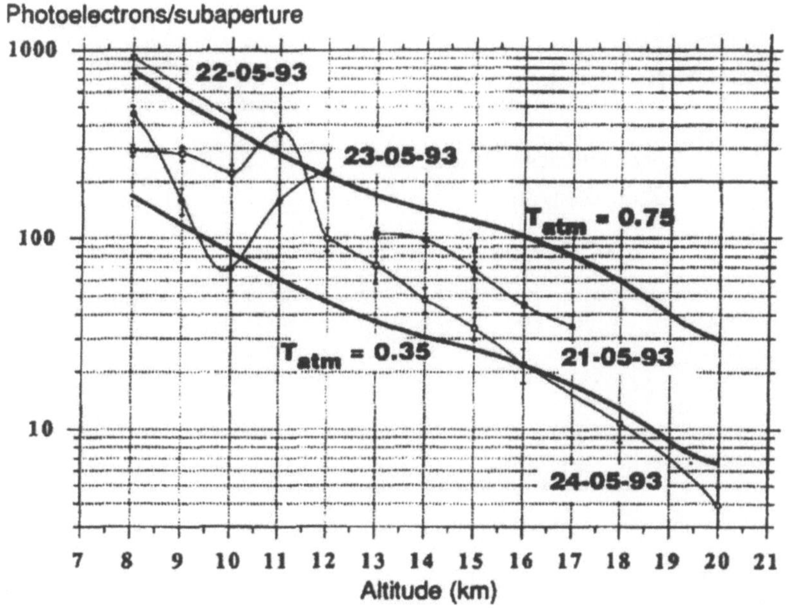

Figure 5. Flux detected per subaperture of the Shack-Hartmann wavefront sensor with the ATLAS experiment [32].

TABLE 2. Cross section of resonant lines σ and column density of sodium, potassium and iron.

	Na	K	Fe
$\lambda\,(nm)$	589	769	372
$\sigma\,(cm^2)$	1.1 10^{-9}	1.0 10^{-9}	1.0 10^{-10}
$N\,(cm^2)$	5 10^9	3 10^8	1 10^{10}

return more flux than iron atoms. In addition, the resonant potassium line at 769.9 nm is rather embedded in telluric O_2lines. Therefore, it seems that sodium provides the most efficient backscattering mechanism in the mesosphere. It has been implemented e.g.: at the Starfire Optical Range (SOR) ([22]), by the Lawrence Livermore National Laboratory (LLNL) at the LLNL and at Lick Observatory ([37]).

- Polychromatic radiative decay also uses mesospheric sodium atoms. It involves not two but at least five energy levels to generate a radiative decay. It aims at providing tilt measurement as well ([16]); first experiments on the sky have been made at the LLNL ([20]).

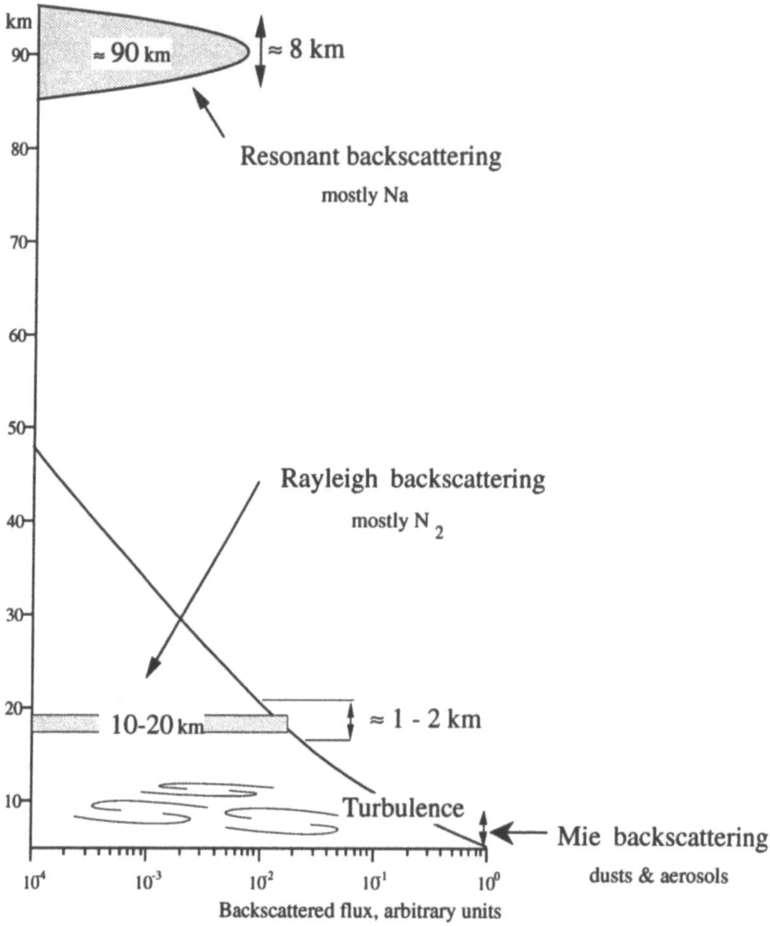

Figure 6. Backscattering processes. Solid countours show the relative efficiency of backscattering as a function of altitude ([56]).

With the exception of the Mt Wilson experiment, all new developments in the field of the LGS involve mesospheric processes.

4. Some properties of the mesospheric sodium layer

The so-called *sodium layer* in the mesosphere is located from 75 to 110 *km* above sea level. There are two theories to explain the sodium concentration at this altitude. One possibility is that sodium atoms are produced by the dissociation of NaCl molecules. These molecules would come from the innermost atmosperic layers. It is rather difficult to understand what process could carry them to the mesosphere through the non convective stratosphere. A more popular explanation of mesospheric sodium atoms is ablation of meteorites. When meteorites enter Earth's atmosphere, their average temperature at $\approx 100\,km$ allows sodium atoms to be freed([53], [13], [29], [30]).

The average altitude of the sodium layer varies with season ([9], [34]). In winter, it is at $\approx 89\,km$, and the temperature is $210 \pm 25\,K$; in summer, it is at $\approx 91\,km$ and the temperature is $160 \pm 25\,K$. The sodium column density varies from 2 to 10 $10^9\,atoms.cm^{-2}$ from summer to winter ([8], [23]). For comparison the density of molecules at $90\,km$ is $\approx 7\ 10^{13} cm^{-3}$. Therefore the total mass of sodium in the layer amounts to $\approx 1 metric\ ton$; this is quite small, so that one could think about feeding it using rockets, but ...

The influx of sodium atoms is $\approx 1.6\ 10^4 cm^{-2}.s^{-1}$. The mean residence time of sodium in the layer is $\approx 6\,days$: ... such a short time precludes feeding the sodium layer with rockets. Two loss mechanisms have been proposed. In the first one, Na transforms to NaO_2 through the reaction

$$Na + O_2 + M \rightarrow NaO_2 + M,$$

where M stands for a neutral atom ([54]). The mass loss rate is proportional to the squared metal atoms density, and is an inverse function of temperature. The other loss mechanism is the formation of cluster ions $Na^+.(H_2O)_n$ ([45], [46], [29]). The mass loss rate is again an inverse function of temperature, and also a function of zonal winds. Both mechanisms account for the observed seasonal variations of the sodium column density because of the correlated temperature variations. Molecules or particles which are formed settle slowly gravitationaly and leave the sodium layer.

The time between collisions of a sodium atom, mostly with nitrogen molecules, is $\approx 100\,\mu s$.

The reader interested in more information about the properties of the sodium layer should refer e.g. to the paper [26]¿

5. Excitation of mesospheric sodium

The resonant excitation of sodium atoms involves absorption of photons from a laser beam at $589\,nm$ which raises the valence electron of sodium from the ground energy level ($3S_{1/2}$) to the sublevel 5/2 of the first excited level ($J = 3P$). This is an electric dipole transition. Due to spontaneous emission, atoms return to the fundamental state, emitting the D_2 line. Until now, backscattering from the D_1 line at $589.6\,nm$ (transition $3P_{1/2} \rightarrow 3S_{1/2}$)

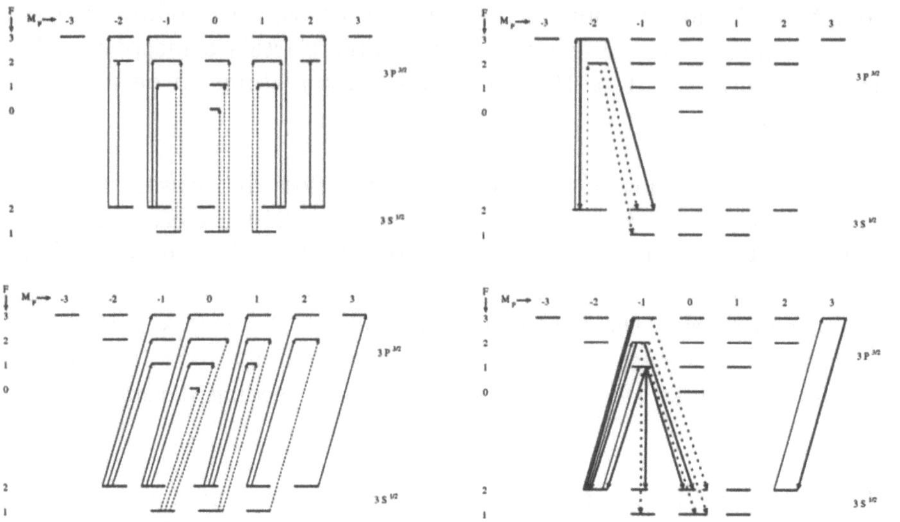

Figure 7. Energy level diagram of the sodium atom for the D_2 transition.
Up: linearly polarized excitation ($\Delta M_F = 0$); on left: all the allowed transitions; on right, it is shown from some $\Delta M_F = 0, -1$ and $+1$ radiative decays that the population of the level $3S_{1/2}(F = 1; M_F = 0)$ is enhanced with respect to the other levels; transitions from this level have the lowest probabilies; therefore after a few excitation-desexcitation cycles, the absorption in the mesosphere decreases.
Bottom: left circularly polarized excitation; on left: all the allowed transitions; on right it is shown that in this case the population of the $3S_{1/2}(F = 2; M_F = 2)$ is enhanced; the D_2 transition from this level has the highest probability; therefore after a few cycles, the D_2 absorption increases.

has never been considered for the LGS because its probability is half that of the D_2 one. This scattering process slightly favours the emission towards the direction of propagation of the exciting field, i.e.: backscattering.

Figure 7 shows the detailed structure of the energy levels $3S_{1/2}$ and $3P_{3/2}$. Both levels are characterized by the principal quantum number $n = 3$ of the valence electron. For the ground state, the electron orbital angular momentum is 0, so that, given the spin angular momentum quantum number $S = 1/2$, the total electron angular momentum quantum number is $J = 1/2$. The nuclear spin angular momentum quantum number is $I = 3/2$. Therefore the total angular momentum quantum number of the ground state has two values $F = I \pm J = 1$ or 2. These two sublevels have different energies because of the magnetic interaction between the electron and the nuclear dipoles; this results in the hyperfine structure of the D_2 line (Fig 7), of which the two components are $1.772\,GHz$ apart.

The first excited level has an electron orbital angular momentum quantum number 1; hence the total electron angular quantum number has two values: $J = 1/2$ and $J = 3/2$, which gives the two levels $3P_{1/2}$ and $3P_{3/2}$.

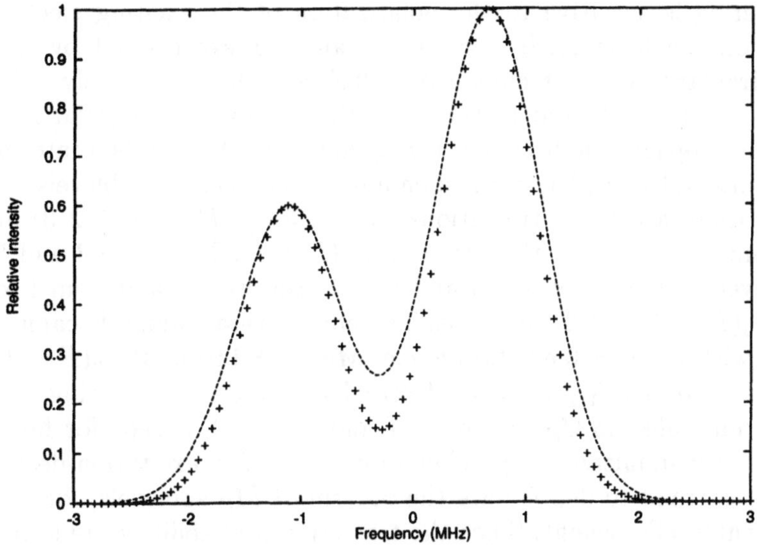

Figure 8. The profile of the mesospheric D_2 line showing the hyperfine structure. Solid line: winter time; dotted line: summer time.

Their energies differ by $\Delta E = 520\,GHz$, because of the spin orbital interaction which is the magnetic interaction between the electronic spin and the orbital momenta. This causes the splitting of the D line into the D_1 and the D_2 lines. Then a further splitting occurs because of the interaction between the total electron magnetic momentum and the nuclear magnetic momentum. The $3P_{1/2}$ level splits into $F = 1$ and $F = 2$; $3P_{3/2}$ splits into $F = 0, 1, 2$, and 3. The ratio of the probabilities of the two hyperfine transitions is equal to 3/5.

The lifetime of the $3P_{3/2}$ level is $\tau = 16\,ns$, and the corresponding homogeneous linewidth of the D_2 transition is given by

$$\Delta\nu_H = 2/(4\,\pi\,\tau) = 10\,MHz \tag{1}$$

This is much smaller than the Doppler linewidth:

$$\Delta\nu_D = 2\,\nu_0/c\sqrt{\ln 2}\sqrt{2\,RT/A} \tag{2}$$

which ranges from 0.96 to $1.1\,GHz$ depending on the temperature T which varies with the season. As usual, R is the gas constant, A the atomic mass and ν_0 the central frequency of the line. Figure 8 shows the profile of the hyperfine structure of the D_2 mesospheric line.

The peak cross section of the D_2 transition for homogeneous linewidth is $1.14\ 10^{-9} cm^2$. Therefore it is $1.14\ 10^{-9}\ .\ \Delta\nu_H\ /\ \Delta\nu_D\ \approx\ 10^{-11} cm^2$ for the Doppler width.

Backscattering is sensitive to the polarization of the exciting electric field. Transitions such as $\Delta M_F = 0, -1$ and $+1$ are excited by linearly, right circularly and left circularly polarized light respectively. The M_F quantum number is bound to the projection of the total angular momentum on the propagation direction; it may have any of the $2F+1$ integer values in the interval $(-F; F)$, corrresponding to the Zeeman sublevels.

Therefore there are more transitions toward $3P_{3/2}(F = 0,1,2,3)$ at $\Delta M_F = 0$ from $3S_{1/2}(F = 2)$ than from $3S_{1/2}(F = 1)$. There are 3 and 5 Zeeman sublevels in the levels $F = 1$ and $F = 2$ respectively, and from 1 to 7 for the $3P_{3/2}(F = 0,1,2,3)$ sublevels. Therefore several sublevels cannot be populated with a $\Delta F = 0$ excitation, e.g. $M_F = \pm 3$ from $3S_{1/2}(F = 2)$ or $F = 3, \forall M_F$ from $3S_{1/2}(F = 1)$. Excited atoms decay following the classical selection rules: $\Delta M_F = 0$ or $\pm$ 1 and $\Delta F = 0$, except for both $\Delta M_F = 0 \Delta F = 0$ simultaneously. Therefore after a few excitation-decay cycles, there is a trend to depopulate the outermost M_F sublevels, for the $3S_{1/2}(F = 1)$ sublevel's benefit. Since the transition probabilities are lower from $3S_{1/2}(F = 1)$ than from $3S_{1/2}(F = 2)$, the absorption in the sodium layer decreases; this holds as long as a collision (most probably with a N_2 molecule) resets the sublevels population distribution. Therefore linearly polarized laser beams are not efficient.

On the other hand, with circularly polarized light, transitions ΔM_F =-1 or = 1 are involved, which allows the highest M_F sublevels, which have the highest probabilities, to be populated. Then the absorption from both $3S_{1/2}(F = 1)$ and $(F = 2)$ is more efficient than with linearly polarized light, and an indirect optical pumping from $F = 1$ to $F = 2$ occurs. Therefore the emitting laser should be circularly polarized.

This holds as long as the same volume of sodium atoms is concerned; this process is partly balanced:

- by winds in the mesosphere, which renew this volume typically every $0.03\,s$ for a $1''$ seeing,
- and by diurnal motion of the astronomical source, which forces the laser beam to scan the mesosphere to track it. The scan velocity is a function of the latitude of the observatory, and of the coordinates of the source; it is $\approx 460 \cos L\, m/s$ for a source at zenith observed from a site at latitude L.

In spite of the comparatively high cross section of the D_2 transition, saturation, i.e.: maximum absorption, occurs at low laser power. A single sodium atom in the ground state cannot absorb more than one photon per deexcitation time τ per cross section unit. This corresponds to a saturation intensity of:

$$I_s = h\nu/\sigma\tau = 185\, W/m^2 or \Phi_s = 1/\sigma\tau = 5.5\, 10^{16} photons/s/cm^2, \qquad (3)$$

or $\approx$ $5W/cm^2$ taking into account that the full D_2 bandwith spans $\approx$ $300\Delta\,\nu_H$ intervals or velocity groups of sodium atoms. This yields the absorption:

$$A = \frac{\rho_h/\tau}{\Phi_s}\frac{\Delta\nu_H}{\Delta\nu_T} \approx 2\% \qquad (4)$$

where $\Delta\,\nu_T = 3\,GHz$ stands for the total bandwidth of the D_2 line.
For laser power at or above saturation, the amount of absorption decreases and there is no increase in the return power: emitting more energy is wasting energy.

6. The LGS wavefront

Several effects make the wavefront emitted by the LGS different from the perfect reference source for the adaptive optics. Let us mention:

the cone effect , or focus anisoplanatism, due to the finite distance of the LGS,

the tilt , which cannot be determined in a simple way due to the principle of inverse return of light, and

the pointing error , due to the different direction of the LGS with respect to the programme object direction. Pointing errors may be caused either by the uncompensated tilt, or by pointing or tracking differential errors in case of a LGS generated by an auxiliary telescope or device, or to uncompensated differential atmospheric refraction when observing a programme source quite far from zenith at a wavelength significantly different from the LGS wavelength. Since these errors results in biases or errors which are not specific to the LGS, but to anisoplanatism, I will not discuss them in this lecture (see Rigaut's lesson in this book).

the defocus due to the altitude of the centre of gravity of the sodium layer which may vary significantly over a time scale as short as a few minutes. Sporadic high density narrow peaks in the vertical sodium density distribution in the mesosphere have been reported ([12]). Typical duration are $1hour$. These variations deserve further investigations. I suggest that the average altitude is monitored by measuring the time of light round trip to the sodium layer using either an additional lidar device or a slight modulation of the LGS repetition rate.

off-axis LGS produce elongated images. The signal to noise ratio in the direction of the elongation is decreased proportionally to the image aspect ratio.

The laser power stability is not critical. Of course errors in the laser wavelength which are not negligible with respect to the inhomogeneous width of the transitions decrease the LGS flux and increase errors in phase measurements.

6.1. THE CONE EFFECT

Figure 9 shows the cone effect. The conical volume subtended by the LGS at the telescope pupil from the LGS does not allow us to sample the whole cylindrical volume subtended by the programme source, which can be supposed at infinity. An area at the edge of the pupil is illuminated by the LGS through turbulent cells which are different from those crossed by the light from the natural guide star (NGS). Phase disturbance measurements derived by any wavefront sensor from the LGS for this pupil area are wrong and the correction applied to the adaptive optics deformable mirror results in degradation of the image quality. The effect increases with the telescope diameter and decreases with the coherence length: it is smaller in a good astronomical site and in the infrared. The d_0 parameter ([18]) plays for the cone effect the same role as the r_0 parameter for the turbulence; d_0 is the diameter of the telescope for which the rms phase error due to the cone effect is less than $\lambda/6$. The phase error due to the cone effect is written as a function of the telescope diameter D as:

$$\sigma_\phi = (D/d_0)^{5/6} \tag{5}$$

Figure 10 shows the variation of d_0 with wavelength for two altitudes of the LGS, 20 and 90 km, and in both cases for an optimistic and a pessimistic Hufnagel-Valley model of $C_n^2(h)$ ([56]).

Figure 11 ([32]) shows the correlation, as a function of the Zernike radial degree (see Rigaut's lesson in this book), between the wavefront measured from a NGS and that measured from a Rayleigh LGS formed at an altitude of 15 km. The loss in correlation due to the cone effect alone is shown by the upper curve: in this case both the LGS and the NGS are superimposed on the sky. The lower curve show the additional affect of anisoplanatism: the LGS is located 20" away from the NGS. The observations were made by the ONERA team with the 1.52 m telescope of the Lunar Laser Ranging Station of Observatoire de la Côte d'Azur in France.

Diaphragming the telescope pupil would decrease the intensity of the cone effect to reject the poorly compensated outer annulus; of course this is not acceptable. The way to avoid the cone effect is to sample the sky with several LGS close to the programme object. The first work in this field has been done by the Lincoln Lab team at MIT ([65] and [36]). They used the laser beam of their SWAT experiment at Maui, separating the pulses in time faster than the atmosphere coherence time. Then they split the pupil into two areas, each of them looking at that spot facing it (see Fig 12a).

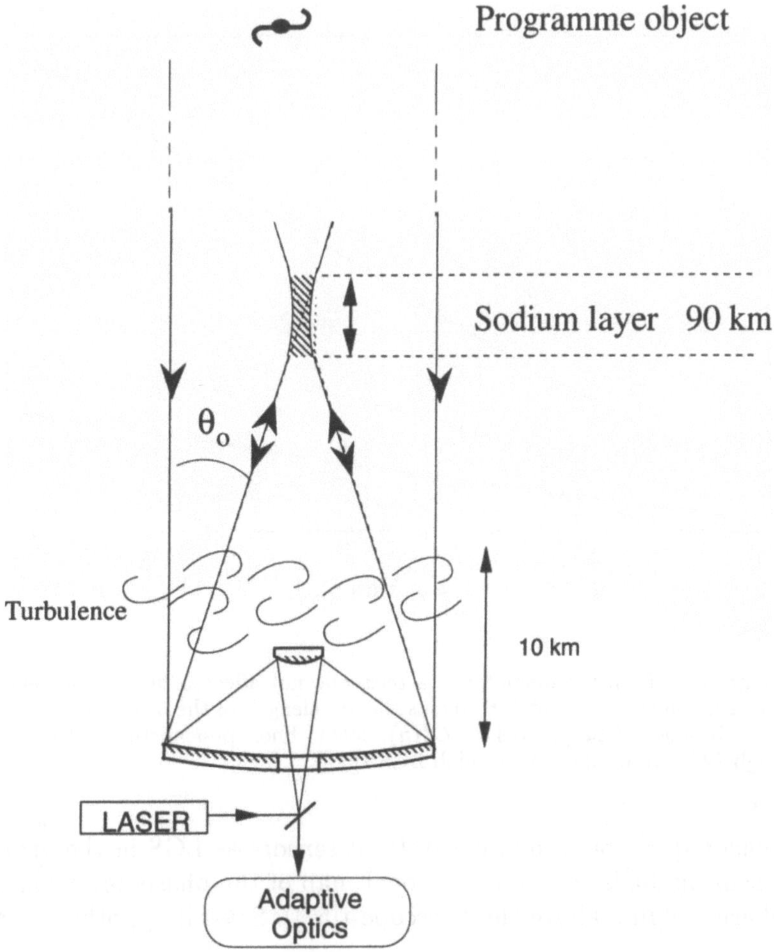

Figure 9. **Due to the finite distance of the LGS, a conical volume is sampled in the turbulent layers, instead of a cylindrical volume for the programme source ([56]).**

In the SWAT case, the pupil was 60 *cm* in diameter, and there were two spots. Stitching was performed: the SWAT adaptive optics device corrected the two parts of the incoming wavefront independently of each other. A comparison of the uncorrected image of the star Pollux with the image corrected using a single LGS and two LGS is given in Fig.13([51]).

The gain in Strehl ratio is marginal, which is explained by the lack of tilt correction for the two areas of the pupil. Indeed, the two areas produce two almost diffraction limited images at two different random locations on the detector, which results in an image degradation.

Another method is to observe the whole spot array from every cell of the wavefront sensor (see Fig.12 b). Then one needs either a wide field

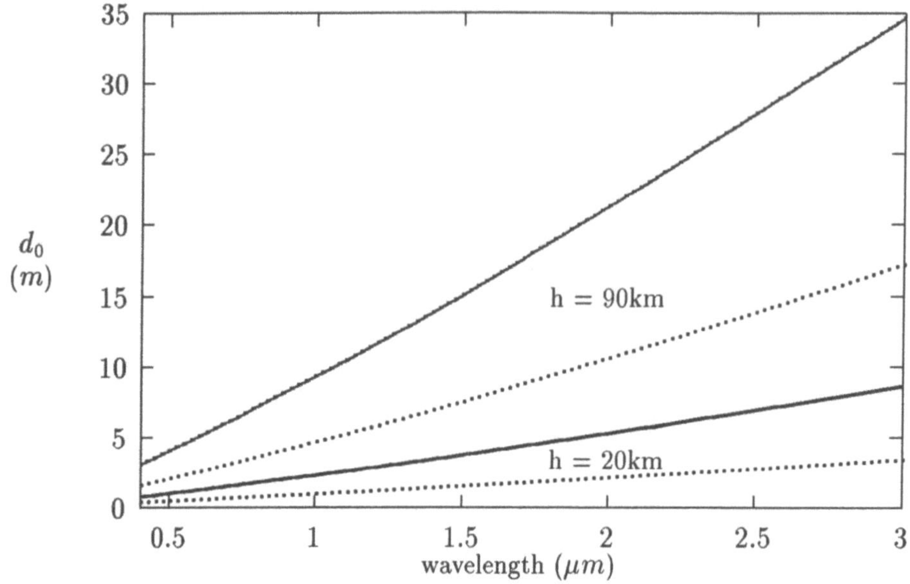

Figure 10. Variation of d_0, the diameter of a telescope not affected by the cone effect to an rms error level on the phase of $\lambda/6$, versus the wavelength of the observation. Solid lines: optimistic Hufnagel-Valley model for $C_n^2(h)$; dotted lines: pessimistic model; thick lines: 90 km high LGS; thin lines: 20 km high LGS ([56]).

wavefront sensor ([55]) or as many wavefront sensors as LGS in the array. This global sensing leads to a 3-dimensional map of the phase deviation in the atmospheric volume above the telescope ([57]). Several hypotheses are assumed:

- most of the turbulence strength $\int_0^\infty C_N^2(h)\,dh$ is distributed among a few discrete turbulent layers. This hypothesis is supported by soundings of $C_N^2\,(h)$ profiles either by the SCIDAR technique [62] or balloon [10] [63], as shown in Fig.14 and 15. Fig.14 shows a typical $C_N^2(h)$ profile, which is high mostly within three thin layers. From Fig.15, they contribute for $\frac{\int_{lower\ boundary}^{upper\ boundary} C_N^2(h)\,dh}{\int_0^\infty C_N^2(h)\,dh} = 33\%, 32\%,$ and 13% respectively to the total seeing, i.e.: $\approx 80\%$ of the seeing is produced by only three thin layers.
- small perturbation approximation; this is a good approximation as long as the zenith angle $z \leq 60^o$ ([64]). Then, following Roddier [48], the phase perturbations due to each turbulent layer simply add to give the final perturbations.
- phase perturbations are small enough that diffraction effects are negligible. This approximation can be relaxed at the expense of heavier

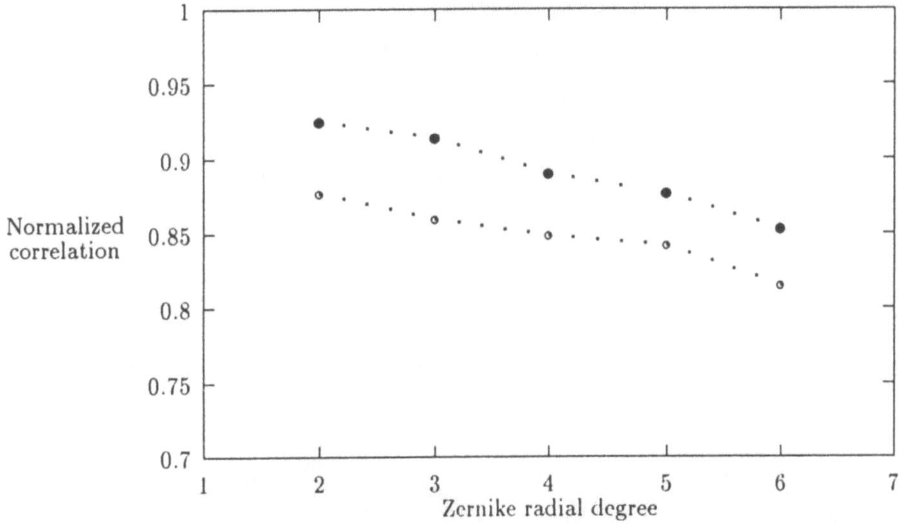

Figure 11. Loss in correlation as a function of the degree of the radial Zernike polynomials in case of a laser guide star in the direction of the programme object (full circles) or 20" apart (open circles) ([32]) .

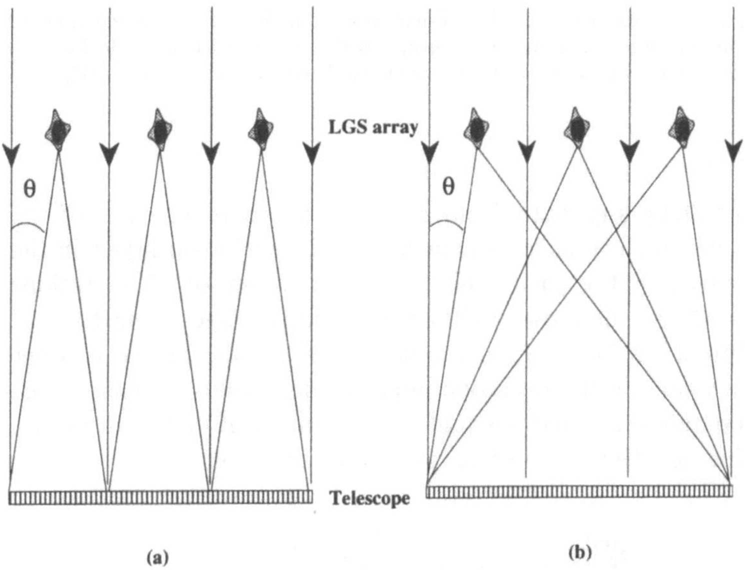

Figure 12. Two ways to tackle a LGS array. (a) the pupil is divided into sections. Each observes the LGS in the array facing it, and the signal is processed independently of the other areas. (b) Each of the pupil sections measures the whole array and the signal is processed globally ([56]).

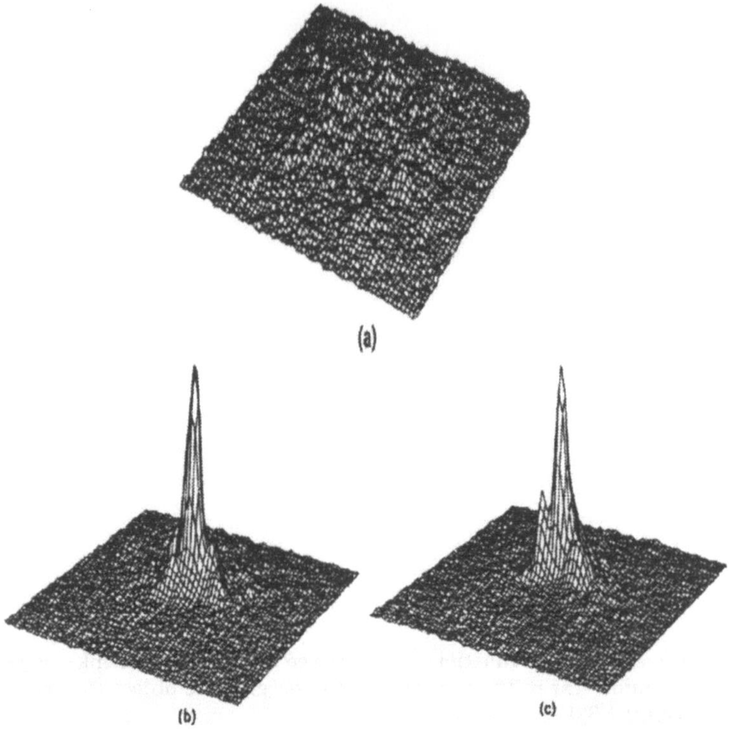

Figure 13. Images of Pollux with the SWAT experiment at Maui. a) uncompensated image; b) image compensated with an adaptive optics device fed with a single LGS; c) image compensensated with an adaptive optics device fed with a pair of LGS ([65]).

computations.

The phase error along any path from each of the spots to each of the subpupils is the sum of the phase shifts in each of the turbulent layers in the coherent cells crossed by the path. This leads to a system of $2N$ equations per laser spot, with $2N$ unknown shifts per turbulent layer (one tip and one tilt per subpupil projected onto the layer); N stands for the number of subpupils which sample the telescope pupil in the wavefront sensor. The number of spots required is roughly equal to the number of turbulent layers. Figure 16 explains notation in the final system of equations:

$$\theta_X^{(m)} - a_0^{(m)}(\underline{X}_i) = \sum_{k=1}^{n} (1 - \frac{h_k}{H}) a_k^{(m)} \underline{X}_{ik}^{(m)} \tag{6}$$

$$\theta_Y^{(m)} - b_0^{(m)}(\underline{X}_i) = \sum_{k=1}^{n} (1 - \frac{h_k}{H}) b_k^{(m)} \underline{X}_{ik}^{(m)} \tag{7}$$

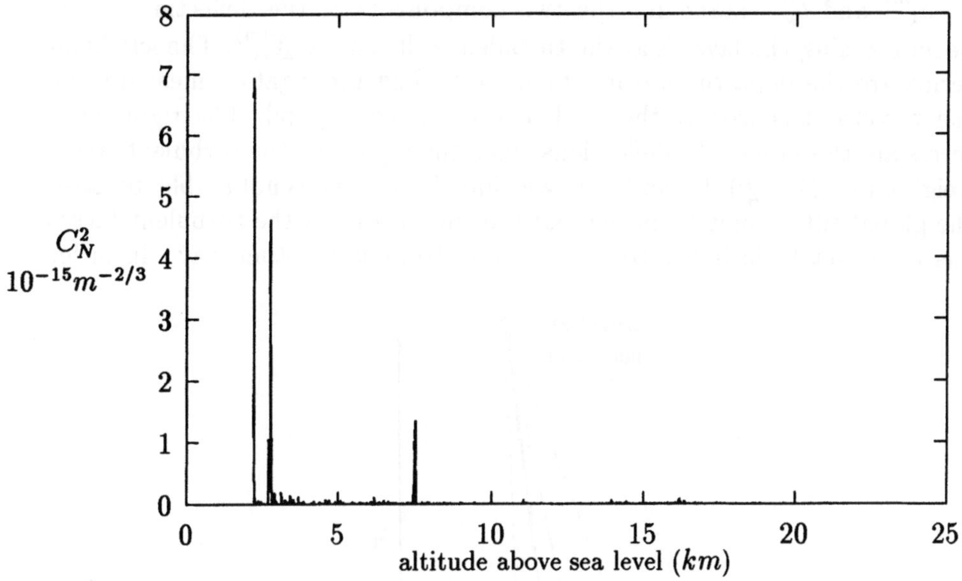

Figure 14. **Distribution of C_N^2 with altitude on July 1991 above the Observatorio del Roque de los Muchachos, located at 2200 m above sea level at the top of La Palma island (Canaria Islands) ([63]).**

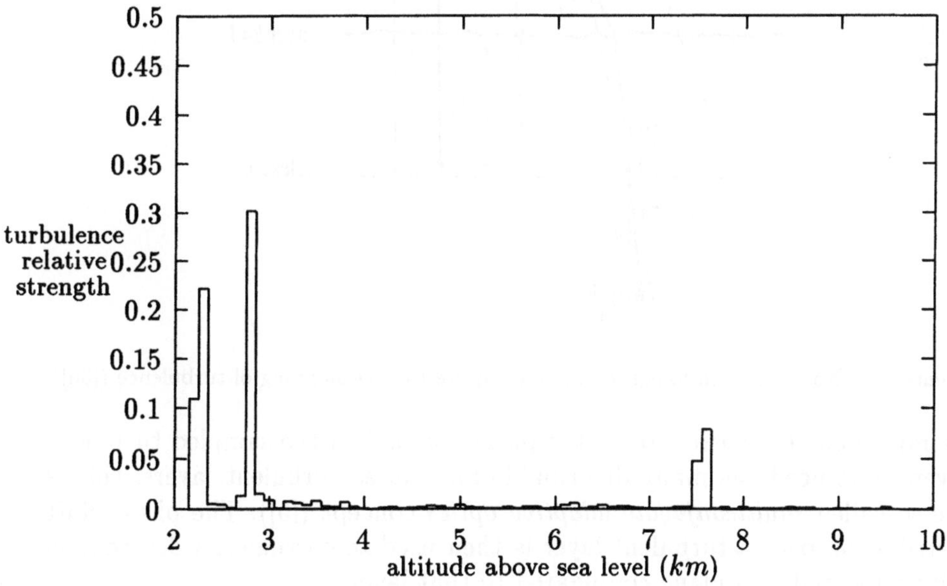

Figure 15. **Same as Fig.14, but for the relative strength of the turbulence ([63]).**

$a_k^{(m)}$ and $b_k^{(m)}$ stand for the two components of the deflection of the beam crossing the layer k at the turbulent cell sample $\underline{X}_{ik}^{(m)}$. The left hand terms are the departures from the unperturbed propagation measured by the wavefront sensor at the level of the telescope pupil. The right hand terms are the sum of the deflections when the rays cross the turbulent layers weighted by $(1-\frac{h_k}{H})$. It can be shown that this system is not stable, because the global tilt cannot be measured; the tilt for one of the turbulent layers has to be set to 0; it has to be measured from some other way. In order

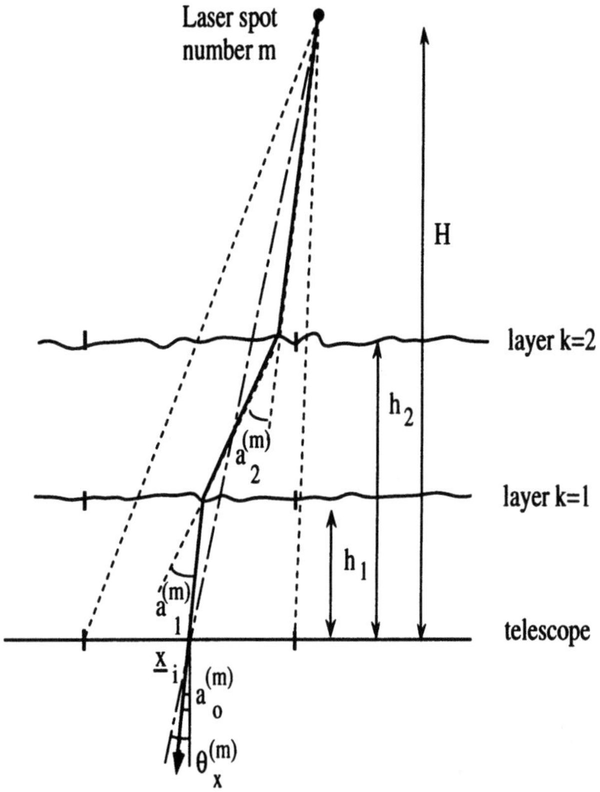

Figure 16. Notation used in eqs. 6 & 7 to compute the 3D-mapping of turbulence ([56]).

to implement the correction of the phase for each of the sampled turbulent layers, one needs as many deformable mirrors as turbulent layers; this is the so-called *multiconjugate adaptive optics* concept ([5]). The phase shift distribution over a turbulent layer is then used to corrugate a deformable mirror located in a plane conjugated to that layer.

The geometry of the wavefront sensor have to be specified in such a way that the images of the individual LGSs of the array given by each of the subpupils (in the case of a Shack-Hartmann) never overlap, whatever the local slope of the wavefront. Then identification of the components of the array is derived from the array distribution on the sky and from the wavefront sensor geometry.

A major by-product of the 3D mapping of the phase shifts in the turbulent volume above the telescope is the widening of the corrected field. From the distribution of shifts, one can compute the total phase shift for each of the telescope subpupils in directions different from that of the reference source; therefore the size of the corrected field is independent of the anisoplanatic patch. The typical field in which all rays to an 8 m telescope are corrected by a 3-spot and a 4-spot array are 7"and 50" respectively. This is significantly larger than the isoplanatic patch of $2 - 4''$ at short wavelengths. Beyond this field, correction is not possible over the whole pupil: the corrected field may be much larger, but it is vignetted. Finally it is worthwhile to note that the spot array is distributed quite far away from the center of the corrected field, e.g.: on a circle of radius of 24" and 41" in the 3-spot and the 4-spot arrays respectively. This will make it much easier to reject the laser light from the astrophysical field with a simple spatial filter. There are geometrical constraints which set the minimum number of spots to 3 for two turbulent layers when widening the fied of view beyond the isoplanatic patch ([58]).

Similar approaches have been proposed (e.g.: [28]). No experimental test has been carried out of the 3D mapping of turbulence, either in the lab or at the telescope .

6.2. THE TILT

The tilt of the incoming wavefront at the scale of the whole telescope pupil is modeled by a prism with a small angle which varies with time. The time of round trip for the laser light to and from the mesosphere exceeds 1 ms only for zenith distances $z > 53^o$; this is much shorter than the coherence time of the atmosphere. Therefore light on the way to the mesosphere is deflected in the opposite direction of its deflection on its way back from the mesosphere to the telescope, because of the principle of inverse return of light. As a result, the apparent direction of the LGS as view from the telescope is constant, whatever the instantaneous prism angle, i.e.: whatever the tilt angle. The absolute position of the LGS cannot be determined since it wanders on the uplink portion of its propagation and there is no way, at least with a single receiving telescope, to distinguish an undetermined position of a guide star from tilt in the wavefront. Then adaptive optics fed with a LGS provides instantaneous images corrected up to the diffraction limit, but these images continuously move with a characteristic time equal to the coherence time for the tilt: diffraction limited images are blurred when integrated on the detector. The effort of running an adaptive optics system plus a LGS is not worthwhile if the tilt cannot be measured ([39], [52]).

It has been argued that the tilt could easily be measured using a NGS in the field. The NGS could be faint, since the tilt is measured with the whole pupil. The probability of finding a suitable NGS should be large since the tilt isoplanatic patch is much larger than the isoplanatic patch for higher orders of the wavefront. Unfortunately this is wrong ([41]). Indeed, since ≈ 90% of the variance of the phase error lies in the tilt component, the accuracy required to measure the tilt is ≈ 10 times higher than for higher orders. Then the probability of finding a suitably bright NGS for tilt correction is not significantly higher than for higher order correction.

Since the wavefront from the reference source departs from that of the programme object as soon as it is located at a distance $\alpha \neq 0$, using a NGS to sense the tilt adds an error in the restored wavefront. The phase variance of the corrected wavefront is ([41]):

$$\sigma_r^2 es = \sum_{i=J_{max}+1}^{\infty} < a_i^2 > + \sum_{i=2}^{J_{max}} < b_i^2 > + \sum_{i=2}^{J_{max}} < (a_i(0) - a_i(\alpha))^2 > \quad (8)$$

where a_i is the coefficient of the Zernike polynomial of rank i, b_i is the noise variance propagation on the ith mode, and J_{max} is the highest corrected Zernike mode. In the case of the LGS, assuming that it provides perfect correction for orders higher than the tilt, the first term, which is the turbulence residual error, cancels, and the other terms, the noise and the contribution of anisoplanatism, reduce to tip and tilt Zernike orders :

$$\sigma_{res}^2 = \sum_{i=2}^{3} < b_i^2 > + \sum_{i=2}^{3} < (a_i(0) - a_i(\alpha))^2 > \quad (9)$$

Following [41], the noise term is:

$$\sum_{i=2}^{3} < b_i^2 > = (\frac{\lambda_{ws}}{\lambda})^2 r_{0_{ws}}^{-2} \frac{2\pi}{N_{ph}} \sum_{i=2}^{3} p_i \quad (10)$$

where p_i stands for the noise variance propagation coefficient on the ith mode and the ws index refers to the wavefront sensor observation wavelength. The anisoplanatism term is given by:

$$< (a_i(0) - a_i(\alpha))^2 > = 2(C_{n(i)n(i)}(0) - C_{n(i)n(i)}(\alpha)) \quad (11)$$

where the covariance $C_{ii}(\alpha)$ can be written:

$$C_{ii}(\alpha) = < a_i(0) a_i(\alpha) > = (\frac{D}{r_0})^{5/3} \frac{\int_0^L C_n^2(h) \sigma_n(\frac{2\alpha h}{D}) dh}{\int_0^L C_n^2(h) dh} \quad (12)$$

([11]), where L is the maximum altitude of integration, and where σ_n for the radial degree n is given by:

$$\sigma_n(x) = 3.90(n+1)\int_0^\infty k^{-11/3} J_{n+1}^2(k) J_0(xk)dk. \qquad (13)$$

The sky coverage given in Table 3 ([41]) is derived from these equations, assuming:

- a Hufnagel atmosphere model with $r_{0_{w.}} = 20\,cm$
- a wavefront spatial sampling of $r_{0_{w.}}$
- a wavefront time sampling of $1\,ms$
- $\sigma^2_{res} \leq 1\,rd^2$.

It emphasizes how marginal is the gain in sky coverage percentage which one gets when using a LGS, because of the tilt problem.

A few ideas have been proposed to overcome the problem of the tilt determination with te LGS, which are reviewed below.

6.2.1. *Dual Adaptive Optics concept*

The Dual Adaptive Optics (DAO) concept ([41]) aims at increasing the probability of finding a reference star for the tilt measurement by allowing us to use fainter stars. Two lasers and two adaptive optics systems are used simultaneously. The first LSG-AO system is used to improve the Strehl ratio for the programme object, possibly providing a diffraction limited image. The second system is used to shrink the image of a neighbouring star. This star may be too faint to make tilt measurements. But shrinking its image improves the determination of its centre of gravity, and therefore the accuracy of the tilt measurement. Since tilt measurement with DAO can use fainter stars, then the probability of finding a suitable star within the isoplanatic patch is larger. Table 3 gives this probability as a function of the galactic latitude ([41]).

"The larger the telescope, the larger the sky coverage, as the gain in resolution brought by the increase of optics diameter is fully expoited" ([41]). It is also emphasized that real conditions will lead to lower sky coverage, or to lower Strehl ratios.

6.2.2. *Polychromatic Artificial Star*

The Polychromatic Artificial Star (PAS) concept aims at providing a 100% sky coverage, by overcoming the principle of the reversability of light. The basic idea ([16]) is to create in the mesosphere a physical process which emits a wide spectrum from a quasi-monochromatic upgoing beam. Backscattered beams emitted at wavelengths different from that of the ongoing beam have different optical paths on their way to the telescope, because of the differential atmosphere refraction (see Fig. 17). The shorter the

TABLE 3. Probability of finding a reference source as a function of wavelength for a seeing of 0.8" at a 2 m and a 6 m telescopes, using a LGS and using a Dual Adaptive Optics device ([41]).

mode	*wavelength*(μm) D(m)	1	0.6
LGS	2	5%	1.5%
	6	5%	1.5%
DAO	2	40%	20%
	6	80%	50%

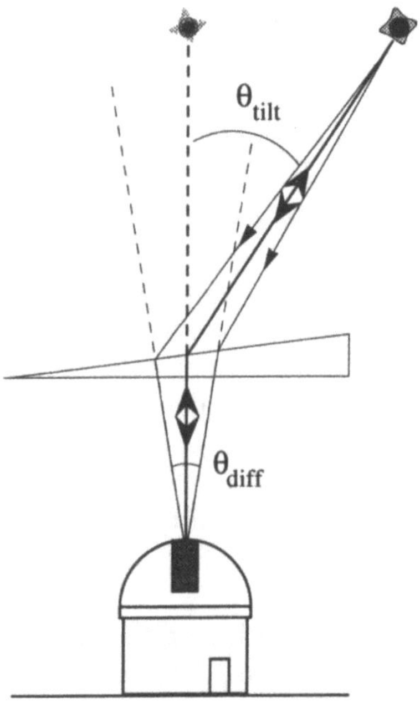

Figure 17. Principle of the Polychromatic Artificial Star: a quasi-monochromatic laser beam excites an emission at several wavelenths, which follow different optical paths to the telescope ([56]).

short wavelength, the larger the differential tilt between two wavelenghts since the refractive index of air changes rapidly with wavelength in the UltraViolet (UV), as shown in Fig. 18.

In the case of small perturbation approximation, the amplitude of any Zernike mode (except the piston mode) is proportional to its differential

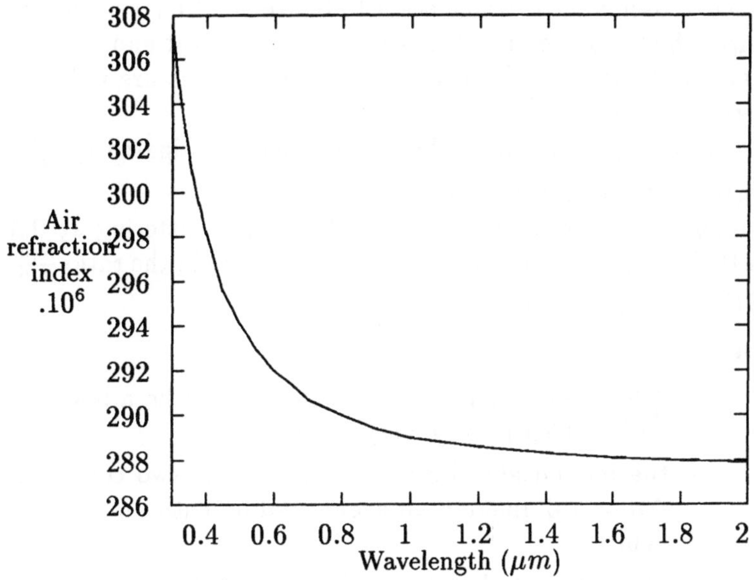

Figure 18. Variation of air refraction index with wavelength ([1]).

measurement at two wavelengths with a differential air refractive index Δn ([16]):

$$\theta = \Delta\theta \frac{n-1}{\Delta n} \tag{14}$$

where in our case θ is the tilt. It is worthwhile to note that this ratio is wavelength dependent. Therefore the tilt can be measured from the differential tilt between two wavelengths. The maximum error allowed in $\Delta\theta$ is :

$$\sigma(\Delta\theta) = \frac{1.22\lambda_c}{fD} \frac{\Delta n}{n-1} \tag{15}$$

where λ_c is the correction wavelength, and f is the accuracy of the tilt measurement in fractional Airy disk unit.

The standard deviation of the angles of arrival of rays onto the pupil is:

$$\alpha = 0.062(Dr_0^5)^{-1/6} cos^2 z \tag{16}$$

For an $8\,m$ telescope, with $r_0 = 15\,cm$, $\alpha = 0.21''$ and $0.43''$ respectively at the zenith and at $z = 45^o$. Hence $\sigma(\Delta\theta)$ is small, ranging from a few milliarcseconds for a $4\,m$ telescope in the infrared to a fraction of a milliarcsecond for an $8\,m$ telescope in the visible. Therefore the PAS has to be bright, and it is highly deasirable to have a component emitting in the UV in order to get the largest possible $\frac{\Delta n}{n-1}$ ratio.

Telescope oscillations, which seem to be frequent ([7]) are not measured by a PAS. They could be corrected by any star bright enough in the field, whatever its distance to the PAS; indeed oscillations are independent on the anisoplanatism.

A few processes to create a PAS have been investigated ([16]).

Rayleigh + Raman scattering .

Raman scattering associated with Rayleigh scattering at $0.35\,\mu m$ is emitted at $0.38\,\mu m$. The number of photons are in the ratio $\frac{N(0.38\,\mu m)}{N(0.35\,\mu m)} = 0.9 10^{-3}$. Pros and cons are :

Cons:

- $\frac{\Delta n}{n-1} = \frac{1}{125}$. Then in the visible $\Delta\theta$ will be a few microarseconds, which is very, very small!
- the magnitude difference between the two colours is huge : $\Delta m = 7.6$, making an accurate differential measurement difficult;
- LGS generated from Rayleigh scattering have low altitude, which makes the correction of the cone effect more difficult
- fluorescence in the UV of the lower atmosphere could add light pollution.

Pros:

- speckle correlation between the images at $0.35\,\mu m$ and at $0.38\,\mu m$ should give improved accuracy in the differential centre of gravity of the images.
- light pollution is only in the UV, making it easy to filter it out.
- excimer laser can deliver high powers in the UV at comparatively low cost and without saturation effects.

It is expected that the drawbacks are stronger than the advantages.

Excitation of the $4P_{3/2}$ level of Na .

The basic concept is to excite the $4P_{3/2}$ atomic level of sodium, from which a radiative decay occurs which emits a line spectrum from $0.33\,nm$ in the UV very close to the ozone transmission cutoff, to $2.207\,\mu m$ in the near infrared. It provides the maximum wavelength range for the variation of the refractive index of air. Figure 19 shows the energy level diagram of the sodium atom, with the possible ways to excite the $4P_{3/2}$ and the most probable decay pathways. Several excitation processes are possible ([16]):

$3S_{1/2} \rightarrow 4P_{3/2}$ **transition.** The $4P_{3/2}$ energy level can be excited directly from the ground state level. But the probability of the

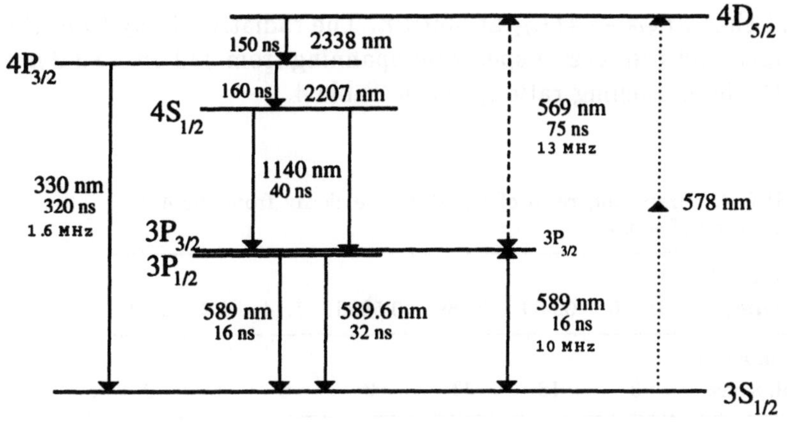

Figure 19. Energy level diagram of the sodium atom with the most probable relaxation pathways from the $4D_{5/2}$ level. Wavelengths in nm, lifetimes in ns and homogeneous linewidths in MHz of transitions are given ([16]).

resonant absorption process is very low :

$$A \leq \sigma \frac{\Delta\nu_H}{\Delta\nu_T} \rho_h \approx 1.610^{-3} \tag{17}$$

where $\sigma = 410^{-10}\, cm^2$ is the cross section, $\Delta\nu_H = 1.6\, MHz$ is the homogeneous linewidth and $\Delta\nu_T = 3\, GHz$ is the total linewidth due to the hyperfine structure and the Doppler width. In addition, the energy per laser pulse is proportional to $(\Delta\nu_D/\Delta\nu_H)/(\sigma\lambda)$ where $\Delta\nu_D = 1.8\, GHz$ is the Doppler linewidth: 50 times more energy is required for this laser than to excite the $3P_{3/2}$ energy level from the same population of sodium atoms. This factor is increased more if one takes into account the atmospheric extinction on the way to the mesosphere, which is $\approx 50\%$ at $330\, nm$ at good astronomical sites. Consequently it is not worthwhile considering this excitation process.

Non-resonant two-photon excitation of $4D_{5/2}$. The energy of the $4D_{5/2}$ level is $E = 34\,520\, cm^{-1}$. Then two photons at $578\, nm$, which carry half that energy, will be able to populate the $4D_{5/2}$ level through a virtual intermediate energy level. It involves a second-order non-linear process, of which the efficiency is much too low to produce enough backscattered photons from the mesosphere.

$3S_{1/2} \rightarrow 3P_{3/2} \rightarrow 4D_{5/2}$ transitions. The most efficient way to populate the $4P_{3/2}$ level is to populate the upper level $4D_{5/2}$ via the two transitions $3S_{1/2} \rightarrow 3P_{3/2}$, which is the wellknown D_2 line, and then $3P_{3/2} \rightarrow 4D_{5/2}$ at 569 nm. The radiative decay from the $4D_{5/2}$ level produces a spectrum spanning from 330 nm to 2.3 μm with the branching ratio given in Table 4.

TABLE 4. Branching ratio of the radiative decay from the $4D_{5/2}$ atomic level of sodium.

wavelength (μm)	0.330	0.569	0.589	0.5896	1.14	2.2	2.34
Number of photons	3	18	22	2	6	6	9

In the case of an incoherent sum of the pumping beams over the modes, the saturation intensity of the $3P_{3/2} \rightarrow 4D_{5/2}$ transitions is found to be given by:

$$I_{s_{569}} = h\nu_{569}/\sigma_{569}\tau_{4D_{5/2}} = 475\,W/m^2 \tag{18}$$

or

$$\Phi_{s_{569}} = 1/\sigma_{569}\tau_{4D_{5/2}} = 1.36\,10^{17} photons/s/cm^2, \tag{19}$$

with $\sigma_{569}\tau_{4D_{5/2}} = 1.4710^{-10} cm^{-2}$ taking into account hyperfine structures.

The rate of resonant two-photon absorption ([25]) is given by:

$$\gamma \approx (\tau_{4D_{5/2}} + \tau_{3P_{3/2}})\sigma_{3P_{3/2}}\Phi_{589}\sigma_{4D_{5/2}}\Phi_{569} \tag{20}$$

per sodium atom, where Φ_{589} and Φ_{569} are the intensities of the two pumping beams in units of $photons/s/cm^2$. This rate is at least one magnitude higher in the case of a coherent sum.

At the moment, the excitation of the $4D_{5/2}$ level is the most efficient process known for the polychromatic LGS.

The bias of the tilt due to the cone effect in the case of the polychromatic LGS has been estimated ([44]). I suspect that solving the problem of the cone effect, e.g.: with the 3-D mapping of the turbulence, should provide as a byproduct the correction for the tilt as well.

At the beginning of this section we have emphasized that the polychromatic LGS has to be bright, because the tilt is measured from a differential effect. Let us check whether this condition can be fulfilled. One can compute the population of the levels with simplifying assumptions (laser pulses long enough so that steady state applies, and small Rabi frequencies so that rate equations model is valid)(([16])). As a result, it turns out that the total laser output is optimized when an equal power is emitted through the two transitions, which leads to $\Phi_{589}/\Phi_{sat} \approx 0.45$ and $\Phi_{569}/\Phi_{sat} \approx 0.18$. Then the flux per transition per mode is $85\,W\,m^{-2}$. Taking into account the inhomogeneous widths, the required total laser *peak* power is:

$$P_{peak} = 170\,\frac{\Delta\,\nu_T}{\Delta\,\nu_H} W\,m^{-2} = 7.4\,kW \tag{21}$$

for $r_0 = 0.15\,m$ in the laser spot area $S = \frac{\pi}{4}(\frac{1.22\,\lambda\,H}{r_0})^2$. The *average* power is derived by multiplying by the duty cycle $T_i/t_c(\lambda_c)$, where t_c is the atmosphere coherence time at the wavelength of the wavefront correction λ_c. T_i is the minimum integration time to get the number N of photons for the center of gravity measurement at $330\,nm$. From Eq. 15 and with the assumption that the image of the laser spot is a gaussian, one gets:

$$N \geq 9.43\,10^{-17}\,f^2\,\frac{\lambda_r^{-2/5}}{\lambda_c^2}(\frac{D}{r_0})^2\,(\frac{n-1}{\Delta\,n})^2 \tag{22}$$

where f stands for the accuracy of the centre of gravity measurement in Airy disc units at the wavelength λ_c. Since the power emitted at $330\,nm$ is $a\,\rho_h\,S/\tau_{4P-3S}$, the limiting integration time is found to be:

$$T_i \geq 1.3\,10^{-15}\,\frac{\tau_{4P-3S}}{a\,\eta}\,\frac{f^2}{\rho_h}\,\frac{\lambda_r^{-2/5}\,\lambda_{laser}^{-2}}{\lambda_c^2}\,(\frac{n-1}{\Delta\,n})^2 \tag{23}$$

Hence there are enough photons from the polychromatic LGS as long as $T_i \leq t_c$. Figure 20 displays T_i and t_c versus wavelength for two values of the atmosphere coherence time $t_{0.5}$ for the tilt at $\lambda = 0.5\,\mu m$: $t_{0.5} = 2\,ms$ (pessimistic case) and $t_{0.5} = 60\,ms$ (optimistic case). It shows that under rather good seeing conditions, correction for the tilt is feasible down to visible wavelengths.

The radiative decay has been observed in the laboratory using a stream of sodium atoms and two lasers ([3]). The first experiment on the sky used the AVLIS laser of the LLNL ([20]). Preliminary processing of the data confirms that the $330\,nm$ laser spot provides enough photons for tilt correction, at least in the near infrared.

6.2.3. *Observation of the mesospheric beam with an auxiliary telescope*

Until now we have considered light backscattered from the LGS downward to the large telescope. But sodium atoms excited at the $3P_{3/2}$ level in the mesosphere with a monochromatic beam emit scatter light in every direction. In the next two sections, we will discuss the observation of the backscattered beam toward or from an auxiliary telescope $d\,m$ away. It

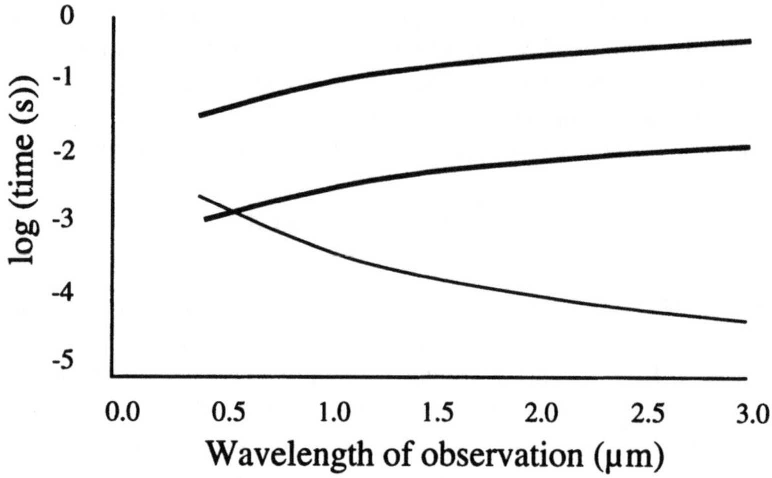

Figure 20. The wavelength variation of the minimum integration time of the 330 nm laser spot required to get enough photons for tilt measurement based on the 330 nm and 569 nm laser spots. The bold lines show the wavelength variation of the tilt coherence time for coherence times at 0.5 μm of 60 ms (upper) and 2 ms (lower) ([16]).

looks like a strip with an angular length $\phi \approx d\tau / h^2$, where τ stands for the thickness of the mesospheric sodium layer.

The tilt can be measured from the deviation of the mesospheric beam with respect to a foreground NGS at the focus of an auxiliary telescope located in such a way that the NGS lies within the isoplanatic patch of any portion of the laser strip (see Fig. 21)([42]). The tilt of the wavefront incoming onto the auxiliary telescope equally affects the laser strip and the NGS, as well as any image displacement due to telescope vibrations or tracking errors. Of course, the auxiliary telescope can measure only the projection of the tilt in the direction perpendicular to the laser strip; indeed the elongated beam does not allow us to derive accurate distance along that axis. Therefore two auxiliary telescopes in two perpendicular directions as seen from the main telescopes are necessary to measure both components of the tilt.

Only the fraction ϕ/ϵ of the laser strip (where ϵ stands for the isoplanatic patch size) can be used to measure one component of the tilt. From preliminary studies, the telescope diameter of the auxiliary telescopes should range around $\approx 25\,cm$ ([43]).

This concept should allow full sky coverage, provided that the auxiliary telescopes are movable to set the alignment with both the laser strip and the NGS, and to keep the NGS close to the laser strip as it tracks the programme source observed from the main telescope.

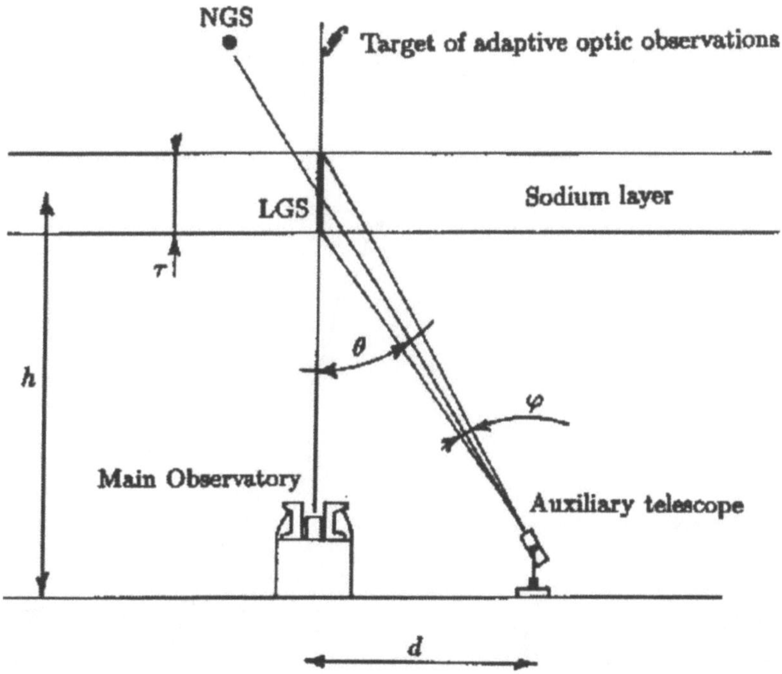

Figure 21. Principle of tilt measurement from a movable auxiliary telescope looking at a foreground NGS tracked within the isoplanatic patch of the mesospheric laser strip ([43]).

6.2.4. *Observation of mesospheric beams emitted by auxiliary telescopes*

Another concept is to use two auxiliary telescopes to emit two laser beams aiming at the programme source observed by the main telescope ([42]). Figure 22 shows what the image at the focus of the main telescope looks like. The two laser strips from the auxiliary telescopes cross close to the programme source; their images are distorted due to seeing effect over a field significantly larger than the isoplanatic patch. Both components of the tilt at the programme source location can be derived from the departure of the laser strips from the average straight lines which fit the distorted laser strips. N independent measurements of these departures can be obtained over the length ϕ of the laser strip: $N \approx \phi \overline{h} / (0.3\,D)$, where $\overline{h}$ stands for the average altitude of the turbulent layers weighted with their relative turbulence strength ([11]), and $\psi = 0.3\,D / \overline{h}$ is the isokinetic angle ([50]). The isokinetic angle is the isoplanatic angle for the tilt. From the standard

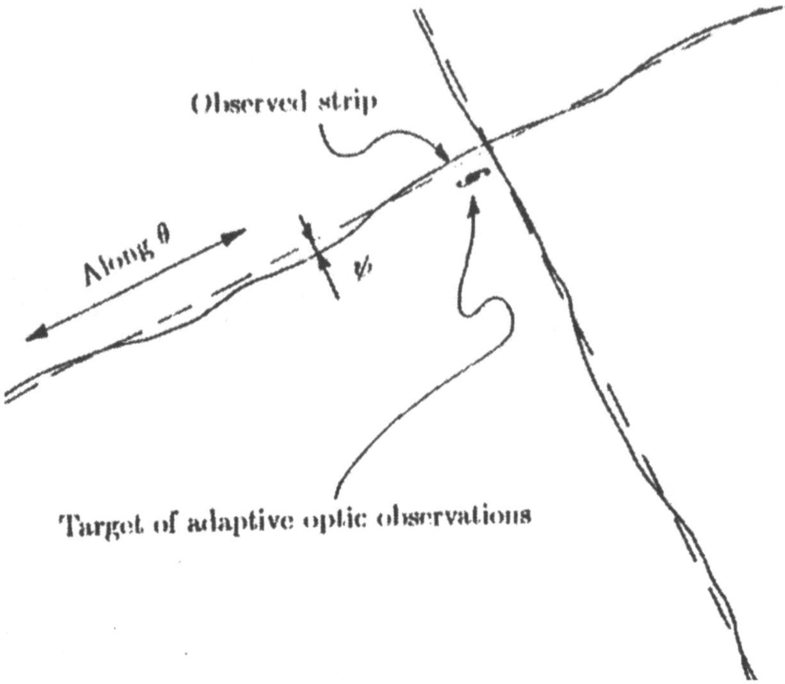

Figure 22. Principle of tilt measurement from two distant auxiliary telescopes shooting at the sky with a laser beam. The two laser strips cross in the field of the programme source at the focus of the main telescope ([42]).

deviation of the tilt measurement:

$$\sigma_\psi \approx \frac{0.4}{\sqrt{N}} \left(\frac{\lambda}{r_0}\right)^{5/6} \left(\frac{\lambda}{D}\right)^{1/6} \tag{24}$$

[42] derive the required length of the laser strip:

$$\phi \approx \frac{0.2\, D^{8/3}}{\overline{h}\, r_0^{5/3}} \tag{25}$$

and the distance between the auxiliary telescopes and the main telescope:

$$d \approx 0.2 \frac{D^{8/3}\, h^2}{\tau\, \overline{h}\, r_0^{5/3}} \tag{26}$$

Values of d are large, ranging from $\approx\ 10\,km$ for $4\,m$ telescope to $d\ \approx 60\,km$ for $8\,m$ telescope, assuming $r_0\ =\ 20\,cm$ and an error in the tilt determination of $0.5\,\lambda\,/\,D$. These high values raise a lot of difficulties, either about the process itself (e.g.: the homogeneity of the sodium layer) or about the setup of the experiment.

7. Operational issues

A lot of problems have to be addressed to operate a LGS on an astronomical site. Let us mention:

light pollution

The laser light is backscattered not only from the atmospheric layer where the LGS is produced, but also from lower altitudes, due to Mie and Rayleigh scattering. In order to protect detectors against this stray light, it has to be filtered either by time filtering (e.g.: for Rayleigh scattered LGS, which is required in any case to select the altitude of the LGS), or spatial fitering (when the laser is emitted through an auxiliary telescope), or spectral filtering (e.g.: in case of the polychromatic LGS). Light is scattered not only back to the telescope which uses the LGS, but also to all telescopes on the same site. This deserves special attention mostly because of scattering at low altitude above the site, which produces a lot of light. Quantitative estimates of the effect remain to be undertaken. Nevertheless, it is worthwhile to note that the presence on the site of Observatoire de Haute-Provence of a lidar station for more than 30 years did never raise severe difficulties. Coordination of telescopes settings and spectral filtering should help to solve the problem.

safety

The operation of the laser has to fulfill general safety regulation for the use of power lasers. In addition, special protection has to be implemented to avoid to harm airplane pilots or damage satellites in case of the use of powerful devices. This has been done very carefully at the LLNL, where powers as high as $1.2\,kW$ have been used; for airplane detection, they use a three level safety procedure: a wide field radar, a narrow field radar and an observer. In addition, shooting is forbidden during time windows for satellite passing over. It is planned that safety for the Calar-Alto LGS device will rely on an infrared camera.

laser type

Studies have to be continued to optimize the most efficient laser type, which depends on the LGS process. Are continuous wave lasers better matching the LGS requirements than pulsed lasers? Among the pulsed lasers, it is not clear whether copper vapor or Nd:YAG lasers are the best suited. It seems that sum frequency lasers are more sophisticated; let us recall that they used the chance coincidence that the sum of the frequencies of the two lines at 1.06 and $1.32\,\mu m$ exactly equals that of the $NaI\,D_2$ line; these frequencies can be added in a non-linear crystal ([31]).

Wall plug efficiency is a major concern for astronomical sites in particular because heat has to be evacuated without spoiling image quality! It is worthwhile to recall here the wall plug efficiency of some lasers (Table 5); in the case of dye lasers, efficiency of the pump is not included.

TABLE 5. Plug efficiency of different kinds of lasers.

laser	efficiency (%)
flash lamp pumped NdYAG	0.15
diode pumped NdYAG	5
Copper vapor	0.5
CuHBr	1.5
y Argon ion	0.05
dye laser, low flux	15-20
dye laser, high flux	40-50

mode

Two case have been considered to shoot aı the sky: either from the main telescope itself (monostatic mode), or from an auxiliary telescope (bistatic mode). Theoretically, shooting from the main telescope is better, since the LGS elongation is minimum; but operation is much more delicate, because of problems of light pollution. The balance between the pros and the cons is not completely studied.

8. Conclusion: state of the art

Table 6 summarizes most of the experiments underway or already running across the world at the time of this conference. It shows the big effort undertaken to master the LGS technics.

Several major problems have to be investigated before the LGS becomes an operational device allowing astronomers to observe faint objects with adaptive optics systems whatever the direction in the sky. Namely, it is mandatory to to solve the cone effect to work at very large telescopes, and to solve the tilt determination from the laser (without any natural guide star) to be able to integrate diffraction limited long exposure images. The optimisation of the lasers (pump laser and dye laser) is also very important; indeed it leads to the plug efficiency of the laser, which is an essential parameter for astronomical sites. This optimidation raises problems in the fields of laser physics *and* of atomic physics *and* of detectors and data processing. The long term goal of these researches should contribute to

TABLE 6. LGS experiments in the world. In column "Process", symbols refer to the physical backscattering process used to create the artificial star; R: Rayleigh, F: fluorescence of sodium in the mesosphere, P: polychromatic. λ_{obs} is the wavelength of the observations for which the experiment is optimized. SOR: Starfire Optical Range, MMT: Multi-Mirror Telescope, ONERA: Office National d'Etudes et de Recherches Aérospatiales, "France": French consortium of public research labs and industries, MPE-MPA: Max Planck Extraterrische Physik-Astronomie, UIUC: University of Illinois at Urbana Champaign, LLNL/Lick: Lawrence Livermore National Lab/Lick Observatory, IC: Imperial College.

Institute	Telescope diameter (m)	Date of first shoot	Laser type	power (W)	rep. rate (kHz)	process	λ_{obs} (μm)
SOR	1.50	5/89	cvl	75-200	5	R	0.88
			freq. sum	20	0.84	F	0.88
	3.5	1997	Nd:YAG	100-200	≈ 10	F	
MMT	6*1.80	1993	cw Ar^+	2.5		F	
ONERA	1.50	1988	Nd:YAG	1	0.1	R	
France	1.50	1999	CuHBr	2*25	tbd	P	
MPE-MPA	3.50	1996	cw Ar^+	2.5		F	≥ 1.2
		≥ 1997	tbd	30		F	≥ 0.7
UIUC	2.50		excimer	50	0.3	R	
LLNL/ Lick	3.0	1995	Nd:YAG	22	10	F	2.2
IC	4.2	≥ 1999	cvl?	10-20		F	≥ 0.82
Univ. Chicago	3.5						

clarify the specific domains of interests of space with respect to ground-based observatories in the field of high angular resolution.

In Europe, teams working about the LGS are coordinating their works within the framework of the network "LGS for 8 m class telescopes" funded by the Training and Mobility for Researchers Programme of the European Union. ESO will organize in 1997 a workshop dedicated to the lasers for the LGS, and this European network has proposed a NATO Summer School which will address the basic problems of the LGS, which have been skimmed over in this lesson, namely aeronomy, physical and optical properties of the atmosphere and of the mesosphere, atomic physics, laser physics, adaptive optics, instrumentation, and on another hand astrophysical throughtput expected from the equipment of forthcoming giant telescopes with the LGS.

Acknowledgements I wish to thank all those who have helped me in the preparation of this lesson, and in particular F. Biraben, H. Friedman, G. Massacrier, M. Séchaud and M. Tallon. I thank also N. Devaney for a critical reading of the manuscript.

References

1. Allen C.W. 1973, *Astrophysical quantities* ed. The Athlone Press Univ. of London, p.125
2. Arsenault R, Salmon D.A., Kerr J.M., Rigaut F.J., Crampton D., Grundmann W.A. 1994, *SPIE* **2201**, 833
3. Aussel, H., Petit, A., Sjöström, S., Weulersse, J.-M. 1993, in ICO 16 Satellite Conference
4. Bahcal U.N., Soneira R.M. 1980, *Ap.J.Suppl.Ser.* **44**, 73
5. Beckers, J.M. 1988, in "Very Large Telescopes and their Instrumentation" ESO Conf. ed. M.-H. Ulrich, p. 693
6. Belen'kii M.S. 1995, *Opt. Lett.* **20**, 1359
7. Belkine, I. 1996, PhD thesis, Université Joseph Fourier, Grenoble
8. Blamont J.E., Donahue T.M. 1961, *J. Geophys. Res.* **66**, 1407
9. Bowman M.R., Gibson A.J., Sandford M.C.W. 1969 *Nature* **221**, 456
10. Bufton J.L., Minott P.O., Fitzmaurice M.W., Titterton P.J. 1972, *J. Opt. Soc. Am.* **62**, 1068
11. Chassat F. 1989, *J. Optics (Paris)* **20**, 13
12. Clemesha B.R. 1995, *J. Atmos. Terr. Phys.* **57**, 725
13. Clemesha B.R., Kirchhoff W.J.H., Simonich D.M., Takahashi H. 1978, *Geophys. Res. Lett.* **5**, 873
14. Doel A.P., Dunlop C.N., Major J.V., Myers R.M., Sharples R.M. 1991,*SPIE* **1542**, 319
15. Foy R., Labeyrie A. 1985, *A &A* **152**, L29
16. Foy R., Migus A., Biraben F., Grynberg G., McCullough P.R., Tallon M. 1995, *A & A Suppl.Ser.* **111**, 569
17. Foy R., Tallon M., Séchaud M., Hubin N., Brixon L. 1989, In: Active Optical Systems, Conf. SPIE **1114**, 174
18. Fried D.L., Belscher J.F. 1994, *J.O.S.A. A* **11**,277
19. Friedman H., Erbert G., Kuklo T., Salmon T., Smauley D., Thompson G., Malik J., Wong N., Kanz K., Neeb K. 1996, In: *Adaptive Optics* ESO Conf **54**, ed. M. Cullum, p.207
20. Friedman H., Foy R., Tallon M., Migus A. 1996, OSA meeting Maui, in press
21. Fugate R.Q., Ellerbroek B.L., Higgins C.H., Jelonek M.P., Lange W.J., Slavin A.C., Wild W.J., Winker D.M., Wynia J.M., Spinhirn J.M., Boeke B.R., Ruane R.E., Moroney J.F., Oliker M.D., Swindle D.W., Cleis R.A. 1994, *J.O.S.A. A***11**, 310
22. Fugate R.Q., Fried D.L., Ameer G.A., Boeke B.R., Browne S.L., Roberts P.H., Ruane R.E., Tyler G.A., Wopat L.M. 1991, *Nature* **353**, 141
23. Gardner C.S., Voelz D.G., Sechrist C.F., Segal A.C. 1986, *J. Geophys.Res.* **91**,

13659
24. Greenaway A. H. 1992, in: *Laser guide star adaptive optics workshop*, Albuquerque (NM), ed. Robert Q. Fugate, p. 663
25. Grynberg, G., Cagnac, B., Biraben, F. 1980, in "Coherent non-linear optics" (Spinger-Verlag) p. 111
26. Happer W., MacDonald G.J., Max C.E., Dyson F.J. 1994, *J.O.S.A. A* **11**, 263
27. Hubin N., Beuzit J.-L, Gigan P., Léna P., Madec P.-Y., Rousset G., Boyer C., Gaffard J.-P., Richard J.-C., Vittot M., Rigaut F., Gendron E., Merkle F. 1993, *SPIE* **1780**, 850
28. Jankevics A.J., Wirth A. 1991, *SPIE* **1543**, 438
29. Jegou J.-P., Granier C., Chanin M.-L., Megie G. 1985, *Ann. Geophys.* **3**, 163
30. Jegou J.-P., Granier C., Chanin M.-L., Megie G. 1985, *Ann. Geophys.* **3**, 299
31. Jeys T.H., Brailove A.A., Mooradian A. 1989, *Appl. Opt.* **28**, 2588
32. Laurent S. 1996, Thèse de Doctorat Université Paris VII
33. Le Louarn, M. 1996, private communication
34. Megie G., Blamont J.E. 1977, *Planet. Space Sci.* **25**, 1093
35. Morris J.B. 1994, *J.O.S.A. A* **11**, 832
36. Murphy D.V., Primmerman C.A., Zollars B.G., Barclay H.T. 1991, *Opt. Letters* **15**, 1797
37. Olivier S.S., An A., Avicola K., Bissinger H.D., Brase J.M., Friedman H.W., Gavel D.T., Max C.E., J.T. Salmon, Waltjen K.E. 1995, *ESO Conf.* **54**, 75
38. Papen G.C. 1995, in *1995 Technical Digest Series "Adaptive Optics"* **23**, 247
39. Pilkington J.D.H. 1987, *Nature* **330**, 116
40. Primmerman C.A., Murphy D.V., Page D.A., Zollars B.G., Barclay H.T. 1991, *Nature* **353**, 141
41. Rigaut F., Gendron E. 1992, A. & A. **261**, 677
42. Ragazzoni, R. *Astron. Astrophys.* **305**, L13
43. Ragazzoni, R., Esposito, S., Marchetti, E. 1995, *Mon. Not. R. Astron. Soc.* **276**, L76
44. Ragazzoni, R., Marchetti, E., G. Brusa G. 1996, *Astron. Astrophys.* in press
45. Richter E.S., Sechrist C.F. 1979, *J. atmos. Terr. Phys.* **41**, 579
46. Richter E.S., Sechrist C.F. 1979, *Geophys. Res. Lett.* **6**, 183
47. Robin A., Crézé M. 1990, *Astro. Astrophys.* **157**, 71
48. Roddier F. 1981, *Progress in Optics* **XIX**, 281
49. Roddier F., Anuskiiewicz J., Graves J.E., Northcott M.J., Roddier C. 1994, *SPIE* **2201**, 2
50. Roddier F., Northcott M.J., Graves J.E., McKenna D.L., Roddier C. 1993, *J.O.S.A. A* **10**, 957
51. Sasiela R.J. 1994, *J.O.S.A. A* **11**, 379
52. Séchaud M., Hubin N., Brixon L., Jalin R., Foy, R., Tallon, M. 1988, in ESO Conf. *Very Large Telescopes and their Instrumentation*, ed M.-H. Ulrich (ESO, Garching) p. 705
53. Slipher V.M. 1929, *Publ. Astron. Soc. Pacific* **41**, 262
54. Swider W. 1985, *Geophys. Res. Lett.* **12**, 589
55. Tallon, M. 1989, PhD thesis, Université de Nice
56. Tallon, M. 1996, private communication
57. Tallon M., Foy R. 1990, Astron.Astrophys. **235**, 549
58. Tallon M., Foy R., Vernin, J. 1992, in: *Laser guide star adaptive optics workshop*, Albuquerque (NM), ed. Robert Q. Fugate, p. 555
59. Théodore B., Petitjean P., Hubin N. 1994, The Messenger **77**, 20
60. Thompson L.A., Gardner C.A. 1987, *Nature* **328**, 229
61. Thompson L.A., Xiong Y.-H. 1995, *SPIE***2534** 38
62. Vernin J., Azouit M. 1983, J. Optics (Paris) **14**, 131
63. Vernin J., Muoz-Tun C. 1993, Astron.Astrophys. **257**, 811
64. Young A.T. 1970, J. Opt. Soc. Am. **60**, 248
65. Zollars B.G. 1992, Lincoln Lab. J. **5**, 67

ASTROPHYSICS AND HIGH ANGULAR RESOLUTION

SOLAR-SYSTEM STUDIES AT HIGH ANGULAR RESOLUTION

TH.ENCRENAZ
DESPA, Observatoire de Paris
F-92190 Meudon

1. Introduction

Solar-system objects are characterized by an extreme variety of physical and orbital conditions. They include planets, satellites and ring systems, asteroids, comets and interplanetary dust. The dimensions of Sun-orbiting objects range from micron-size (in the case of interplanetary and cometary dust) to several ten thousand km (in the case of the giant planets). Their distances to the Sun range from less than an AU (0.3 AU in the case of Mercury, less for Sun-grazing comets) to several tens of AU (40 AU in the case of Pluto).

1.1. A LARGE VARIETY OF OBJECTS

The Solar-system planets range into two categories: the terrestrial planets, at less than 2 AU from the Sun, characterized by a relatively low diameter, a large density, a small number of satellites, and (with the exception of Mercury) an atmosphere dominated by relatively heavy gases (CO_2, N_2, with in addition O_2 in the case of the Earth). In contrast the giant planets, beyond 5 AU, have large dimensions, a low density, a large number of satellites, a ring system, and an atmosphere dominated by hydrogen and helium. They probably have a solid central core but they have no observable surface. Pluto, the most remote planet, does not belong to these categories and shows more analogies with the satellites of the giant planets.

Most of the satellites orbit around the giant planets. Most of them have no atmospheres. There are three exceptions: Titan, Saturn's largest satellite, with a N_2 atmosphere and a surface pressure of 1.5 bar; Triton, Neptune's satellite with a tenuous N_2 atmosphere (surface pressure ofa few microbars), and Io, Jupiter closest satellite, with a very tenuous SO_2 atmosphere (surface pressure of a few nanobars).

A-M. Lagrange et al. (eds.), High Angular Resolution in Astrophysics, 233–258.
© 1997 *Kluwer Academic Publishers.*

Asteroids are mostly found in the asteroidal belt, at an heliocentric distance of about 3 AU. Their sizes range between several tens to several hundred km (about a thousand for the biggest asteroid, Ceres). They probably are the remnants of a planet which could not complete its accretion stage due to the strong gravitational field of the neighbouring planet Jupiter. They are too small to retain a permanent atmosphere and show a bare surface. The brightest asteroids have visual magnitudes ranging from 8 to 12.

Comets are very small bodies made of ices and rocks, usually less than 10 km in size. Their orbits are very elliptical, so they spend most of their life at very large heliocentric distances where they remain unaltered. When their trajectories bring them back in the vicinity of the Sun, the ices of the surface sublimate with an outgassing of parent molecules and an ejection of dust, forming the coma and the dust tail. The parent molecules are then photodissociated into radicals, atoms and ions which form the plasma tail. These phenomena become usually observable at heliocentric distances smaller than about 2AU, where water sublimation starts.

1.2. INTEREST OF SOLAR-SYSTEM STUDIES

Solar-system studies provide information on a wide variety of domains. From the dynamics studies (determination of orbits, rotation periods, rotation axes, etc...) information is derived about the dynamical history of the various Solar-system populations; the techniques used are astrometry, imaging and photometry. From the chemical studies (determination of the chemical composition of atmospheres and surfaces), one can derive constraints upon the chemical composition of the primordial solar nebula, and on the early processes of formation and evolution in the Solar system (accretion, differenciation...); the methods used are basically spectroscopy and imaging spectroscopy. From the morphology of planetary disks observed through imaging and spectro-imaging, one can determine either the dynamics and climatology of the atmospheres, or the age and formation processes of bare surfaces.

As an example, two major scientific objectives of Solar-system studies are listed below:

1) We need to better understand the early stages of the Solar-system history. This can be done through the chemical analysis of the giant planets and comets, the morphological study of the asteroids and the outer satellites, in order to better understand the link between comets and asteroids, between comets and the recently detected trans-neptunian objects, and the connection betwen cometary and interstellar matter. Considering the Solar system as a specific example of a stellar system, this information can be

used as an input for stellar formation models, and can help addressing the question: is the Solar system unique in the Galaxy?

2) A fascinating problem is the comparative study of the terrestrial planets, Venus, the Earth and Mars. Starting from reasonably close conditions of heliocentric distance, size, composition and temperature, these three planets have evolved toward the extreme conditions of today: for Venus and Mars respectively, surface pressures of 93 bar and 0.007 bar, and surface temperatures of 730 K and 230 K, with intermediate values in the case of the Earth. All atmospheres exhibit CO2 and N2, with , in the case of the Earth, the oxygen coming from the apparition of life. A better understanding of the complete processes which led to such extreme evolutions are needed in order to address the question: is the Earth unique?

Dealing with these fundamental problems, more specific questions can be addressed:

-The history of the Mars and Venus atmospheres can be studied through the study of the history of water on these two planets, the determination of their chemical composition, and (in the case of Mars) the surface mineralogy and morphology. This study requires spectroscopy and high-resolution mapping of these planetary disks.

-The formation processes of the giant planets and their evolution can be studied through the study of their chemical abundances, their spatio-temporal variations, and the determination of elemental and isotopic ratios. This requires spectroscopy and mapping of the planetary disks.

-The connection between the outer objects (comets, asteroids, trans-neptunian objects) and the interstellar medium can be studied through the determination of the orbits of the objects, their dynamical parameters, their chemical composition, the measurement of isotopic ratios, the study of their surfaces and the possible irradiation processes. This requires astrometry, photometry, spectroscopy and high-resolution imaging.

Solar-system objects exhibit a wide variety of angular sizes, ranging (apart from the Moon) from several tens of arcsecs to a few milliarcsecs (see Table 1). In Section 2, high-resolution imaging studies are reviewed, from the largest to the smallest sizes, in the visible, IR and radio range. In Section 3, these studies are extended to imaging spectroscopy, mostly in the infrared and millimeter range. In Section 4, expected future studies are discussed, in particular with 8-m class telescopes and with the VLTI.

TABLE 1
ANGULAR SIZES OF SOLAR-SYSTEM OBJECTS

OBJECT	ANGULAR DIAM.(")	DIAMETER (km)	GEOCENTRIC DISTANCE (AU)	RES.(km) in 0.5"
MOON	1900	3500	0.0025	1
MERCURY (quadrature)	6.5	4900	1	380
VENUS (quadrature)	23	12100	0.7	263
MARS (opposition)	18	6800	0.5	190
JUPITER	40	143000	5	1875
IO	1.0	3600	5	1875
SATURN	16	120000	10	3750
TITAN	0.8	5100	10	3750
URANUS	3.5	50800	19	7125
NEPTUNE	2.16	48600	30	11250
TRITON	0.12	2700	30	11250
PLUTO	0.08	2300	40	15000
CERES	0.7	1000	1.8	675
TYPICAL ASTEROID	0.06	100	2	750
TYPICAL COMET D = 1 AU	0.013	10	1	375
TYPICAL COMET D = 5 AU	0.003	10	5	1875
TYPICAL TRANS-NEPT.OBJ	0.004	100	35	13125

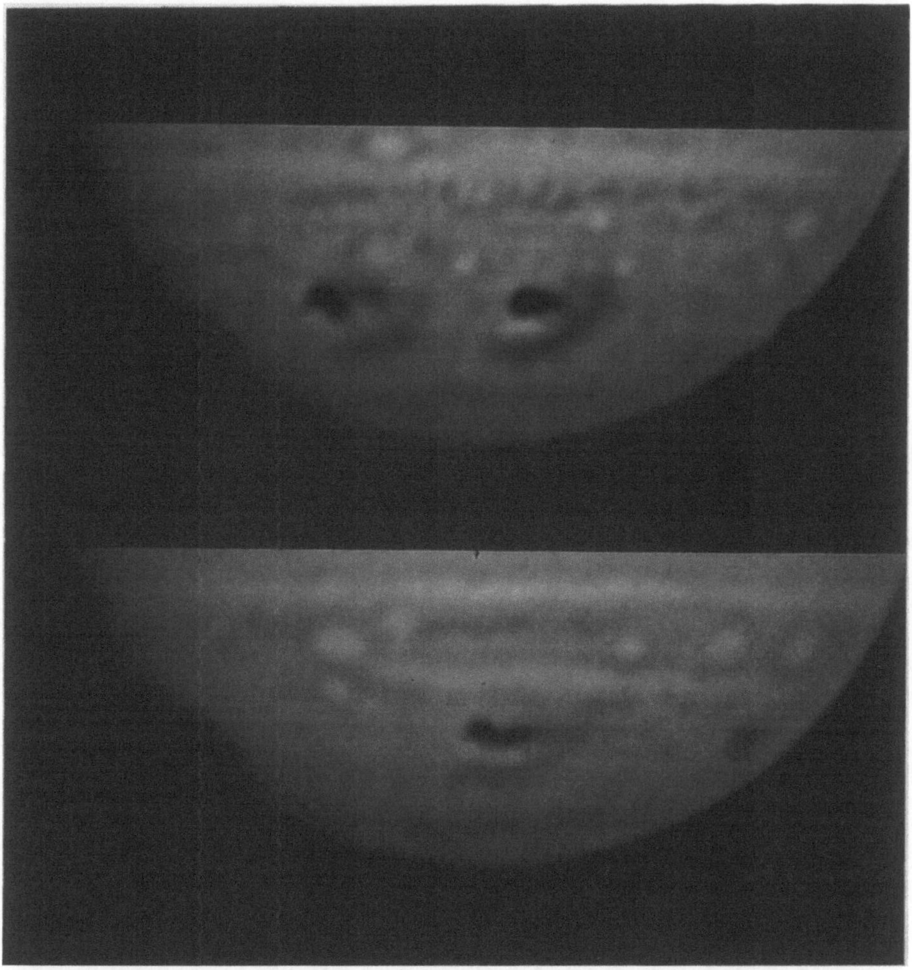

Figure 1. Impact sites on the Jovian disk, at a latitude of -44¡ , at the time of the collision of comet Shoemaker-Levy 9 with Jupiter, in July 1994. The full disk equatorial diameter is 38 arcsec. The image has been taken at the 1-meter telescope of Pic du Midi Observatory (Lecacheux, 1996).

2. High-resolution imaging of Solar-system studies

2.1. MORPHOLOGY OF PLANETARY DISKS

2.1.1. *Large planetary disks (Venus, Mars, Jupiter, Saturn)*

Visible and infrared imaging. With a diameter larger than 15 arcsec at opposition, these planets are large enough to be mapped in the visible and in the infrared from the Earth under good seeing conditions (0.5 arcsec) with several tens of pixels along a diameter (Fig. 1). This allows a study of spatio-

Figure 2. The disk of Jupiter as observed with the wide-field camera (WFPC2) of the Hubble Space Telescope, before the collision with comet Shoemaker-Levy 9 (NASA-STScI).

temporal variations of morphological structures (clouds and evolution of anticyclonic features on the giant planets, dust storms on Mars...).

Using the HST with the WFPC2 camera, the spatial resolution is improved by a factor 10 in the visible range. As a result, the pixel size is about 200 km on Jupiter. It can be noticed that this spatial resolution is better than the one achieved by the Voyager probes for the global images of the planet (Smith et al, 1979). These high-quality images have been used, in particular, for studying the cloud morphology of Jupiter and Saturn, and for monitoring the Jovian dust haze after the collision of comet Shoemaker-Levy 9 in July 1994 (Fig. 2).

In the infrared range, the use of adaptive optics has also allowed a gain by a factor 3 or 4 in angular resolution. Using a 3.6m telescope at 2 microns, a resolution of 0.15 arcsec is actually achieved with the ADONIS adaptive optics system at ESO (La Silla, Chile). Because a satellite is needed as a point-like source for reference, such studies are currently limited to the regions of the planets which are close to the limb. The method, of course, cannot be used for Venus which has no satellite. In the case of Mars, previous attempts to use Mars' satellites Phobos and Deimos have failed in the past, but this technique should become possible in the future with the use of more sensitive wavefront sensors; it will then allow a detailed study of the Martian polar caps. Using adaptive optics, galilean satellites are especially useful to measure the evolution of the PSF versus the angular distance between the reference and the object. This technique has been very successfully used on Jupiter in the 2-micron region for the study of the polar haze in the high latitude regions, allowing a determination of its scale height (Drossart, 1996). In the future, observations at higher wavelength (3-4 microns) with adaptive optics will be essential for an accurate mapping of the auroral regions in the H3+ emissions.

Radio observations. Radio mapping of large planets has been achieved for several years using radio interferometers. The largest one is the VLA (Very Large Array) which provides an angular resolution of 0.3 arcsec at a wavelength of 7 mm (43 GHz frequency). Jupiter and Saturn have been mapped at various wavelengths in the centimeter range (de Pater et al, 1982; de Pater and Dickel, 1982). The Jupiter images distinctly show the two components of the jovian radio emission: the thermal component of the disk which dominates at lower wavelengths, and the synchrotron radiation associated with the radiation belts, which increases in intensity with the wavelength (de Pater, 1982; Fig. 3).

Another remarkable example of planetary radio imaging has been the study of the manetospheric activity of Jupiter at the time of the SL9 collision, through a monitoring of its radio emission at 13 and 22 cm wavelength at the Australian telescope. This instrument consists in six 22-m antennae along a 6-km EW line. A significant increase of the radio emission was measured after the first impacts, implying the presence of newly accelerated electrons in the radiation belts (Leblanc and Dulk, 1995).

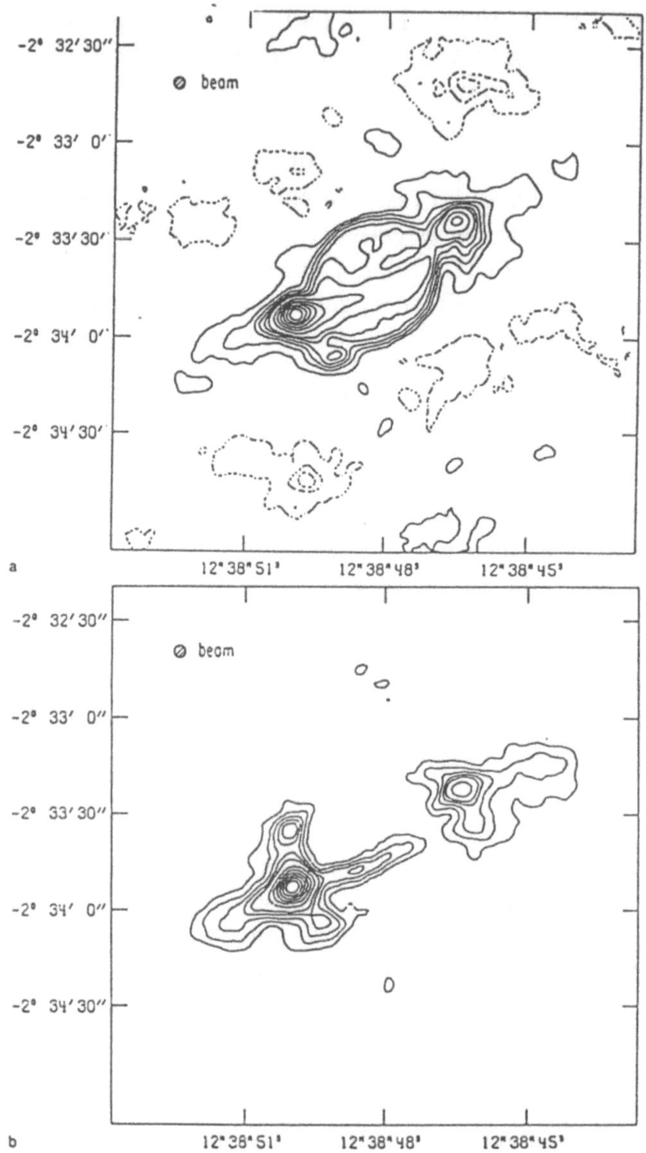

Figure 3. VLA images of Jupiter at l = 22 cm. Upper: intensity (thermal emission and synchrotron radiation); lower: polarized intensity. The peak intensity is 182 mJy/beam in the upper image, 107 mJy/beam in the lower image. The figure is taken from de Pater et al (1982).

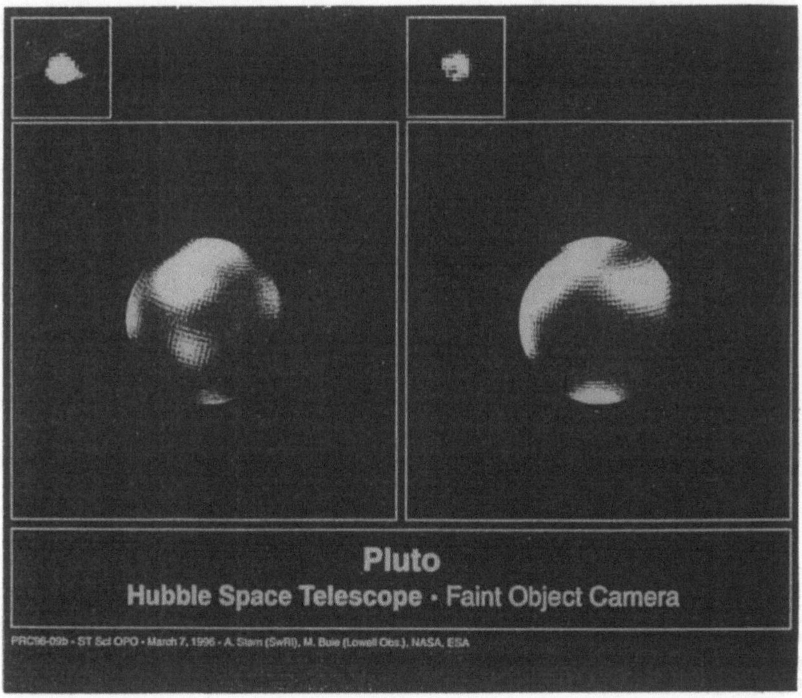

Figure 4. **First image of Pluto recorded with the Faint Object Camera (FOC) of the Hubble Space Telescope. The disk of Pluto is 0.08 arcsec in diameter (A. Stern and M. Buie, NASA, ESA).**

2.1.2. *Small disks*

We consider here planets and satellites which angular diameters between about one tenth and a few arcsec or less: Uranus, Neptune, the Galilean satellites, Titan, Triton, Pluto and the biggest asteroids. These objects are poorly resolved or unresolved from the ground under normal seeing conditions.

With the HST, these objects can be mapped using WFPC2; the number of pixels along a diameter is in the range of 10 to 70. This is sufficient for identifying morphological structures, associated to atmospheres (Uranus, Neptune), or to surfaces (Io, Titan, Vesta). Spectacular results have been achieved with the detection of large climatologic changes in Neptune, the

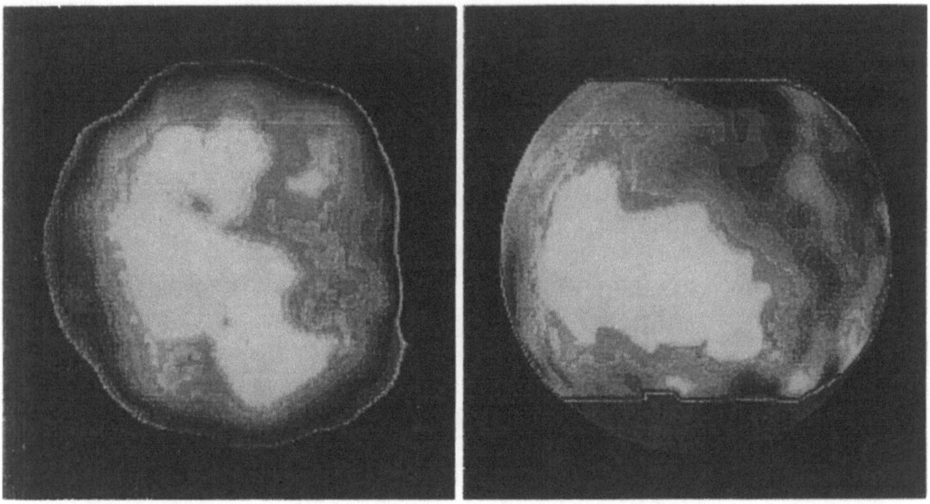

Figure 5. Titan's surface at eastern elongation as observed with (left) the COME-ON+ adaptive optics system (ESO, La Silla) at 2.0 mm and (right) with the Hubble Space Telescope at about 1 mm. The spatial resolution is 0.13 arcsec for the ESO data, 0.10 arcsec for the HST data. The ESO data are from Combes et al (1996); the HST data are from Smith et al (1995).

identification of bright spots on Io and Pluto (Fig. 4).

In the infrared range, significant progress has been achieved with the ADONIS adaptive optics system at ESO. In the case of Titan, in particular, following the pioneering work of Saint-P et al (1993), new images have been taken in the 2 micron region, in two selected filters K1 and K2. These filters were selected to be sensitive to radiation coming respectively from the upper atmosphere (K1, within a CH4 band) and from the lower atmosphere and the surface (K2, outside the CH4 band). Using a proper combination of these two sets of images, it was possible to retrieve a mapping of the surface which shows evidence for a bright spot (Combes et al, 1996; Fig. 5).

Another successful example is provided by the ADONIS images of the Galilean satellites Galymede and Callisto (Forni, 1996). The morphological and mineralogic mapping was derived from images obtained at different wavelengths (1.7 and 2.0 microns). In particular, a bright spot on Callisto is associated with the Valhalla impact basin (Forni, 1996).

Figure 6. **Observation of Saturn's satellites with the Hubble Space Telescope (WFPC2) at the time of the ring-plane crossing in August 1995 (P. Nicholson, NASA)**

In the case of asteroids, the first image taken with adaptive optics was obtained on Ceres in the K filter (Saint-P et al, 1993) using the COME-ON system at ESO, and was used for deriving a thermal structure model of the asteroid. Images of Vesta were later obtained with ADONIS at ESO (Hainaut, 1994; Dumas, 1996); they showed variations of the albedo and were also used to constrain Vesta's thermal model and physical properties.

2.2. DETECTION OF SMALL OBJECTS

2.2.1. *Search for planetary satellites*

High-angular imaging has been used for several decades to search for new satellites or identify them visually. In 1980, the development of visible bidi-

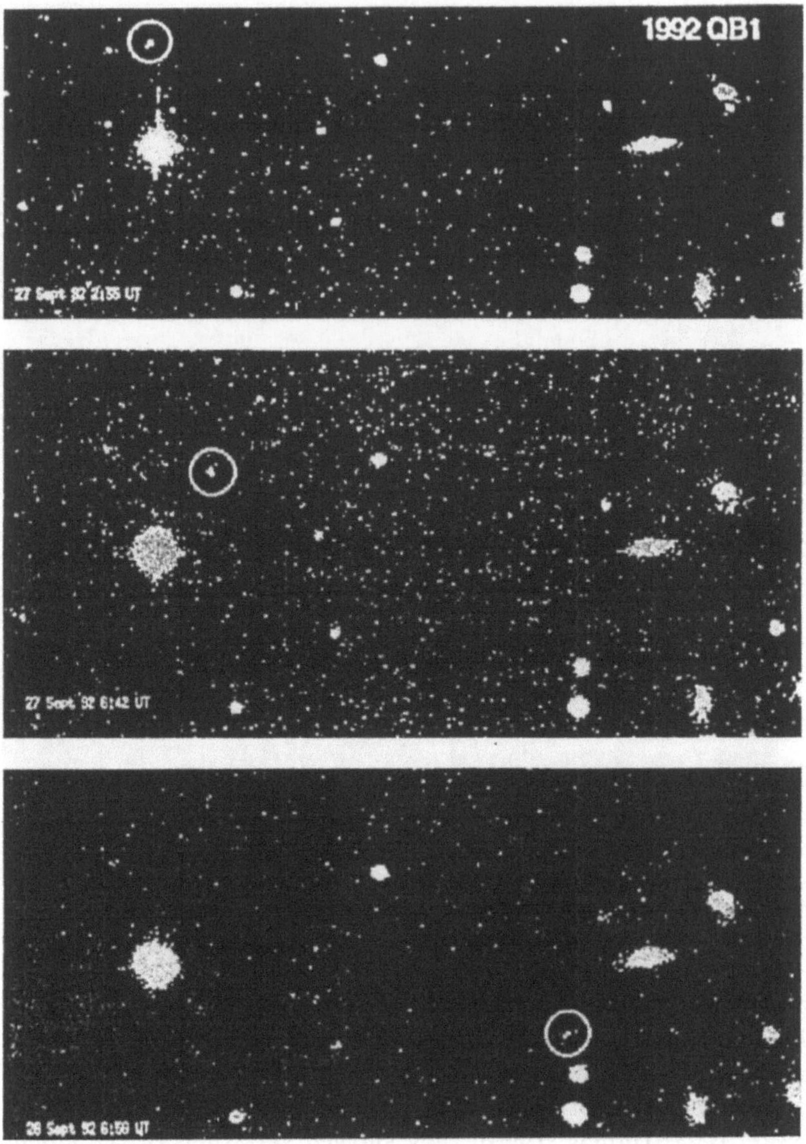

Figure 7. Detection of a Trans-Neptunian object (1992 QB1) using the 3.5-m telescope (NTT) at ESO (La Silla). The exposure time is 5 minutes for each image. The two upper images are taken on Sept. 27, 1992; the lower image is taken on Sept. 28, 1992. Observers: A. Smette and C. Vanderriest. The detected object has a visual magnitude of 23.8, a heliocentric distance of 41 AU and an estimated diameter of 100 km. The figure is taken from Hainaut (1994).

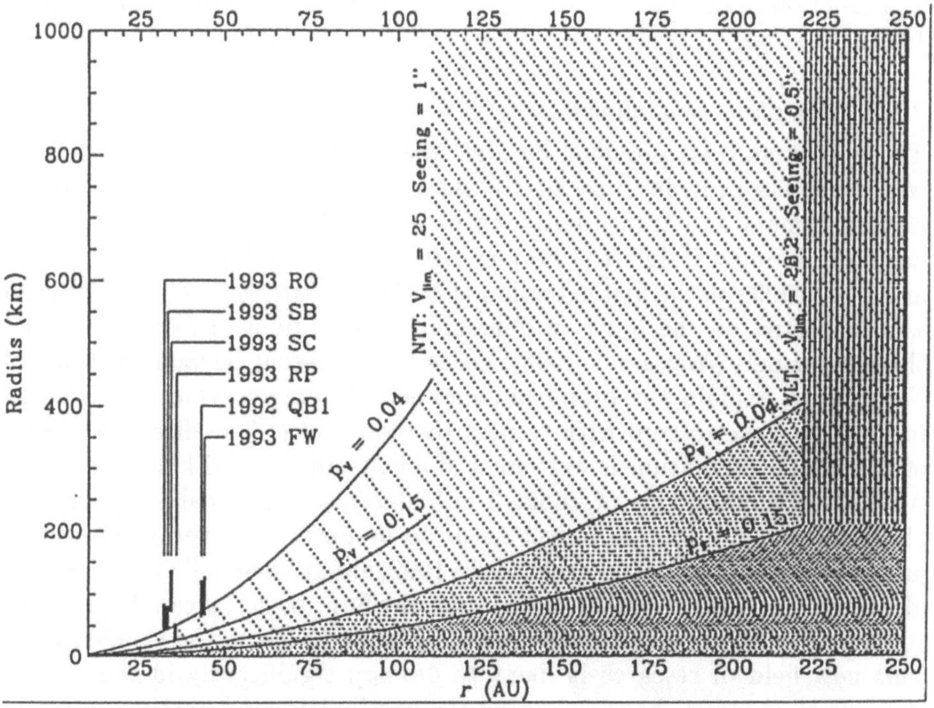

Figure 8. Possibilities of detection of Trans-Neptunian Objects using a 8-m telescope. The curves show the radius (in km) of an object at opposition and at magnitudes V=25 (NTT limit) and V=28.2 (expected limit for one VLT unit), as a function of heliocentric distance, for two geometric albedoes. The figure is taken from Hainaut et al (1994).

mensional detectors with improved sensitivity allowed the first image of the Pluto-Charon system, separated by 0.6 arcsec at maximum elongation (Bonneau and Foy, 1980).

Many attempts have been made to search for Saturn's small satellites, taking advantage of the edge-on geometry which occurs every 16 years when the Earth crosses the plane of Saturn's rings. During the 1980 campaign, only ground-based observations were made, in particular at Pic du Midi Observatory where the seeing may be as good as 0.3 arcsec. A new satellite, Helene, was detected and the main thickness of the ring was measured (1 km). These observations were repeated during the 1995-96 campaign; in addition, HST and ground-based adaptive optics images (in particular with ADONIS) were obtained. As a result of this campaign, 3 new satellites were detected and 3 transient objects, close to the Roche limit, were also identified. These have a limited lifetime due to the strong tidal forces which prevent their permanent cohesion. The structure of the rings and its width as a function of the distance to Saturn were also analysed in detail (Sicardy, 1996).

2.2.2. *Detection of weak objects*

The better contrast provided by high angular resolution in the search for weaks objects provides a sensitivity gain which can be used for searching distant comets and trans-neptunian objects (TNO's). A pioneering work has been done by Hainaut (1994) on distant comets (Fig. 7) and by Jewitt and Luu (1993) on the first detection of TNOs. The presence of these objects was suspected on the basis on dynamical observations: a reservoir, called the Kuiper belt (Weissman 1995), needs to be present beyond the orbit of Neptune to explain the origin of short-period comets, which have all an inclination close to zero with respect to the ecliptic. The direct detection of the TNO population is extremely difficult since a 100-km body at 40 AU has an angular size of about 3 milliarcsec and a visual magnitude of about 24. In addition, an object needs to be observed several times for a proper determination of its orbit. Since 1992, about twenty TNOs have been identified, at distances of 30-45 km from the Sun, with a mean diameter of 100 km. The observations have been done using CCDs in the visible range. This new field of research is likely to develop significantly in the future. Hainaut et al (1994) have estimated the possibilities of detection of TNOs with a 8-m telescope. Using the VLT and assuming an albedo of 0.04, it is expected that 100-km objects will be detected up to heliocentric distances in the range of 200 AU (Fig. 8).

2.3. ACTIVITY OF COMETS

2.3.1. *Distant comets*

Comets spend most of their life at large heliocentric distances where they remain inactive. They are then extremely difficult to detect, as the angular diameter of a 10-km body at 5 AU is only 5 milliarcsec; in addition, the albedo is expected to be very low (0.04 for P/Halley) and the temperature very low too (equilibrium temperature below 200 K at 5 AU). As the comet approaches the Sun, its surface is heated and ices start to sublimate, indicating the beginning of cometary activity . The heliocentric distance at which the activity begins depends upon the nature of the cometary ice. Water sublimates at about 2 AU, while other organic ices (CO, CO2, H2CO) sublimate at lower temperatures, i.e. higher heliocentric distance. Although water is believed to be the main cometary constituent, cometary activity has been reported in several cases at heliocentric distances much larger than 2 AU, which implies the presence of other ices, probably organics. The search for activity on distant comets is thus a powerful technique to

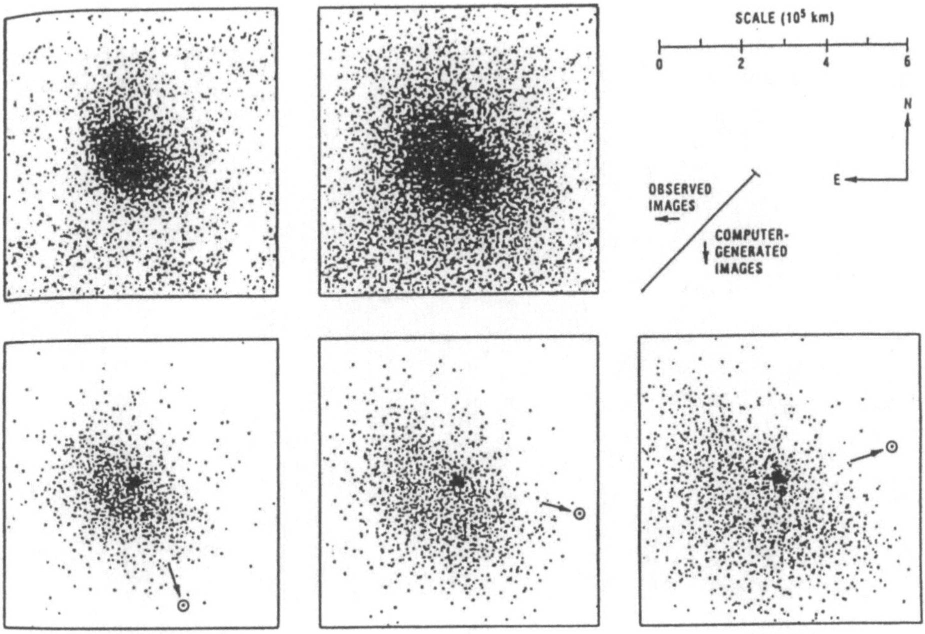

Figure 9. Modelling of an outburst of P/Halley, observed in February-March 1991. Upper images: observations; lower images: computer-generated images. Information is retrieved upon the date of the outburst, the particle velocity and the size distribution of the particles. The figure is taken from Sekanina et al (1992).

learn about their composition at distances where direct spectroscopic evidence is not possible. As a remarkable example, an outburst was detected on comet Halley in 1991, when the comet was at 14 AU from the Sun (Hainaut, 1994). From two sets of images recorded in February and March 1991 rspectively, the outburst was modelled, and the particle sizes and velocities were inferred (Sekanina et al, 1992; Fig. 9).

2.3.2. *Nearby comets*

Nearby comets provide the astronomers with unique opportunities to study their near-nucleus region or even their nucleus, usually hidden behind the coma. Two occasions have occurred over the past fifteen years. In 1983, a new comet discovered by the infrared satellite IRAS, IRAS-Araki-Alcock, approached the Earth at a distance of 0.03 AU. Ground-based observations

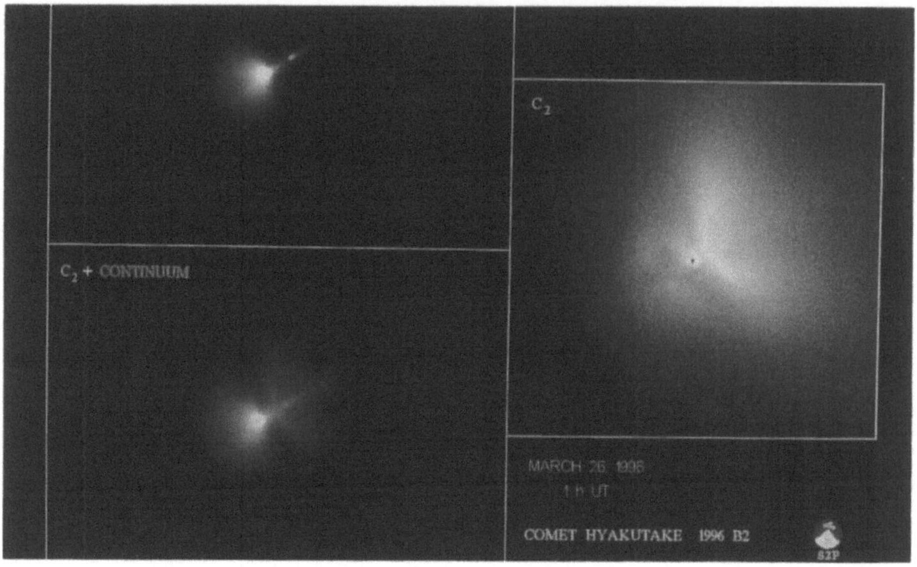

Figure 10. Comet Hyakutake, as observed on March 26, 1996, from the 1-meter telescope of Pic du Midi observatory, in different visible filters (Lecacheux, 1996).

with a seeing of 0.5 arcsec provided images with 10 km spatial resolution. This was still higher than the size of the nucleus which was not directly observed. In March 1996, a new comet, Hyakutake, approached the Earth at a distance of 0.1 AU. This comet was much more active than the previous one; ground-based images were taken, in particular at Pic du Midi Observatory (Fig. 10), giving unprecedented information upon the near-nucleus structure. Near-infrared images were recorded in the J, H and K filters using the ADONIS system at ESO. The comet was then at 0.6 AU from the Earth and the angular resolution was 0.1 arcsec, corresponding to a spatial resolution of 40 km in the J filter. For the first time, icy grains (identified from their color index) were clearly detected in the solar direction (Marco et al, 1996). This observation also demonstrated the possibility of observing comets with adaptive optics, using the cometary nucleus itself as a reference.

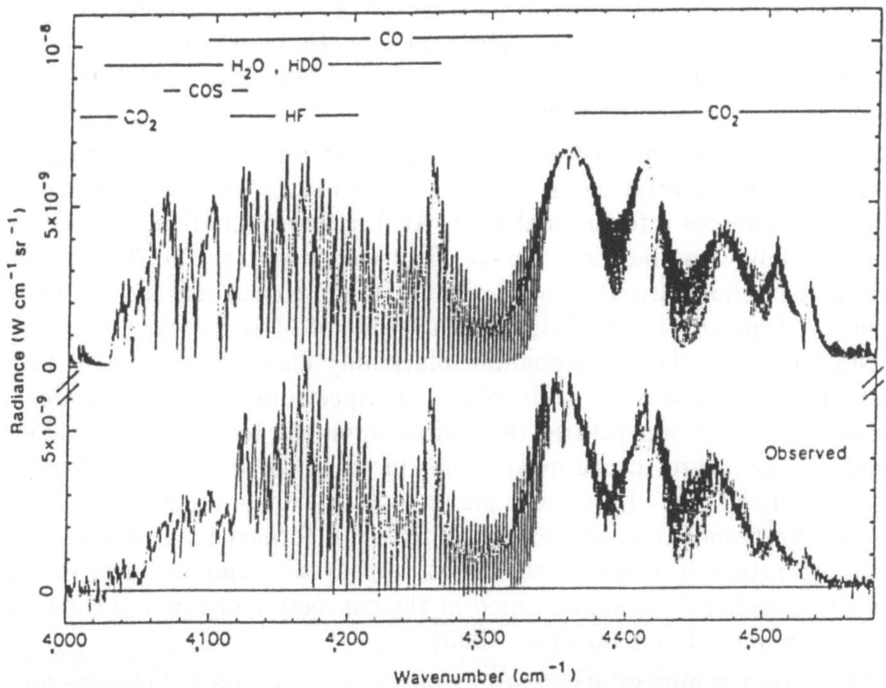

Figure 11. Thermal emission from the lower atmosphere of Venus, observed at high spectral resolution in the near-infrared range, using the CFH telescope (Mauna Kea, Hawaii). Upper curve: synthetic spectrum; lower curve: observations. The spectral signatures are due to CO2, CO, H2O, HDO, HF and OCS. The figure is taken from Bézard et al (1990).

3. Imaging spectroscopy of Solar-system objects

3.1. SCIENTIFIC OBJECTIVES OF SPECTROSCOPIC STUDIES

The main purposes of spectroscopy of Solar-system objects is (1) the study of the chemical composition of the gaseous phase (planetary and satellite atmospheres, comets) and (2) the mineralogic study of surfaces (Mars, asteroids, satellites), for a determination of their chemical composition.

All Solar-system objects show a spectrum which characterized by two components, a reflected solar spectrum and a thermal component. The reflected component shows a solar blackbody continuum and thus peaks in the visible range. Planetary signatures appear as absorption features in front of this continuum (molecular bands in the case of an atmosphere, or mineralogic signatures in the case of a surface). From the depth of the molecular absorption bands, an information is derived upon the abundance of the absorber. As an example, in the case of the giant planets, the main absorber of the solar reflected spectrum is methane which dominates the whole spectrum in the visible and near infrared range; for the telluric planets, CO2 is the main absorber.

The thermal component,which extends in the infrared range, coresponds to the part of the solar energy which is absorbed by the Solar-system object and converted into thermal heat. At first approximation, the thermal spectrum follows a blackbody at the effective temperature Te. Te is a function of the heliocentric distance and the albedo (mean reflectivity) of the object; it typically ranges from 400 K (orbit of Mercury) to 50 K (orbit of Neptune) and the corresponding blackbody curves peak from 7 to 50 microns. In the case of an atmosphere, the spectrum is mostly function of the atmospheric temperature. Depending upon the temperature gradient, molecular signatures can appear either in emission or in absorption. For instance, in the case of the giant planets which show a temperature inversion at the tropopause, stratospheric features, formed in a region where the temperature increases with height, appear as emission lines; in contrast, tropospheric features, formed in the convective region, are shown as absorption lines (Encrenaz et al, 1995).

For studying mineralogic features on the surface of a Solar-system object, a moderate spectral resolution is sufficient (R = 50 - 100), at least where no ice is expected (Mars, asteroids). In the case of the outer icy bodies, for which organic ices can be involved, a resolving power of 300-1000 is more appropriate. In the case of planetary and satellite atmospheres, where lines are usually formed at pressure levels ranging from a few bars to a few millibars, a resolving power of 1000 to 100000 is best suited. The most demanding observations are those of cometary molecular emissions, which are Doppler-broadened and require, to be resolved, a resolving power in the range of 106. Heterodyne spectroscopy is the best tool for this research and has already led to the discovery of a large number of cometary parent molecules.

3.2. LARGE PLANETARY DISKS

In the case of Jupiter, Saturn, Venus and Mars, the flux per pixel is sufficient to allow imaging spectroscopy in the near-infrared range (1-5 microns) with both high spectral and spatial resolution. A Fourier Transform spectrometer equipped with a bidimensional camera is ideally suited for this type of study. Such an instrument is presently operating at CFHT (Maillard, 1996); with a typical spatial resolution of about 0.5", it provides a resolving power of 1000 to 10000 over the 1-5 micron range. A large number of scientific programs are going to benefit from this instrument which offers an important complement to previous space observations.

In the case of Mars, an imaging spectrometer had been flown aboard the PHOBOS spacecraft (Bibring et al, 1989). Part of the Martian surface was mapped with a spatial resolution of a few km. The spectral resolution was moderate (60) but allowed a mineralogic analysis of the Martian surface and a study of minor atmospheric constituents (CO and H2O). In particular, anomalies were found in the CO distribution over the Martian disk (Rosenqvist et al, 1992). Ground-based observations of Mars in the (2-0) CO band at 2.3 microns with a FTS-2D will be best suited to solve this problem.

Imaging spectroscopy of Venus has been achieved in the near infrared range by the NIMS instrument (Near Infrared Mapping Spectrometer) aboard the Galileo probe, during its flyby of Venus in February 1990. A spatial resolution of 25 km was obtained on the Venus disk. The near infrared range allows, between the CO2 absorption bands, to probe the deep atmosphere of Venus hidden behind a thick cloud deck. Spectral images were recorded in various points of the disk. They were used for an analysis of the motions of the lower cloud deck and a study of the minor constituents (CO, H2O) in the lower atmosphere (Carlson et al, 1991). In parallel, ground-based FTS observations recorded in various positions of the Venus disk allowed a refined analysis of the spectra in some specific wavelength windows, leading to an abundance determination of several minor constituents (Bzard et al, 1990; Fig. 11) and a measurement of the D/H ratio (de Bergh et al, 1991). In the future, observations of Venus with a FTS-2D should allow to combine the advantages of both spatial and spectral resolutions.

Other scientific programs which should benefit from this instrument include the study of the auroral zones of Jupiter in the H2 and H3+ emissions, at 2 and 3.5 microns (Drossart et al, 1993), and the study of the hot spots of Jupiter in the 5 micron window. This last program is especially promising in view of the recent results of the Galileo probe, which might indicate that the atmospheric composition at the entry site of the probe was significantly different from the mean values inferred from previous measurements.

3.3. SMALL DISKS

This class of objects includes diameter sizes ranging from about 0.5 to a few arcsec: Titan, the galilean satellites and the largest asteroids. A significant progress has been achieved recently in the near infrared range using adaptive optics combined with a low-resolution dispersing system (CVF, R= 50-100). As mentioned above, this spectral resolution is adequate for studying the mineralogy of surfaces.

Titan spectral images have been obtained in the J, H and K filters and will be used to refine the subtraction of the atmospheric contribution through a complete modelling of the CH4 absorption. As a result, it should be possible to get information upon the nature of the surface, and, in particular, the origin of the bright spot detected in the broad-band filters. Near-infrared spectroscopic data at high spectral resolution suggest the possible presence of water ice over the surface of Titan (Coustenis et al, 1996). Imaging spectroscopy of the Titan disk will allow to assign the chemical nature of the bright spot and to discriminate between an altitude effect (mountain) or a chemical differenciation (Combes et al, 1996).

Similar analyses are in progress on the Galilean satellites. On Io, the hot spots have been associated to SO2 frost. On Ganymede and Callisto, spots of H2O ice have been identified on Ganymede, while water ice seems more uniformely distributed on Callisto (Forni, 1996).

Images of Ceres have been obtained at 10 wavelengths in the 1.5-2.4 micron range using the COME-ON+ system at ESO (Hainaut, 1994), clearly showing evidence for albedo and geological variations. The asteroid program is continuing with observations of Vesta using ADONIS at ESO and the adaptive optics system at CFHT (Dumas, 1996).

3.4. UNRESOLVED POINT-LIKE SOURCES

Triton, Pluto, small asteroids and trans-neptunian objects cannot benefit from imaging spectroscopy, but high angular resolution is still necessary to improve the contrast and increase the sensitivity. Another important consequence of the use of adaptive optics is that it allows high-resolution spectroscopy. Indeed, in the case of a point-like source, if the flux is concentrated in 0.1" instead of the typical 0.5" corresponding to good seeing without adaptive optics, the entrance slit of the spectrograph can be narrowed accordingly, which allows an increase in spectral resolution by a factor 5.

This advantage is especially interesting for Triton and Pluto which both exhibit tenuous atmospheres, with surface pressures in the range of a few microbars. High spectral resolution is thus essential for searching for spectral features of minor atmospheric species like CH4 and CO with maximum contrast (Encrenaz et al, 1992).

3.5. COMETS

Comets are exciting targets for imaging spectroscopy in the visible and near-infrared range. This spectral domain is rich of signatures of parent molecules (in the IR), and radicals and ions (in the visible). The individual lines are Doppler-broadened, so the highest achievable resolving power (to be optimized as a function of the cometary flux) is needed. A spectral

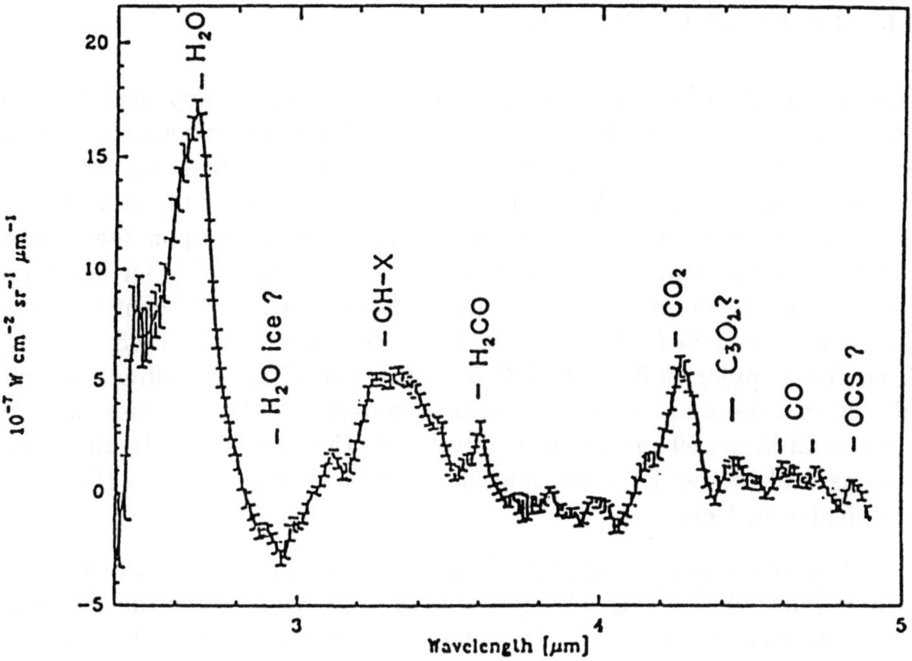

Figure 12. The infrared spectrum of comet Halley, as observed from the IKS spectrometer aboard the VEGA probe. Resonant fluorescence emissions of parent molecules are detected in the 2.5-5 mm range. The figure is taken from Combes et al (1988).

range of specific interest is the 3-4 micron region where various kinds of hydrocarbons have been detected (Combes et al, 1988; Fig. 12). Apart from the detection of methanol (Bockele-Morvan et al, 1995), the exact nature of these hydrocarbons is not completely clear and seems to vary from comet to comet.

At a distance of 1 AU from the Earth, the spatial resolution obtained on a comet is 375 km without adaptive optics (seeing of 0.5“), 75 km with AO (0.1“ in the J filter). Observations with ADONIS with a CVF can provide a mapping of the near-nuclear region in the 3-4 micron range, especially important for the observation of hydrocarbons and H2O ice. Such observations will hopefully be performed on comet Hale-Bopp in 1996 and 1997. In the case of very bright comets, FTS-2D observations would be of unprecedented quality, as the high spectral resolution would allow to separate the individual cometary lines and possible identify the various hydrocarbons which contribute to the 3.2-3.6 cometary emission.

3.6. MILLIMETER SPECTROSCOPY

The advantage of heterodyne spectroscopy lies in its very high spectral resolving power (106) which is crucial for analysing narrow emission or absorption lines in the atmospheres of Solar-system objects: Mars, Io, stratospheres of Venus, giant planets and Titan, and comets. In the case of planetary and satellite atmospheres, information is derived upon the vertical structure of the atmospheric species (Encrenaz et al, 1995); for Mars and Venus; velocity maps are derived from observations at different points of the planetary disks (Lellouch et al, 1991; Rosenqvist et al, 1995; Fig. 13), from the Doppler shift of the CO lines (at 230 GHz, the diffraction limit of a 30-m antenna is about 12", to be compared to the maximum sizes of Venus and Mars, 60 and 18" respectively). In the case of comets, the observations lead to the determination of abundances and velocities (Crovisier and Schloerb, 1991).

Using the VLA, a mapping of H2O on Mars has been achieved at 22 GHz (Clancy et al, 1992). The spatial resolution was 0.5" and the planetary diameter was about 15" at the time of observation. The H2O line was detected with highest intensity on the limb, due to the strong airmass effect which increases the line of sight. From the observations, information was inferred on latitudinal variations, North-South asymmetries and local time effects. These data provide important constraints for climatology models of the Martian atmosphere.

Millimeter interferometers are now able to provide these measurements with spatial resolution. The IRAM interferometer at Plateau de Bure (France) is presently equipped with four 15-m antennae, with about 1 km baseline; at 230 GHz the spatial resolution is about 1.5". The first planetary program to benefit from this instrument should be the determination of winds on Mars and Venus. As a future program, the mapping of bright comets in some strong transitions (HCN, CH3OH, CH3CN...) might become possible.

4. Future programs

High angular resolution observing program are expected to develop significantly in the future, with the on-going improvement of adaptive optics techniques, the future operation of a near-infrared imaging spectrometer aboard the HST, the future use of 8-m class telescopes and, as a next step, the development of space interferometry at the VLT (VLTI).

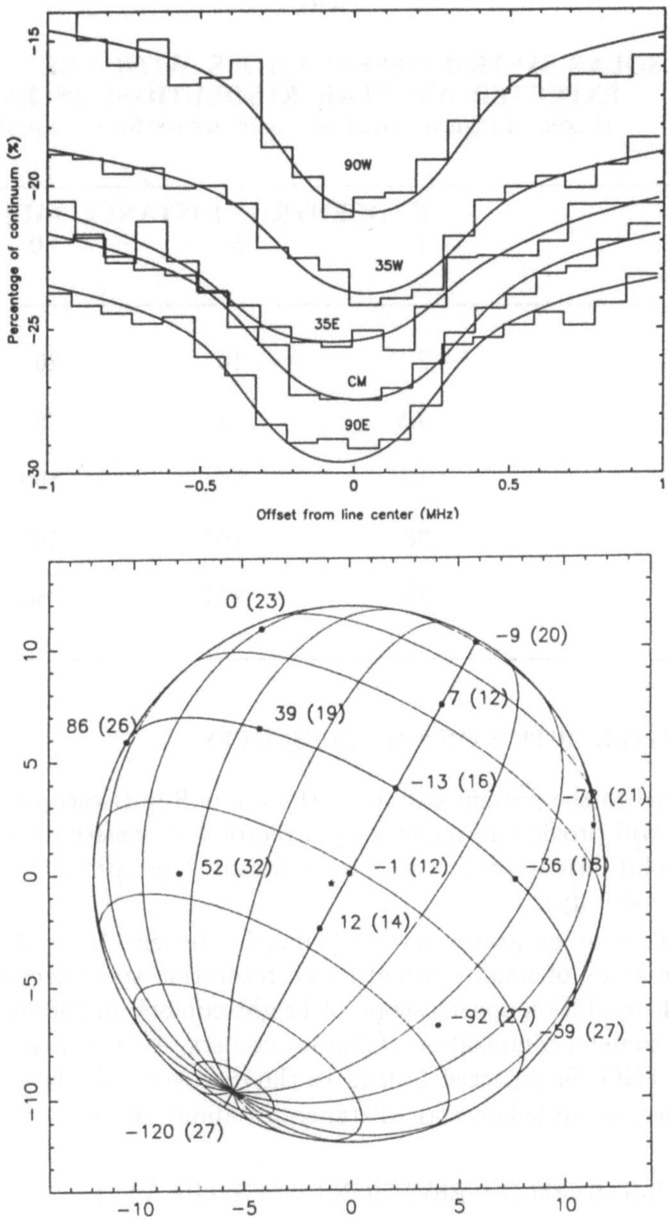

Figure 13. Determination of the winds in the atmosphere of Mars (z= 40-70 km) from a measurement of the Doppler shifts of the J= 2-1 line of CO at 230 GHz. Observations were recorded with the IRAM 30-m antenna. The field of view was 12 arcsec and the total diameter of Mars was 24 arcsec. The measurements, performed during southern summer solstice, clearly indicate a global easterlies flow in the southern hemisphere. The figure is taken from Lellouch et al (1991).

TABLE 2

SOLAR-SYSTEM OBSERVATIONS WITH THE VLTI: EXPECTED ANGULAR RESOLUTIONS IN KM
(Expected angular resolution: 10 milliarcsec for $\lambda = 2$ μm)

λ (μm)	GEOCENTRIC DISTANCE (AU) 1	5	10	50
0.5	2	10	20	100
2	7.5	37	75	375
5	19	95	190	950
10	38	190	380	1900
20	75	375	750	3750

4.1. ADAPTIVE OPTICS ON 8-M TELESCOPES

As compared to the present situation, the use of 8-m telescopes with adaptive optics will provide an increase by a factor 4 in sensitivity (at constant spatial resolution) or an increase by a factor 2 in spatial resolution (at constant sensitivity).

The following programs should primarily benefit from this improvement: mineralogy of small asteroids; high-resolution spectroscopy of Io and Titan; high-resolution spectroscopy of bright comets; mineralogy of trans-neptunian comets; composition of distant comets. As mentioned above, the search for TNOs is expected to lead to the discovery of a large number of new samples, up to heliocentric distances of about 200 AU.

4.2. SOLAR-SYSTEM OBSERVATIONS WITH THE VLTI

At 2 microns, the angular resolution provided by the VLTI is expected to be about 10 milliarcsec. Table 2 shows the corresponding spatial resolution in km for various wavelengths and various heliocentric distances.

VLTI observations will be devoted to the observation of small objects, in the range of an arcsec in angular diameter: asteroids, satellites of the outer Solar system, Pluto and Charon, cometary nuclei and trans-neptunian objects. In the near infrared, at 2 micron, the VLTI will resolve a cometary nucleus at 1 AU, all asteroids larger than 50 km in the asteroidal belt, all outer satellites with a diameter larger than 300 km, Pluto and Charon. In the visible, at 0.5 micron, the VLTI will even resolve a cometary nucleus at 5 AU (i.e. all Jupiter-family comets) and all trans-neptunian objects larger than 100 km in diameter.

The following programs should greatly benefit from the VLTI: the study of TNOs and distant comets; the search for asteroids companions (the unexpected detection of Ida's companion, Dactyl, by the Galileo probe suggests that the event may not be exceptional in the Solar system); the morphology of Triton, Pluto and Charon and the morphology of small asteroids. The VLTI should thus enable the study of small Solar-system bodies to make a significant progress in the coming decade.

References

Bézard B, de Bergh C, Crisp D and Maillard J P (1990) The deep atmosphere of Venus revealed by high-resolution night-side spectra, Nature, 345, 508-511.

Bibring J P et al (1989) Results from the ISM experiment, Nature, 341, 591-593.

Bockele-Morvan D, Brooke T Y and Crovisier J (1995), On the origin of the 3.2-3.6 micron features in comets, Icarus, 116, 18-39.

Bonneau D and Foy R (1980) Interfromtrie au 3.60m CFH. I. Rsolution du systme Pluton-Charon, Astron. Astrophys. 92, L1-L4.

Carlson R W et al (1991) Galileo infrared imaging spectroscopy measurements at Venus, Science, 253, 1541-1548.

Clancy R T, Grossman A W and Muhleman D O (1992) Mapping Mars water vapor with the Very Large Array, Icarus, 100, 48-59.

Combes M et al (1988) The 2.5-12 mm spectrum of Comet Halley from the IKS-VEGA experiment, Icarus, 76, 404-436.

Combes M, Vapillon L, Gendron E, Coustenis A and Lai O (1996) 2-micron images of Titan by means of adaptive optics, The Messenger, 83, 40-42.

Coustenis A, Lellouch E, Maillard J P, and Mc Kay C P (1995) Titan's surface: composition and variability from the near-infrared albedo, Icarus, 118, 87-104.

Crovisier J and Schloerb F P (1991) The study of comets at radio wavelengths, in "Comets in the post-Halley era", R L Newburn, M Neugebauer and J. Rahe editors, Kluwer, pp. 149-174.

de Bergh C, Bézard B, Owen T, Crisp D and Maillard J P (1991) Deuterium on Venus: observations from Earth, Science, 251, 547-549.

de Pater I, Kendernine S and Dickel J (1982) Comparison of the thermal and non-thermal radiation charcteristics of Jupiter at 6, 11 and 21 cm with model calculations, Icarus, 51, 25-38.

de Pater I and Dickel J (1982) VLA observations of Saturn at 1.3, 2 and 6 cm, Icarus, 50, 88-102.

Drossart P (1996) Private communication.

Drossart P, Maillard J P, Caldwell J and Rosenqvist J (1993) Line resolved spectroscopy of the Jovian H3+ auroral emissions at 3.5 micrometers, Astrophys. J., 402, L25-L28.

Dumas C (1996) Private communication.

Encrenaz T, Combes M and Saint-P O (1992) Solar-system infrared observations at high

spectral resolution, ESO CWP-40, 115-126.
Encrenaz T, Bibring J P and Blanc M (1995) The Solar System, Springer-Verlag, 2nd edition.
Encrenaz T et al (1995) Detectability of molecular species in planetary and satellite atmospheres from their rotational transitions, Plan. Space Sci., 43, 1485-1516.
Forni O (1996) Private communication.
Hainaut O R (1994) Selected observational studies of minor bodies in the solar system, PhD dissertation, Institut d'Astrophysique de Lige and European Southern Observatory.
Hainaut O R, West R M, Smette A and Marsden B G (1994) Imaging of distant comets: current and future limits, Astron. Astrophys. 289, 311-325.
Jewitt D C and Luu J X (1993) Discovery of the candidate Kuiper belt object 1992 QB1, Nature , 362 730-732.
Leblanc Y and Dulk (1995) Changes in brightness of Jupiter's radiation belts at 13 and 22 cm during and after the impacts of comet SL9, Geophys. Res. letters, 22, 1793-1796.
Lellouch E, Goldstein J J, Bougher S W, Paubert G and Rosenqvist J (1991) First absolute wind measurements in the middle atmosphere of Mars, Astrophys. J., 383, 401-406.
Maillard J P (1996) Private communication.
Marco O, Encrenaz T and Gendron E (1996) First images of icy dust grains in a comet, submitted to Nature.
Rosenqvist J et al (1992) Minor constituents in the Martian atmosphere from the ISM/Phobos experiment, Icarus, 98, 254-270.
Rosenqvist J, Lellouch E and Encrenaz T (1995) Global circulation in Venus' mesosphere from IRAM CO observations (1991-1994): a tribute to Jan Rosenqvist, Bull. Amer. Astron. Soc., 27, 1080-1080.
Saint-Pé O, Combes M, Rigaut F, Tomasko M and Fulchignoni M (1993) Demonstration of adaptive optics for resolving imagery of solar-system objects: preliminary results on Pallas and Titan, Icarus, 105, 263-270.
Saint-Pé O, Combes M and Rigaut F (1993) Ceres surface properties by high-resolution imaging from Earth, Icarus, 105, 271-281.
Sekanina Z, Larson S M, Hainaut O R, Smette A and West R M (1992) Major outburst of comet Halley at a heliocentric distance of 14 AU, Astron. Astrophys. 263, 367-386.
Sicardy B (1996) Private communication.
Smith B A et al (1979) The Jupiter system through the eyes of Voyager 1, Science, 2O4, 951-957.
Smith P, Lemmon M T, Lorenz R, Sromovsky L A, Caldwell J J and Allison M D (1996) Titan's surface revealed by HST imaging, Icarus, 119, 336-349.
Weissman P R (1995) The Kuiper belt, Ann Rev. Astron. Astrophys., 33, 327-357.

CIRCUMSTELLAR MATERIAL AROUND MAIN SEQUENCE AND EVOLVED STARS

F. PARESCE
European Southern Observatory
Karl Schwarzschild Str. 2
D-85748 Garching b. München, Germany

Abstract. Our current understanding of dust and gas orbiting around main sequence stars as evidenced by IR photometry, optical coronography and high resolution UV spectroscopy is reviewed and placed in the context of protoplanetary system formation and evolution. The case of Beta Pictoris, the brightest known Vega-like disk, is used as an example of the kind of information one might be able to deduce from the highest possible spatial resolution and contrast measurements planned in the near future from the ground and space. The possible existence of planetary-sized bodies in the Beta Pic disk, its evolutionary stage and the incidence of disks around other main sequence stars are also topics of great interest in this field that we can just begin now to clarify with existing and, especially, planned high resolution techniques discussed at this school. Finally, the exciting potential for actually reading the past complex evolutionary history of a star as it winds its way across the HR diagram as impressed on the circumstellar gas and dust structures is explored in some detail. It is shown that much can already be learned from such studies, for example, on supernovae, Miras and luminous blue variables but that only attainment of high contrast milliarcsecond resolution from the optical to the thermal IR will allow a real breakthrough in the quantity and quality of data required for this important endeavour.

1. Introduction

We have known for a long time that appreciable amounts of dust and gas can be found in the immediate vicinity of certain types of stars as diverse as pre main sequence or planetary nebulae and that this circumstellar (CS) material has crucially important information to convey on the evolutionary

A-M. Lagrange et al. (eds.), High Angular Resolution in Astrophysics, 259–283.
© 1997 *Kluwer Academic Publishers.*

history of its parent. The higher resolution and better sensitivity instrumentation that has recently become available has also revealed that this material is present in detectable amounts around stars that had never been, heretofore, expected to retain such envelopes. Today, in fact, the real observational challenge is to find a way to detect and characterize CS material around any astrophysical object as a way to gain a better understanding of the complex processes that create and maintain it.

In this lecture, this rather vast subject will be divided into two arbitrary sections for simplicity: the first dealing with the issue of the CS material recently found around normal main sequence (MS) stars (the so-called Vega phenomenon) while the second will consider briefly what can be learned from detailed studies at very high angular resolution and contrast of the highly structured CS medium around objects losing mass in their voyage across the upper part of the HR diagram. This division should not be interpreted as having any underlying physical significance since it is becoming ever more apparent that there must be a smooth continuum of structural and dynamic properties of the CS envelopes extending from the very early accretion disk from which the star formed to the mass loss episodes in the last stages of its evolution. It is precisely this continuity that makes studies of the CS environments so profitable scientifically especially since it is unlikely that an envelope can very quickly erase memory of the past events that have shaped it.

2. The Vega Phenomenon

2.1. GENERALITIES

This fascinating subject started in 1983 when IRAS first detected a strong IR excess at 25, 60 and 100 microns in the otherwise normal MS A0V star Vega (Aumann et al., 1984). Since then, the subject has blossomed to include a number of candidate Vega-like objects and coverage from the UV to mm wavelengths with many different techniques with sometimes spectacular results. A complete review of the field is not the main objective of this lecture but those who wish to indulge in the details have no shortage of excellent reviews to guide them. These would include Norman and Paresce, 1989; Backman and Paresce, 1993 (BP93); Lagrange-Henri, 1995, Sicardy, 1994 and a whole book devoted to the subject edited by Ferlet and Vidal-Madjar, 1994. Since the field is developing very quickly it is difficult to keep up with the pace of discoveries. On the other hand, there are some fundamentals which underlie the subject and which are not likely to change very drastically in the coming years and it is to these that this lecture is aimed. They need to be grasped and retained, in fact, if one hopes to keep a clear and well balanced perspective on the newest results and the confidence to attach to any implications derived therefrom.

The basic physics and some of the mathematics required for the task are spelled out in considerable detail in BP93; Artymowicz et al., 1989; and Backman et al., 1992 and do not need, therefore, to be repeated here at any great length. The critical parameters in the problem are the grain size and its distribution function, the radiative and absorptive efficiency of the grains in the star's radiation field as a function of wavelength and the grain temperature at any location. The exact way in which the grains absorb and reemit energy is, of course, the main factor driving the equilibrium temperature any grain reaches in the star's CS medium. Unfortunately, this parameter also represents the main uncertainty in the inversion problem that needs to be solved in order to determine the physical structure and composition of the material from the observed emission characteristics and, therefore, the main source of ambiguity in the results. The usual solution is to assume simple power laws for the wavelength dependence of the efficiencies for a restricted set of size categories. For example, large (black body) grains having diameters much larger than the wavelength emit efficiently and so yield the lowest possible equilibrium temperature at a given distance from the star. Very small (ISM) grains, on the other hand, emit inefficiently at all longer wavelengths and, therefore, yield the highest temperature at a given distance. In between lie the intermediate or mid-size grains with sizes that are most interesting in our context because comparable to the wavelengths (1–100 microns) used in most investigations to date whose efficiencies are constant in the optical and linear functions of wavelength in the IR.

These deceptively simple facts have the following important consequences. First, the excess IR emission observed in the Vega-like objects can be attributed to orbiting grains only if the grain temperature T at the distance r from the star follows an appropriate relationship that includes both the stellar luminosity and the radiative and absorptive efficiency of the grains. In the simplest case of BB grains, for example, the expected equilibrium temperature varies as $L_*^{0.25} r^{-0.5}$. Thus, the IR spectrum and the spatial scale of the emission must be consistent with this relation. Second and conversely, a reasonable estimate of the grain size can be immediately obtained from this relation and the two critical observational parameters: emission temperature and size of the emitting region. The most important consequence of the preceding discussion that needs to be emphasized here is that the object must be spatially resolved to have any confidence that the emission originates in a stable, orbiting structure around the star. The IR excess by itself is not a sufficient condition for this to occur. Regrettably, so far only three objects have been resolved and found to be consistent with this hypothesis and, therefore, can strictly be called Vega-like, namely

Vega, Fomalhaut and Beta Pictoris. A fourth object ($BD+31°643$) has recently been resolved by ground-based optical coronography (Kalas and Jewitt,1997) but we should await further study before it is officially inducted in the Vega-type club. ISO (see Habing et al 1996) and, eventually, SIRTF should hopefully enlarge this tantalizing but meager sample.

The main physical parameters for these three stars are shown in Tables I–III of BP93. In the case of Vega itself, a more recent and more accurate determination of the cloud diameter by van der Blick et al., 1994 and recently confirmed by preliminary ISO results (Habing et al., 1996) has yielded a cloud of smaller grains in a slightly larger envelope. In all cases, however, the grains are always considerably larger than those in the ISM or in typical CS mass loss envelopes thereby already indicating that some sort of agglomeration process somewhere in the nebula has already occurred. Although in no case has Keplerian motion been established yet for any of the Vega-like stars, it is nevertheless safe to claim that the grains must be in orbit around these objects because of a) the impossibility for very small ISM grains being found at the observed radii and temperatures and b) the inconsistency of the known mass loss limits of these stars ($\dot{M} < 10^{-10} M_{\odot}/\mathrm{yr}$) with the amount of solid material observed.

One is usually struck by the rather puny amounts of total mass these structures actually involve (see Table III of BP93 and Section A2 of this paper): in IR emitting grains, no more than a lunar mass at most. It is a fundamental, well known but usually unappreciated characteristic of these objects that what counts for their observeability is not the total mass in the CS environment but the effective scattering area. To give a prime example, although our own zodiacal dust cloud contains less than 10^{-10} of the mass of the planets in the solar system, its infrared luminosity is ~ 100 times that of all the planets combined! Similarly, a Vega-like cloud of lunar mass of ~ 10 micron size grains spread over an area of several hundred AU squared around the stars can easily overpower their photospheric flux and be detectable at tens of parsecs! Another peculiar trait of these stars is the fact that the IR emission decreases rapidly with decreasing wavelength (see Figure 1 of BP93) so that, except for Beta Pic that has a residual amount of excess emission there (see section A2 for an explanation), by ~ 10 microns the stellar continuum dominates the spectrum. This implies in the formalism developed briefly above that there is little or no material at $T > 200\mathrm{K}$ or, conversely, there is a clear zone below a cut-off radius r_1 of approximately 20–50 AU (listed in Table IV of BP93). It is probably not fortuitous that r_1 corresponds so closely to the solar system dimensions.

In order to get a grip on the stability and likely evolution of such a dust cloud, one needs to carefully consider the time scales for the various dust destruction processes at work in their CS environment. The ratio of radiation pressure to gravitation is independent of r but inversely proportional to the grain size. This ratio reaches ~ 0.5–1 depending on grain composition at a blowout radius below which the grain is expelled from the system in a free fall time of $\sim 10^4$ yrs and which depends on stellar luminosity and mass. Moreover, grains orbiting in a radiation field will spiral in toward orbit center due to the Poynting-Robertson drag. The orbit decay time scale is proportional to the grain size and the square of the distance. Grains also collide and are presumably lost on a time scale that also depends on r while the ice sublimation time is an extremely sensitive function of grain temperature ($\sim T^{5.5}$). The blowout radius and the typical time scales for these processes are given in Table XI in BP93. A more detailed analysis of these times can be found in Artymowicz, 1988 and 1994 (A94).

A comparison of the grain removal time scales at both the inner and outer edge of the clouds with the likely ages of the stars shows unambiguously that the grains are relatively short lived and that, therefore, they are not primordial but must be continuously replenished by some mechanism. Although subject to considerable observational error that will be certainly reduced with better observations with ISO, the central void radii and temperatures also seem too different at the moment to be clearly ascribed to a phase transition unless the grain composition changes dramatically at the boundary. In any case, the ice sublimation time is so short in the voids that grains there must be refractory as observed in Beta Pic (Knacke et al., 1993). Another interesting result of these calculations is that PR drag is definitely competitive with collisions at the inner edge of the clouds which would imply the existence of some other removal mechanism to prevent grains from moving in and erasing the voids. In summary, the preponderance of evidence now seems directed towards the view that the clouds detected in the IR around the Vega-like systems are not the unprocessed residue of the original accretion disk that formed the central star but rather are the result of rapid and efficient processing by grinding or ablation of relatively large, so far undetected, parent bodies.

2.2. BETA PICTORIS

This last assertion stands the best chance of test and validation in the case of this unique object. It stands apart from the group for one simple reason: its CS cloud can be not only marginally resolved in the IR but also imaged in the visible due to scattering of the stellar light by the orbiting grains. Recent images of this object are shown in Figure 1. I have consciously avoided, up to now in this lecture, the use of the term "disk" for the CS clouds around the Vega-like stars but I cannot continue for long to ignore the obvious fact that, if they are indeed made up of orbiting material, they should be

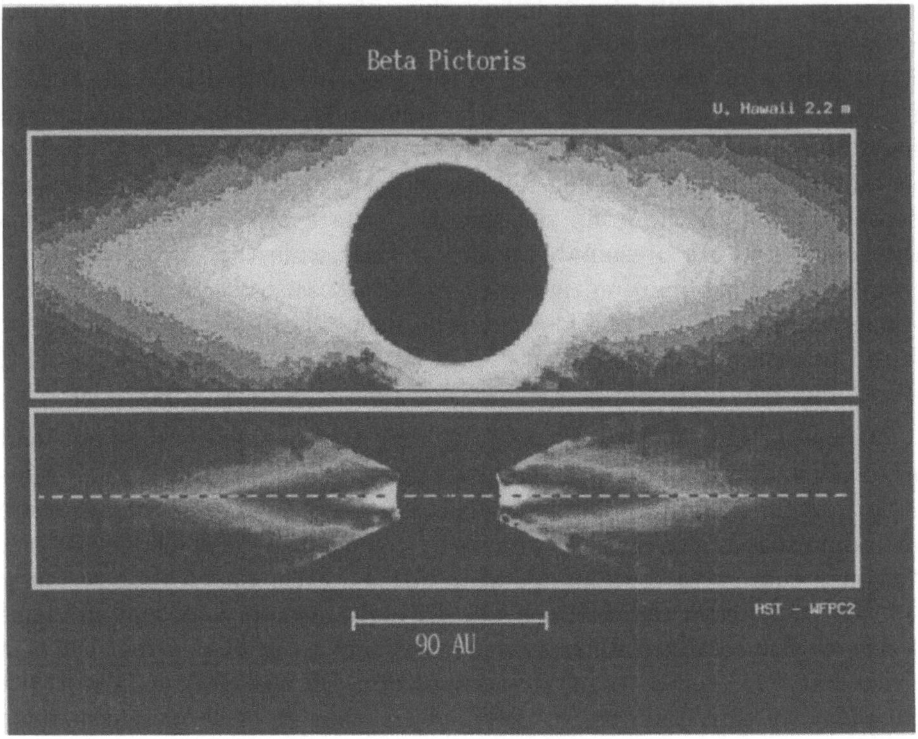

Figure 1. An outstanding example of the diagnostic power of high spatial resolution and high contrast techniques as applied to CS environments around MS stars. a) The Beta Pic disk as seen from the ground with an optical coronograph on the U. of Hawaii 2.2 m (Kalas and Jewitt, 1996 and b) from space with WFPC2 on HST (NASA/ESA) at the same physical scale . The dark areas in the center of the images correspond to the areas covered by the digital masks used to suppress the light from the central star. North is inclined 30 deg to the horizontal axis pointed towards the left with the East pointing downwards.

disk-like structures located in the star's equatorial plane. Although there was some indication already from IRAS scans that Beta Pic and Fomalhaut had spatially asymmetric emission profiles, only the much higher resolution of the optical images of Beta Pic could confirm this hypothesis and, in fact, offer a treasure trove of information on its main physical characteristics which could, then, be reasonably confidently extrapolated to the others.

The scientific importance of Beta Pictoris lies mainly in the fact that the optical and near IR observations with their higher spatial and spectral resolution crucially complement the presently cruder far IR measurements. While the latter provide information on the emitting properties of the grains and a rough idea of where they are located, the former yield the precise spatial distribution of the scattering grains and their composition. Specifically, a model that simultaneously satisfies the IR and optical data with a simple single dust component can now be constructed by inverting the integral flux equation containing the $T(r)$ relationship in order to obtain the fundamental parameters governing the problem, namely the run of grain area with T and the properties of the emitting particles. This exciting prospect has been realized by exploiting two basic inversion techniques: a parametric one in which a specific dust density radial profile is assumed (Backman et al., 1993; Diner and Appleby, 1986; Nakano, 1988) and the parameters are adjusted to obtain a best fit to the data or a non-parametric technique in which no a-priori assumption on the profile is made but where a maximum entropy condition is imposed on the resulting distribution (Artymowicz et al., 1989). These have now yielded a general consensus on such diverse parameters as total scattering area, albedo, disk morphology etc at least within a generous but tolerable range of uncertainties (see BP93 and A94 for more comprehensive reviews).

In view of the fundamental complexity of the problem and the still very limited direct observational overlap between the IR photometric and optical coronographic observing region (see Figure 3 in Norman and Paresce 1992), it is short of amazing that, with such a simple one component dust model, it has been possible to establish unambiguously several important facts concerning this object and, then, by comparison or inference for the other resolved sources. In the next paragraphs, I will briefly review these facts but emphasize what remains to be done.

The consistency between the spatial variation of surface brightness (SB) and optical spectrum of the disk in the optical and the far IR emission spectrum implies that the scattering area in the outer disk ($r > 100$ AU) is composed mainly of mid-size (2–20 micron), high albedo ($A > 0.5$), grains extending out to at least 1000 AU from the central star. These grains are distinguished by a λ^{-1} power law emissivity in the IR and a constant emissivity in the optical and, therefore, relatively easy to handle analytically. The observed morphology is consistent with an essentially edge-on ($i < 5°$) to the line of sight, constant opening angle ($\sim 10°$) flaring disk (see Figure 1). The SB radial profile has a now well established break at $r \sim 80$ AU where it changes from a -3.5 power law index to a shallower -2.4 one (Clampin et al., 1994; Golimowski et al., 1993; Kalas and Jewitt, 1995) extending down to the present limit of reliable detectability ($r \sim 40$ AU). Beyond this point, both the far IR spectrum (BP93) and the recent near IR imaging at 10–20 microns (Lagage and Pantin, 1994) agree on a relatively dust-clear void of 20–30 AU radius surrounding the star.

Returning to the massive outer dust disk for a moment, all models converge on a perfectly scattering cross sectional mid-size grain area in the disk varying with r something like $r^{-1.7}$ so that the total area is of order 10^{29} cm^2 or an astounding 480 AU^2 of high albedo particles. The high albedo itself is a bit of a surprise since most comparable solar system material exhibits low albedos (zodiacal and cometary dust, asteroids, carbonaceous chondrites etc) but may simply be due either to slightly dirty ice or bright silicates (see Norman and Paresce 1992, A94, Greenberg and Li, 1996 for a more complete discussion of this very important issue). Of more immediate concern is to obtain an accurate estimate of the total mass of the disk for comparison with theory. This, unfortunately, is a very difficult task owing to the fact, that what you see may not be necessarily all there is since the most conspicuous material in the particular wavelength band of observation is obviously the one having typical sizes similar to the wavelength itself. Thus, most of the mass of the disk may reside in much smaller or larger objects depending on the particular size distribution function adopted.

Fortunately, we are getting increasingly more precise ideas about this function guided by resonably secure observations and theoretical considerations from a variety of sources including, most recently, our Beta Pictoris analog in the solar system the Kuiper belt (BP93; Weissman, 1995). A lower bound to the integral can, thus, be securely located at or near the blow-out radius which, not surprisingly, corresponds closely with the lower bound set by the gray scattering limit down to the UV and an upper bound, at least for the outer disk, of $\sim$ 1000 km with a $\sim a^{-3.5}$ size distribution in between. The biggest uncertainty is, of course, on this exponent as even small changes can have an enormous influence on the total mass. With these assumptions, one obtains $\sim 0.1\ M_\odot$ for the total mass of the Beta Pic disk. The optical scattering and far IR emitting grain mass alone is probably more of order of a lunar mass or so (10^{26} g; see also Lecavalier des Etangs et al., 1996). The hotter ($T \sim 300$K) mainly $\sim$ 1 micron size silicate dust located in the inner void at $r \sim$ 15–20 AU contributes an insignificant amount to the total.

As noted above,the critical transition region around $r \sim$ 50–100 AU may hold significant clues as to the nature and origin of this enigmatic structure. With current techniques, this area is just barely but unreliably reached by coronographs and near IR spectrographs. Several and presently little understood phenomena tend to occur in the vicinity. First, there is the SB profile break occurring at $\sim$ 80 AU. Second there is the disk flaring that starts nearby (Artymowicz et al., 1989; Kalas and Jewitt, 1995) and, third, a prominent asymmetry in the 10 micron emission between the NW and SE extensions of the inner disk is noticeable at comparable distances (Lagage and Pantin, 1994). Fourth, a possible change in particle type may occur at $r \sim$ 50 AU as a result of an observed (but as yet unconfirmed) drop in the B band disk flux (Lecavalier des Etangs et al., 1993) and, finally, the outer radius of the central void probably falls somewhere in this same area. If and how any of these phenomena are related is not at all clear at the moment and future observations must be directed to this issue as soon as possible. A number of serious possibilities and even wild speculations have centered on the interpretation of these phenomena the most interesting of which are the effect of the ice sublimation boundary and of gravitational perturbation by planets.

On more detailed inspection several items need to be kept in mind in this context. The SB break may not be due to a particle phase change as the ice sublimation radius is located somewhere around 20–30 AU according to recent calculations (A94) but this result is very sensitive to uncertainties associated with the $T(r)$ relation (some MEM models locate the boundary at 75 AU) and the first solid phase transition occurs at temperatures of $\sim$ 1000K that are reached only in the immediate vicinity of the star (Duschl et al., 1996). Depending on whether or not there is a corresponding albedo change (the B band drop in SB might argue for one), this break might rather reflect a change in particle density or type, as suggested by the parametric fits of Backman et al., 1993. A change in particle density does present some difficulties but neither can be strictly ruled out at this point. In the scenario described by Lecavalier des Etangs et al., 1996, a rather arbitrary cut-off in the particle size distribution due to the extinction of orbiting evaporating bodies (OEBs) close to the star is invoked to explain this feature. On the other hand, the effect of the central void would not be felt on the SB profile as a break but as a maximum located at $r \sim$ 20–30 AU. In summary, we are still far from any truly satisfactory explanation for these phenomena and so they remain mysterious at the moment. Almost certainly, apart from the observational uncertainties, a good part of the problem is due to the inherent inadequacy of the simple one-dust-component-does-it-all character of the models used to very good effect beyond $\sim$ 100 AU and much more observational and modelling effort will have to be expended in the near future to unravel the conundrum.

2.3. NATURE AND ORIGIN

Can anything sensible be said about the nature of the disk seen in Figure 1 from all the things said so far and can it be placed into some sort of evolutionary scheme together with the others? First, I believe evidence is slowly but surely mounting that what we are seeing is a young but already well-formed proto-planetary system in the process of clearing out the debris resulting from the grinding down or erosion of planetesimals by mutual collisions. Evidence is also mounting that the outer disk at least is a faithful, albeit more massive, copy of our own Kuiper belt (KB) in many respects. BP93 had already showed that the fractional luminosity of the Beta Pic disk is strikingly similar to that expected from a then only dimly understood KB (see Table XII of BP93). Presently, with a much better idea of its main characteristics (Jewitt et al., 1996, Stern and Campins, 1996) the similarities in size distribution and thickness, for example, are still noticeable. The differences are likely to be found mainly in the total mass of material in the belt that may be much lower for the much older Sun (Backman et al., 1995). This is consistent with the idea that the Beta Pic disk is almost certainly a transient phenomenon seeing its prodigious flow rate estimated by A94 to approach 10^{-12} $M_{\odot}$/yr implying it may be losing almost a solar system mass in a time scale comparable to its expected age.

Parenthetically, there has been some controversy lately concerning this latter parameter. I computed an age of 2×10^8 yrs on the basis of comparisons with theoretical isochrones (Paresce, 1991) but consistency with these was obtained only if a moderate amount of metal deficiency was assumed unless there is considerable absorption of starlight in the disk, something which is not evident in the data. The metal underabundance seemed to be supported by published calibrations of this star's Geneva photometric indices but recent observations of its UV and optical spectrum are consistent with a solar abundance (Lanz et al., 1995, Holweger et al.,1997). Although on the face of it, this may suggest that Beta Pic, therefore, is not a MS star everything else implies otherwise (even a sub-giant identification made on the basis of luminosity indicators!). A simpler solution is that its distance has been underestimated. With the new Hipparchos-derived value of 19.3 pc (Crifo et al.,1997), the star pops back comfortably onto the theoretical tracks at $t \sim 1\text{–}2 \times 10^8$ yrs.

Although this much might seem easy to deduce,the crucial question still hangs in the air: yes, but what else is lurking in the region? This, of course, is a much more difficult question to answer. The simplest answer could lie in carrying just a little further our tantalizing comparison with our own KB and thence extrapolating to the known solar system or relying on the evident similarities with recent planetary and planetesimal formation scenarios (see, for example, Nakano, 1987 and Bailey, 1994). But this is probably too simplistic and it should not mislead us into expecting or even needing planets such as we know them. Let me review here the basic clues that might point to the existence of large (say > 1 km diameter) bodies in the Beta Pictoris and Vega-like systems:

1) grains larger than ISM size indicating that there must have been in the past favorable conditions leading to agglomeration or sticking of grains rather than just erosion;

2) solid material orbiting rather than falling into or flowing away from the star;

3) material arranged in disks in the stellar equatorial planes rather than in spherical clouds or shells (this is probably true even for Vega although we are probably seeing it pole-on);

4) central, relatively dust-free regions of typical inner solar system size;

5) a necessary reservoir of larger bodies to resupply the observed grain population against rapid removal processes;

6) larger bodies may be required to stir the disk to the observed thickness;

7) unexplained asymmetries in the SB distribution at all spatial scales; and

8) transient spectral gaseous absorption features attributed most naturally to star-grazing comets evaporating very close to the star.

These last two arguments need a little more explanation since they may be critical in this context. The first has been exhaustively examined experimentally by Kalas and Jewitt, 1995, and the possible effects of planet-sized bodies on the dust distribution has been explored theoretically by Roques et al., 1994. Barring a serious perturbation due to a passing star or the effect of an unseen companion that have no supporting evidence but cannot be completely ruled out at the moment, a large planet or a system of planets gravitationally upsetting the inherent orbital symmetry of the emitting grains is the only solution our limited imagination can come up with presently. But exactly by what astounding mechanisms such systematic asymmetries can be established over such a large range of distances (from $\sim$ 10 to 800 AU) with planets that are most likely forced to reside within a few tens of AU of the star remains a complete mystery.

The second argument may be more convincing at least at first appearance but contains a two step process namely that 1) star-grazing comets are responsable for the complex and time-varying absorption line shapes of the gaseous component of the shell surrounding Beta Pictoris (see Vidal-Madjar and Ferlet, 1994 for exhaustive reviews) and 2) that these comets are perturbed into suicidal orbits by a planet of the right size and location (Levison et al., 1994). Recently Lecavalier des Etangs et al., 1996 have refined this scenario (first suggested by Weissman, 1984 for Vega) into a largely self-consistent model of the Beta Pic disk that unifies many of the ideas presented here and has, therefore, important significance as a framework in which new observations and theories must be tested. In this model, both the observed dust and gas disks are the result of the evaporation of km sized bodies (comets or asteroids) that are moving in towards the central star (in other words diffusing from a KB belt to a planet- or planetesimal-dominated region). Then, this mechanism would be naturally quite similar to that occurring in our solar system but does depend critically on the perturbing action of a fully formed proto-planetary system located inside ~ 50 AU. As I noted above, it is still not at all clear that the observed large scale asymmetries can be fully and quantitatively explained by this scenario but it is a good start and one that needs more and better observations to buttress it properly.

2.4. CONTEXT AND EVOLUTION

Now that we think we know what Beta Pic is all about (how many surprises are yet in store?), we should ask where do the other Vega-like stars fit in to this scenario? Starting from the other two true Vega-likes, several points can be made. Observationally, it is pretty clear nowadays that Beta Pic's uniqueness i.e. exceptional brightness stems mainly from its very high dust cross-sectional area including albedo and not to the peculiarities of its inclination or distance (see Kalas and Jewitt, 1996 for a complete justification of this assertion). The upper limits to a scattering disk from coronographic observations of Vega and Fomalhaut by these authors are consistent with the estimates of grain geometric cross sections presented in Table III of BP93 obtained for mid-size grain models fitted to the IRAS data. The very short dust destruction time scales, the age and definite MS status, the absence of substantial amounts of gas and any nearby star forming regions all argue unambiguously that a) the disks are not remnants of the primordial CS nebula (see also Zuckerman and Becklin, 1993) and b) all that separates the CS environment of the three stars is the amount of dust generated by the evaporation of comet-like bodies formed in an earlier phase of planetary evolution. Since, clearly, the amount of dust present in the observable CS environment is in a delicate balance between the rapid destruction processes described in section A1 and the comet evaporation replenishment mechanism, the substantial differences must be due either to differences in the primary source, in the mechanisms giving rise to the evaporation, or simply to age. The rate of the destruction processes cannot be too different as all three stars are dwarfs of roughly the same spectral type.

Does Beta Pic then possess a richer cometary/asteroid environment than the other two or is it more efficient in grinding down its store of planetesimals? It is almost certainly younger than the other two although not by very much so an evolutionary effect is still possible but, at the moment, we have no clue as to the real causes of these exceptional differences. More observations, especially of the presently invisible disks will be critical in this regard as will the detection of similar disks in other stars. A healthy number of Vega-like candidates have been identified in the last few years from IRAS data (see Walker and Wolstencroft, 1988 and BP93 for a reasonably complete list). I should emphasize here that we are not necessarily looking for just any object that has an excess IR emission and some indication that it is not a point source but only those whose optical depths or fractional bolometric luminosities are in the same range (10^{-3} - 10^{-5}) as the three true Vega-likes. Those with very high fractional luminosities are more likely to be very interesting, certainly, in their own right but quite different beasts than the ones under discussion here (see Artymowicz, 1996 for a good discussion of this point). From a preliminary investigation of the statistics of this sample, we can come to the following general conclusions: 1) there are at least 50–60 objects in the Yale BSC with the right characteristics; 2) most objects on this list seem to represent true Vega-like analogs as we defined them here namely that the IR emission may be due to orbiting particles (but until they are resolved we cannot be sure); 3) the phenomenon may be the rule rather than the exception among A and F stars at least; and 4) it appears in all spectral types and is not limited to young objects.

2.5. FUTURE POSSIBILITIES

From this brief but intense exploration of the present status of the MS disks, we can naturally extract the wish list for the near future. First, we need to concentrate on the three resolved sources with better imaging and photometry in order to finally detect and precisely characterize the disks around Vega and Fomalhaut at least to the level we have presently reached for Beta Pic. This would ideally include a determination of the inclination angle, thickness and extent of the disks and the inner voids and a proper characterization of the grain population in order to sharpen our dim understanding of what mechanisms drive the CS environment in these stars. In the specific case of Beta Pic, we need to investigate in much greater detail and accuracy the critical 10–100 AU region corresponding to angular distances of 0.6″–5″. Here we need to do good reflectance and emission mineralogy, precise high resolution imaging to clarify the radial and azimuthal symmetries or lack thereof and good photometry to establish unambiguously the nature of the inner void and its boundaries. It will be very exciting if the expected structures impressed on the dust by a planet could be detected. None of this is going to be easy even for such an exceptionally bright source. The real bottom line here is the availability of super smooth optics that would allow a reduction of the instrumental scattered and diffracted light by at least two orders of magnitude over the present best telescope/coronograph combinations. Of course, other factors also play an important role namely a good seeing and low water vapor site, adaptive optics, anti-blooming CCDs etc. but they are not likely by themselves to change the picture substantially in the next few years.

Unfortunately, there is, presently, little motivation in the astronomical community for building such low scattering devices. Several alternatives are being considered such as an actively controlled coronograph for the HST that would compensate for the low to mid spatial frequency ripple in the HST primary and long baseline IR interferometry with the Keck or the VLTI. It is not yet obvious that these techniques would have the required disk/star contrast in the required $0.6''$–$5''$ range of 10^{-5} or better but they are promising in this regard. Paradoxically, our best chance at breaking the contrast ratio barrier lies in NASA's new emphasis on the detection of extrasolar planets since the chances of such a discovery depend critically on an accurate determination of the background against which they will have to be detected namely the CS disks we have been discussing in this lecture. Let's hope this new interest will last long enough and that enough money is invested in the technology to permit a breakthrough even as a fringe benefit.

Second, we need to spatially resolve more IR-excess stars. The candidates are there as we described in the preceding section and all we need is higher resolution IR devices to obtain the diameter and, possibly, the shape of their CS environment together with the IR spectrum. Then, the typical grain sizes and void sizes should be straightforward to obtain. In principle, ISO should improve the situation considerably and the preliminary indications (Habing et al., 1996) are encouraging but it still remains to be established that it can actually perform much better than IRAS in the spatial resolution domain. There is no question that it will provide tremendous improvement in the spectral resolution domain. SIRTF, if it ever gets off the ground, should do even better. Finally, we need to use these instruments to conduct a systematic deep search for more IR excess objects with the Vega-like luminosity fraction in order to get a robust statistical sample and, perhaps to extend the sample to non MS stars such as "old" PMS and some "young" sub-giants. This would give us a better feel for the possible temporal evolution of the disks i.e. their origin and fate.

3. CS Material around Giants and Supergiants

3.1. MASS LOSS IN POST-MS EVOLUTION

In this very brief consideration of a very complex subject, I wish to concentrate on a few specific issues regarding the potential that very high angular resolution techniques should have on our understanding of the physical mechanisms that govern post-MS evolution and give some recent examples of the type of measurement that will be required in the future. We now have detailed theoretical scenarios concerning post-MS evolution with precise predictions as to the evolution and ultimate fate of a star of any mass or chemical composition. What are missing, by and large, are accurate experimental techniques to test these predictions on any single stellar specimen. We have relied so far on observations of small samples of very different stars in widely different evolutionary phases to try to piece together the possible past evolution of that sample. In particular, a star's position on the HR diagram with respect to that of other objects expected or hoped to be of the same kind and to theoretical evolutionary tracks usually yield only a rough idea of the path it has taken so far and that it will take in the future.

Since mass loss seems a natural and inevitable product of most post-MS evolutionary scenarios, it might seem reasonable to explore the possibility that this effect which accompanies almost every twist and turn of the star's life could be exploited to tell us what the star has been through as it winds its way across the HR diagram. This might, then, represent a very powerful way to confirm and sharpen our understanding of the physical mechanisms that affect this critical stage of a star's life. In other words, every significant event in the star's life should be impressed on its CS environment with a characteristic signature whose proper interpretation should allow us to decipher its timing, strength and origin. Conversely and, perhaps, just as importantly this study would shed considerable light on the physics of the mass loss process itself which has, presently, still many uncertainties and ambiguities. Of course, this is all straightforward to assert but is it a practical and realistic goal? My aim here is to show that in many cases it is and that current high resolution observations with HST, for example, clearly suggest that it is a technique that is ripe to be fully and effectively exploited provided the necessary theoretical and technological tools become available.

The key parameters driving the intensity and character of the mass loss process and, therefore, the strength of the CS signature are the stellar luminosity and temperature which determine the mass loss rate and, especially, their rate of change as a function of time. During the time in which the star is blowing a steady thermally or pressure driven wind, we can expect that the CS density varies as r^{-2} out to an outer radius set by the wind speed that can be determined spectroscopically in most cases and by the time interval over which it has been blowing. In the case of red giants, an inner boundary to the dust shell will be located at the condensation radius ($\sim$ 1–1.5 R_*) but when the wind is suppressed for some reason, the inner CS radius would be detached from the star by an amount proportional to its speed and the time since it stopped. In both cases, a precise determination of these boundaries immediately furnishes critical evolutionary time scales. Left to itself, this process would set up a spherically symmetric nebula whose visibility will depend on distance, emitting and scattering area of the gas and dust, stellar luminosity etc. Variations on the theme would certainly include some sort of spatial asymmetry due to rotation, pre-existing CS medium that hampers the initially spherically symmetric flow, presence of companions etc. (Livio, 1994).

For massive stars, this will already give rise to an observable CS structure during MS evolution (Langer, 1994) but for intermediate and low mass stars, the nebula is expected to become consistent and, therefore, detectable only in the later phases when the luminosity increases substantially as the star climbs the RGB and AGB (Iben, 1985). Thus, two radically different wind types can be foreseen depending on the stellar mass or luminosity. The first would be the slow ($\sim$ 10 km s^{-1}), dense wind containing dust grains expected from a red giant or supergiant while the second would be a fast and low density essentially pure gas wind blowing from a blue giant or supergiant. This might be expected to be the typical canvas on which any future activity would express itself. Then, in a very general way, the typical CS environment around post-MS stars might best be described by a colliding wind scenario in which the various excursions across the HR diagram that any star of reasonable mass is expected to be subjected to by the changes in its internal structure (but never observationally confirmed, by the way) give rise to a series of prominent shells of swept-up dust and gas. Their location, morphology, and chemical composition all retain fundamental information on a previous state of the parent star.

To get a general feel of when and how this might happen, one can look for instances in which large and relatively sudden changes to the mass loss rate may occur during a star's evolution. A quick overview would reveal the following possibilities: 1) relatively fast and repeated excursions of low mass stars up and down the RGB and AGB due to shell burning episodes that result in mass loss rate gradients; 2) red-blue-red excursions of massive and intermediate stars that change the temperature of the atmosphere and input spectrum of the radiation field that drive the mass loss; 3) repeated crossings of the stellar instability strip giving rise to radial pulsations that help create and drive dust into the CS environment and form shells of CS material; 4) shell flashes and thermal pulses that modify the star's luminosity on short time scales; 5) excursions across the atmospheric stability boundary giving rise to the LBV phenomenon; 6) instabilities due to an interacting companion that trigger thermonuclear runaways (novae) and, of course; 7) final cataclysmic events such as core collapse that sends matter spewing out into the surrounding CS environment.

At some level, each one of these possibilities has fundamental implications for the CS medium as each event should imprint on it its own spatially and dynamically distinct features and their detailed investigation should reveal much about the events themselves. For example, the chemical composition of each feature (shell, disk, ring, jet etc) should reflect chemical changes due to dredge up events, the Y and Z dependence of horizontal excursions, the products of novae outbursts and the grain chemistry and formation rate during pulsations. The spatial structure should reflect the original atmospheric and CS conditions and the hydrodynamics of the resultant flow and the eventual asymmetries like jets or bipolar outflows, the collimation mechanisms, magnetic fields and possible binary nature of the source region. The locations and speeds of features also provide accurate information on the time scales (time since onset or end, duration of event etc) and together can yield accurate distances. Finally, the interaction with the ISM sheds light on the chemical evolution of the region by its degree of heavy element enrichment. In the next few paragraphs, I will discuss very briefly some practical examples of these general principles.

3.2. CS STRUCTURE AROUND MASSIVE STARS

While still on the MS or as it evolves off the MS to become a blue supergiant, a massive star ($> 10\ M_{\odot}$) will create around itself a hot bubble or cavity of ionized gas contained within a thin, dense shell of swept up wind and ISM material. For typical time scales, the shell will be located at $\sim$ 20–30 pcs from the star (Castor et al., 1975). Although these shells are quite prominent around WR stars, for example, they are not easily detectable in less massive stars. But, if then, the star undergoes a rapid swing across the HR diagram because of this mass loss to become a red giant, the dense wind begins to flow inside the bubble and, for the typical red giant lifetime, may reach out to $\sim$ 10 pcs from the star. In the process, of course, the star will traverse the pulsational instability strip and, therefore, eject one or more shells of gas that will also plow into the red giant wind. All these structures can be detected in X-rays, radio, or IR and optical emission lines provided the sensitivity and resolution is high enough.

Moreover, as most likely occurred to the SN 1987A progenitor Sk202-69 because of its low metallicity (Woosley, 1988), another swing back to the opposite side of the HR diagram occurs, it will restart the fast, hot wind that will now compress the dense gas on the leading edge of the shock into another much denser shell located at 1–2 pcs from the star and enclosing another smaller cavity. It might also cause the shells ejected from each passage through the instability strip to fragment. In the particular case of SN1987A, of course, the brilliant flash of light from the explosion

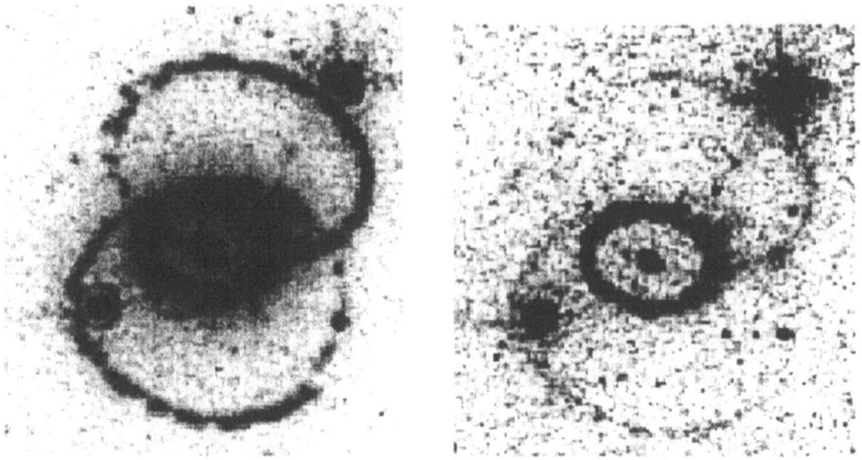

Figure 2. An HST image of the CS environment around the SN1987A as seen through a filter centered at the [NII] 6584A (left panel) and the [OIII] 5007A (right panel) lines (R.Gilmozzi,private communication). The lower [NII]/[OIII] brightness ratio of the outer rings compared to the inner one indicates a smaller N/O abundance in the outer rings.

quickly traversed the CS region exposing its entire life history as it were in one fell swoop (McCray and Wei, 1993) and, in fact, confirming the then surprising conjecture that it had been a red giant in an earlier phase. The spectacular HST picture of the real SN1987A CS environment is shown in Figure 2. From the previous description, one should not be too surprised at its complexity and that it might be rather difficult to interpret. But, although the details still remain to be understood (see Burrows et al., 1996 for a valiant attempt), much has been already learned from the study of the main ring and its spectrum (Panagia et al., 1996). Since there is no particular reason to believe there was anything peculiar about Sk202-69 apart from its metallicity, we should expect that any other similar object would exhibit such complex and revealing CS structure with which to read its past history.

Another important class of objects that exhibit large changes in physical conditions during their lifetime and evidence of episodes of large mass loss is that of the Luminous Blue Variables (LBV, Humphreys and Davidson, 1994) that serve as excellent examples of the concepts sketched above. In this case, the precise cause of the outbursts is still uncertain but dynamical instabilities in the outer layers due to traversal of the empirical temperature-dependent luminosity boundary for hot stars is a prime suspect. This instability may actually prevent these massive ($> 50\ M_\odot$) and most luminous ($L \sim 10^6\ L_\odot$) stars from moving all the way across the HR diagram into the red supergiant region but less luminous LBVs, on the other hand, are most likely to have originated in the red giant region. In any case, these crossings should give rise to some very spectacular mass loss episodes which must be reflected in their CS environment. The exceptional cases of Eta Carinae and P Cyg are well known through many years of high spectral and spatial resolution observations. Even so, because of its greater distance, P Cyg has only recently started to yield its CS secrets (Nota et al., 1995).

Another LBV whose CS environment has been explored in considerable detail in the optical and IR with high resolution imaging and coronography is AG Car. Recent HST images taken in the NII line and continuum are shown in Figure 3. The morphology and possible origin of this nebula were discussed in detail on the basis of earlier ground-based images and spectra by Paresce and Nota (1991) and Nota et al. (1992). The most exciting aspect of these images is the clear and very disturbing difference betwen the gaseous nebula as revealed by the NII emission line image and the dust nebula seen in the continuum image. This suggests that whatever mechanism is responsible for forming the dust cloud (whether it is the effect of dynamical interactions or of source asymmetries is hard to tell yet) must be decoupled from the gaseous component to a surprising degree. Of course, it is quite unlikely that the dust was formed by condensation in a cool region of the lower atmosphere of the star and then blown out into the observed CS region by radiation pressure as is the case in normal red giants since the withering UV emission of such a hot star would destroy it on extremely short time scales. Consequently, the grains must have actually been created in local pockets of very high density CS gas located probably in turbulent regions behind shocks driven through the various shells and subjected to RT type instabilities. Only there would the high opacity of the gas be sufficient to shield them from the destructive effects of the radiation field.

The physical appearance and spatial distribution of the CS dust lend qualitative support to this hypothesis. The disconnected fragments that seem to make up the main part of the structure of the dust clouds seen in the images in Figure 3 and, especially, the limb brightened dust rings or shells embedded in the immediate vicinity of the brightest parts of the gaseous shells may be due to such protected areas, the exposed surfaces of which are being ablated away by the wind and radiation of the star. This is a situation very reminiscent of the famous proplyds recently detected in the Orion Nebula (O'Dell and Wen, 1994). The gaseous nebula morphology is also very instructive since it clearly indicates the existence of several radially and azimuthally asymmetric shells due to three different ejection episodes (Nota et al., 1996). Most of the mass of gas and dust seems concentrated in a roughly NE-SW plane where the nebulae are pinched radially while it

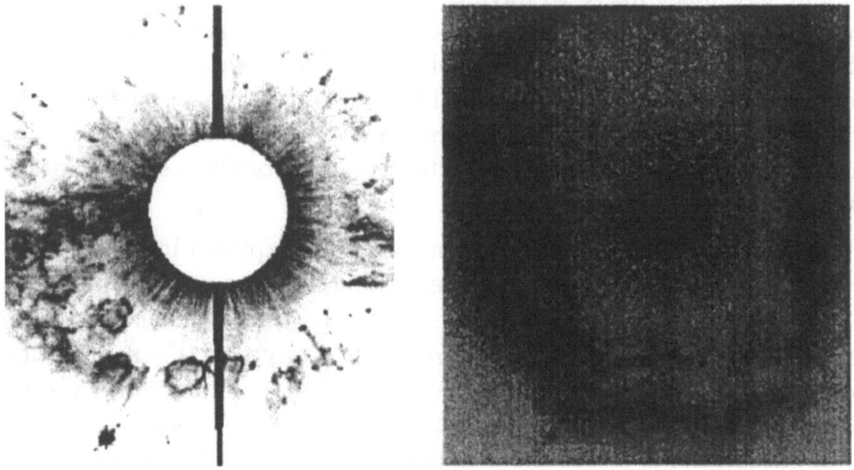

Figure 3. HST images of the LBV AG Car taken in the line of [NII]6584 (right panel) and in the V continuum (left panel). The field of view is 36″ × 36″. Data courtesy A. Nota and M. Clampin.

has expanded more freely in the opposite (polar?) direction, another good example of the effects of the expansion of a wind within a pre-existing density enhancement along an equatorial plane. The extent and position of the shells and the measured velocities are quite consistent with time scales of $\sim 10^4$ yrs, a typical time scale associated with the LBV instability process. From a detailed analysis of images like these, it may be possible to tell whether or not the star needs to be taken back to an earlier cooler evolutionary phase in which it was a red or yellow supergiant and/or if it has a companion that collimates the ejecta and helps form the dust.

3.3. CS STRUCTURE AROUND LATE-TYPE STARS

As the star moves up and down the RGB and AGB, the mass loss rate varies from 10^{-7} to $\sim 10^{-4}$ $M_{\odot}$/yr on time scales of 10^{6-7} yrs. Intermediate mass stars (~ 5 $M_{\odot}$) can also swing across the HR diagram towards the blue and cross even repeatedly the instability strip and become for pulsating Mira variables before they move back up the AGB (see Iben, 1985, for example). As we have amply discussed above, any or all these episodes will dump substantial amount of gas and dust into the star's CS medium but it is especially in the pulsational phases that most of the dust is expected to be produced. In normal conditions, it is very difficult to get the extended atmosphere of a red giant to be both cool and dense enough at the same time to produce dust by the standard process of nucleation and subsequent growth. The only way this could happen is by a combination of the two processes: levitation of the atmosphere by shock dissipation during a pulsation and consequent dust formation in the resulting enhancement followed by radiative acceleration outwards. Frictional gas-grain coupling then entrains the gas along with it. This is all clearly a very tricky and only dimly understood sequence of events that have little or no direct observational support. Thus, perhaps, the biggest challenge in this field today is to explore the inner domain of the dust generating region in these variables in order to understand how dust is produced and what propels it radially outwards.

This will, of course, require very high spatial and spectral resolution in the IR so as to penetrate the absorbing overlying dust. A considerable amount of work in the spectral domain has been accomplished already and will continue with ISO but the critical optical and IR imaging techniques lag far behind. This situation should improve substantially in the near future with NICMOS on HST and the array of IR imagers with adaptive optics planned for the new generation of large ground-based telescopes. In the meantime, we must content ourselves with tantalizing glimpses into the inner workings of the most extended CS envelopes of these objects or the ones that have exposed them to view thanks to repeated eruptions or the effect of mass transfer from a companion. Optical imaging of IRC +10 216 is a good example (Crabtree et al., 1987) where one can start making out radially distinct structures corresponding to several 200–400 yr duration outbursts.

The symbiotic variables which in general consist of a binary system with a Mira LPV as the primary and a hot subdwarf as the secondary represent another useful diagnostic tool in this quest. Quite a few of these objects have been studied by means of direct imaging or bi-dimensional imaging spectroscopy with exciting results. HST, in particular, has allowed considerable progress in this field by direct imaging of a number of them especially of R Aquarii, one of the closest at $\sim$ 200 pcs. Here again all the complex evolutionary history of this object lies exposed in the spectacular CS nebula that surrounds this object (see Burgarella et al., 1992 for a recent review). The crucial issue here is to attempt to peer directly into the wind and jet generating engine, a task which is tantalizingly close to reality due to its proximity (at this distance 1 milliarcsec corresponds to 0.2 AU!). The most recent UV images with $\sim$ 10 AU resolution are shown in Figure 4 taken from a recent paper by Paresce and Hack, 1994. Here, for the first time, one can clearly make out the Mira itself (indicated by M in the Figure) and the beginning of the very narrow jet spewing out from M towards the NE and SW where it interacts with fragments of material at

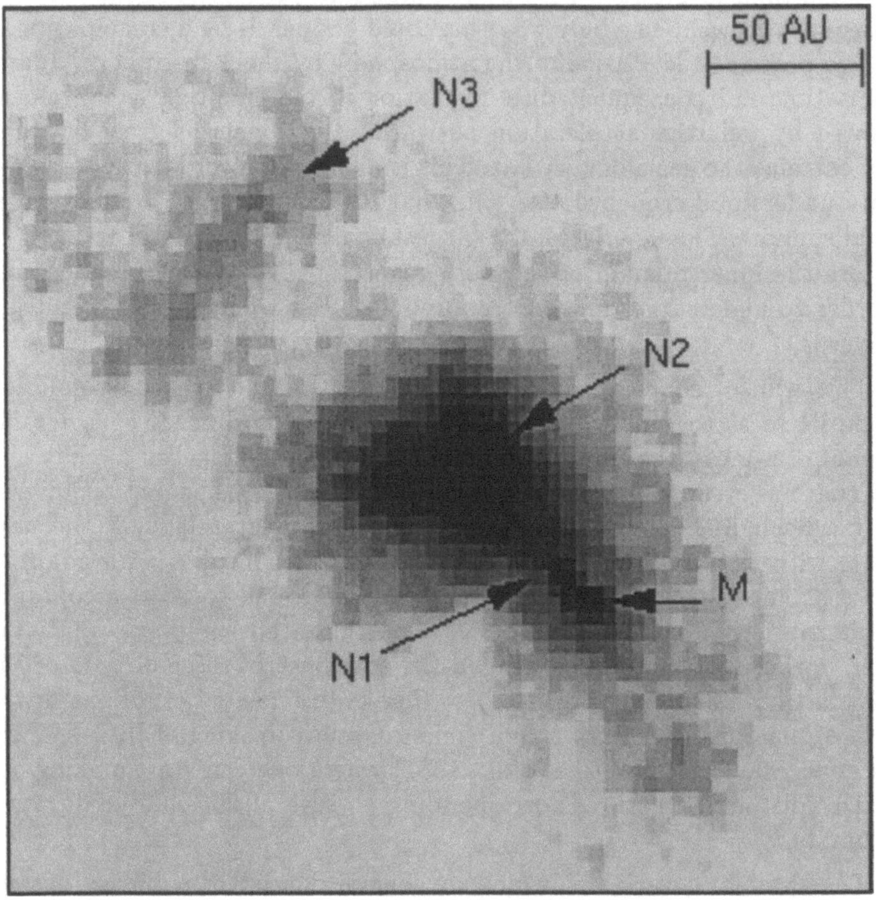

Figure 4. The central 1.4″ × 1.4″ region of the symbiotic variable R Aqr taken in UV light at ~ 1900 A from Paresce and Hack, 1994. North is up and East to the left

N_i and S_i.

The observations shown in Figure 4 taken together with a number of other measurements are starting to delineate the processes by which the Mira loses mass. They imply that the hot companion is presently blowing an energetic wind into the Mira's CS environment at a rate of $\sim 10^{-8}\ M_{\odot}/\mathrm{yr}$. The wind encounters a density gradient where it is collimated and expands supersonically towards N2 in the Figure. The pulsating wind from the Mira probably modulates the physical conditions in this area giving rise to the observed temporal variations in the flow. The wind then undergoes RT-type instabilities to break up into knots and interact with density gradients in the Mira wind. To explain the hour glass morphology and dynamics of the outer regions of the nebula one must invoke the effects of at least two recent outburts from the Mira occurring in succession several hundred years ago (Solf, 1992). To really confirm this scenario in all its details will require another factor of ten improvement in resolution as might be expected for the planned new technology HST and the large interferometric arrays in Hawaii and Chile.

3.4. CONCLUSIONS

The colliding wind scenario has been eminently successful so far in explaining many of the features observed in the CS environment of many different types of stars especially for planetary nebulae (PNe) that are the easiest to detect (Frank et al., 1993). The actual processes that set up and drive the winds in the first place are still shrouded from view, however, and it is here that new techniques such as the ones very briefly described in this lecture will have to be developed and applied. Rather than having to deal with the complex superposition of many effects integrated over the whole lifetime of the star, the objective is to be able to detect and characterize each mass loss event separately or in sequence at the earliest possible time when the action is closest to the star. For objects at 1 kpc distance, this will require resolutions of the order of milliarcseconds in narrow passbands and over several decades of energy. Many space and ground based instruments are planned to exploit this opportunity so the next 10–20 years should be extremely exciting in this regard. The goal of determining observationally the nature of a star's evolutionary path through the HR diagram is well worth the effort. A parallel effort to develop the theoretical tools required to interpret the observations is another very critical need including more sophisticated simulations of what happens when two winds collide or interact with an accretion disk or shell. An example of our ignorance is written all over the image shown in Figure 2 where we still cannot imagine how the object created the inner bright ring and not a shell as one might have expected. Was this how it started expanding or has the top and bottom of the shell been blown out by some mysterious mechanism? These and many others are the intriguing questions we leave for the younger generation at this school to answer.

References

Artymowicz, P., Burrows, C., and Paresce, F.: 1989, ApJ, 337, 494

Artymowicz, P.: 1988, ApJ, 335, L79

Artymowicz, P.: 1994, in "Circumstellar Dust Disks and Planet Formation", R. Ferlet and A. Vidal-Madjar (eds.): 1994, p. 47 (A94)

Artymowicz, P.: 1996, in "The Role of Dust in the Formation of Stars", H.U. Käufl and R. Siebenmorgen (eds.), Springer Verlag, Berlin, in press

Aumann, H.H., et al.: 1984, ApJ, 278, L23

Backman, D.E., and Paresce, F.: 1993, in "Protostars and planets III", U. Arizona Press, p. 1253 (BP93)
Backman, D.E., Gillett, F.C., and Witteborn, F.C.: 1992, ApJ, 385, 670
Backman, D., Dasgupta, A., and Stencel, R.E.: 1995, ApJ, 450, L35
Bailey, M.E.: 1994, in "Asteroids, Comets, Meteors 1993", A. Milani, M. DiMartino and A. Cellino (eds.), Kluwer Academic Publishers, Dordrecht, p. 443
Burgarella, D., Vogel, M., and Paresce, F.: 1992, A&A, 262, 83
Burrows, C., et al.: 1995, ApJ, 452, 680
Castor, J.I., McCray, R., and Weaver, R.: 1975, ApJ, 200, L107
Clampin, M., Robberto, M., Paresce, F., and Maccioni, A.: 1994, in "Circumstellar Dust Disks and Planet Formation", R. Ferlet and A. Vidal-Madjar (eds.): 1994, p. 67
Crabtree, D.R., McLaren, R.A., and Christian, C.A.: 1987, in "Late Stages of Stellar Evolution", S. Kwok and S.R. Pottasch (eds.), Reidel Publishing Co., p. 145
Crifo, F., Vidal-Madjar,A., Lallement,R., Ferlet,R., and Gerbaldi,M.: 1997, A&A, in press
Diner, D.J., and Appleby, J.F.: 1986, Nature, 322, 436
Duschl, W.J., Gail, H.-P., and Tscharnuter, W.M.: 1996, A&A, 312, 624
Ferlet, R., and Vidal-Madjar, A. (eds.): 1994, "Circumstellar Dust Disks and Planet Formation", Editions Frontieres: Gif sur Yvette
Frank, A., Balick, B., Icke, V., and Mellema, G.: 1993, ApJ, 404, L25
Golimowski, D.A., Durrance, S.T., and Clampin, M.: 1993, ApJ, 411, L41
Greenberg, J.M., and Li, A.: 1996, in "The Role of Dust in the Formation of Stars", H.U. Käufl and R. Siebenmorgen (eds.), Springer Verlag, Berlin, in press
Habing, H., et al.: 1996, A&A, 315, L233
Holweger,H., Hempel,M., van Thiel,T., and Kaufer,A.: 1997, A&A, in press
Humphreys, R.M., and Davidson, K.: 1994, PASP, 106, 1025
Iben, I.: 1985, Q.J.R.A.S., 26, 1
Jewitt, D., Luu, J., and Chen, J.: 1996, AJ, in press
Kalas, P., and Jewitt, D.: 1997, Nature, 386, 52
Kalas, P., and Jewitt, D.: 1996, AJ, 111, 1347
Kalas, P., and Jewitt, D.: 1995, AJ, 110, 794
Knacke, R.F., et al.: 1993, ApJ, 418, 440
Lagage, P.O., and Pantin, E.: 1994, Nature, 369, 628
Lagrange-Henri, A.M.: 1995, Ap&SS, 223, 19
Langer, N.: 1994, in "Circumstellar Media in the Late Stages of Stellar Evolution", R.E.S. Clegg, I.R. Stevens and W.P.S. Meikle (eds.), Cambridge U. Press, p. 1
Lanz, R., Heap, S.R., and Hubeny, I.: 1995, ApJ, 447, L41
Lecavalier des Etangs, A., Vidal-Madjar, A., and Ferlet, R.: 1996, A&A, 307, 542
Lecavalier des Etangs, A., et al.: 1993, A&A, 274, 878
Levison, H.F., Duncan, M.J., and Wetherill, G.W.: 1994, Nature, 372, 441
Livio, M.: 1994, in "Circumstellar Media in the Late Stages of Stellar Evolution", R.E.S. Clegg, I.R. Stevens and W.P.S. Meikle (eds.), Cambridge U. Press, p. 35
McCray, R., and Wei, L.H.: 1988, in "Structure and Evolution of Galaxies", Fang Li Zhi (ed.), World Scientific Pub. Co., p. 204
Nakano, T.: 1987, MNRAS, 224, 107
Nakano, T.: 1988, MNRAS, 230, 551
Norman, C., and Paresce, F.: 1989, in "The Formation and Evolution of Planetary Systems", H. Weaver and L. Danly (eds.), Cambridge U. Press, p. 151
Nota, A., Clampin, M., Garcia-Segura, G., Leitherer, C., and Langer, N.: 1996, in "Science with HST-II", P. Benvenuti, F.D. Macchetto and E.J. Schreier (eds.), p. 398
Nota, A., Leitherer, C., Clampin, M., Greenfield, P., and Paresce, F.: 1992, in "Non-isotropic and Variable Outflows from Stars", L. Drissen, C. Leitherer, and A. Nota (eds.), ASP Conf. Ser., 22, p. 377
O'Dell, C.R., and Wen, Z.: 1994, ApJ, 436, 194
Paresce, F.: 1991, A&A, 247, L25
Paresce, F., and Hack, W.: 1994, A&A, 287, 154

Paresce, F., and Nota, A.: 1989, ApJ, 341, L83
Panagia, N., et al.: 1996, ApJ, 459, L17
Roques, F., Scholl, H., Sicardy, B., and Smith, B.: 1994, Icarus, 108, 37
Sicardy, B.: 1994, in "Asteroids, Comets, Meteors 1993", A. Milani, M. DiMartino and A. Cellino (eds.), Kluwer Academic Publishers, Dordrecht, p. 429
Solf, J.: 1992, A&A, 257, 228
Stern, A., and Campins, H.: 1996, Nature, 382, 507
van der Blick, N.S., Prusti, T., and Waters, L.B.F.M.: 1994, A&A, 285, 229
Walker, H.J., and Wolstencroft, R.D.: 1988, PASP, 100, 1509
Weissman, P.: 1984, Science, 224, 987
Weissman, P.: 1995, ARAA, 33, 327
Woosley, S.A.: 1988, ApJ, 330, 218
Zuckerman, B., and Becklin, E.E.: 1993, ApJ, 414, 793

LIFE ON EXTRA-SOLAR PLANETS

A. LEGER
Institut d'Astrophysique Spatiale, CNRS, bat 121
Université Paris-Sud - F-91405 Orsay, France
e-mail: leger@iaslab.ias.fr

1. What Could be Alien Life?

The search for a good tracer of alien life requires first a general definition of what is life. Biophysicists and biochemists agree to define a living being as a system that:

(1) contains information (negentropy),

(2) is able to replicate itself and

(3) undergoes few random changes in its information package that allow a Darwinian selection of the most performing.

Searching for such beings several lightyears away from us seems, a priori, an hopeless task. We show that, fortunately enough, this is not the case.

Trying to avoid being biased by the nature of Life on Earth, one can consider different possibilities for the support of this information. Organised physical structures as those of the magneto-optic or semiconductor memories of our computers could be thought of, but the building of such systems by natural processes appears unlikely. A more attractive possibility is a chemical support: the coding by a sequence of chemical entities (chain cells) in a linear macromolecule which behaves as a message written with letters. A replication process can be imagined provided that cells have homo or hetero affinities - of course, we have in mind the example of DNA but this can happen with quite different macromolecules - .

2. When is a Planet Habitable?

If the search for exo-lives is restricted to chemistry based ones, the concept of habitable planet results as well as a possibility of evidencing it from distance.

A-M. Lagrange et al. (eds.), High Angular Resolution in Astrophysics, 285–292.
© 1997 *Kluwer Academic Publishers.*

Macromolecules can react rapidly, only if they are in solution . Solid state reactions are too slow and large molecules cannot stand the temperatures required for them to be in the gas phase. As pointed out by A. Brack [1], among the different possible solvents, liquid water is a special one. It has a very high dielectric constant, e = 80, that allows salt ionisation and, most important, has the capability of building H-bonds with dissolved molecules. The latter property allows very specific conformation of macromolecules by attraction of their hydrophilic groups (OH, CO, COOH...) and repulsion of the hydrophobic ones (CH, CH3, aliphatic chains...). These specific conformations of macromolecules are favourable to specific chemical reactions as "key-keyhole" ones which are valuable to build reproducible complex structures with a rich information content. Alternative solvents as liquid hydrocarbons, alcohols, liquid NH3 are less favourable than water because of a lower e or/and lack of H-bonding.

Chemically, H2O has some activity, i.e. hydrolysis, that can be important to select between different chemical pathways by destroying the products of some of them [1]. However, its activity remains moderate as opposed for instance to liquid NH3 that attacks essentially any organic compounds.

Last but not least, T. Owen [2] has pointed out that water is also appropriated for a subtle reason: it is indirectly more resistant to UV because some of its photolysis products, O2 and O3, protect it from further attack.

The definition of the habitable zone (HZ) around a star results: it is a region where a planet can sustain liquid water at its surface, at least during some time of its local year.

J.Kasting et al [3] have further proposed the concept of continuously habitable zone (CHZ) requiring that such a situation lasts long "enough" for life to evolve towards elaborated structures. This excludes massive stars which have a short life and possibly small stars where habitable planets are so close to them that their spinning and orbital rotations are phase locked by tidal forces which may lead to unfavourable conditions for life appearance and development, although detailed climate models have to settle this point [4]. The authors concluded that mid to early K stars and G ones (0.78 Mo $<$ M* $<$ 1.1 Mo) were optimal candidates for having a CHZ.

3. How to Prove Life by Remote Sensing?

Fortunately, operational tools for remote detection of life appear if we consider the possible nature of the macromolecules supporting the living being's information [2]. To allow replication, labile bonds are favourable and possibly necessary. Then, carbon based chemistry, organic chemistry, appears to be the most powerful. Carbon can be easily oxidised (CO2) or reduced (CH4) which allows a great variety of chemical species. This property is unique. Even Si, the element most similar to C in Mendeleev periodic table, makes much stronger bonds with O than with H and has a poorer chemistry.

This richness of C chemistry is confirmed by the study of a medium where physical conditions (pressure, temperature...) are quite different from those in the laboratory, the Interstellar one: among 112 species identified in 1996, 84 contain carbon and only 8 silicon.

Consequently, restricting our quest for life to carbon chemistry based ones is probably not too severe and T. Owen [2] concluded that our present understanding of life requirements and planetary conditions provides us with some real support for carbon-water chauvinism.

It is thought that carbon was mostly fully oxidised (CO2) in the primitive atmosphere of terrestrial planets, 108 yrs after their formation [5]. If a carbon based life has developed at a large scale on a planet, it will have required a large amount of organic molecules and made it from the only available raw material, CO2. This implies the reduction of the latter by some process, as photosynthesis or the use of planet internal heat, according to a reaction scheme:

CO2 + H2O + energy —> (CH2O) + O2,

that releases free oxygen. As this gas is very reactive and oxidises iron or sulphurs contained in planet rocks, if not continuously replenished, it would disappear rapidly from an atmosphere, specially if plaque tectonic and volcanism are present which bring fresh reducing material from the planet interior to its surface. On Earth, this would happen in 2 107 yrs [6].

The massive presence of free O2 ($P > 10$ mbar) and H2O in an exoplanet atmosphere appears then as a criterion for the presence of C based life.

To qualify this criterion, it must be shown that abiotic processes cannot provide free oxygen as well. This is a key question that has already been addressed but requires additional studies in view of its first importance. Photodissociation of CO2 has been considered [7] but only a small amount of O2 (< 5 mbar) can be produced under the most favourable conditions. A more plausible possibility is H2O photodissociation by the star UV, followed by H escape from the atmosphere. On terrestrial planets, i.e. rocky ones with temperature close to 300 K, this process is strongly reduced by the presence of a cold trap at the tropopause which blocks most of the ascending water vapour and makes it fall back to the planet surface as rain or snow before it has reached the UV rich upper regions. In addition, the escape of atomic H may be not as easy as initially expected [9]. As a result, 99.9999% of breathable oxygen on Earth is produced by photosynthesis and only 1 ppm comes from H2O photodissociation although the latter species is quite abundant [8]. However, an efficient production of abiotic 02 seems possible when a planet undergoes a runaway greenhouse effect, as Venus did, because the tropopause cold trap no longer exists and the H escape is favoured, the more as the planet has no magnetic field to protect it from its star's wind [9]. This process has be considered [2] [10] with the conclusion that the detection of 02 in the atmosphere of a planet located close to its stars would be of little relevance but, if the planet is located at the outer part of its HZ, such a detection, simultaneously with that of H2O, would be meaningful.

As early as 1975, J. Lovelock [11] has proposed that a good criterion for life on an exo-planet is the presence of atmospheric gases very far from a thermodynamical equilibrium, e.g. 02 and CH4, as they exist on Earth. This idea has also be used by C. Sagan [12] when showing that an observer, using Galileo probe instruments would have evidenced life on Earth. We do not think that, this is an "absolute " criterion either, because a planet is a thermodynamically open system and purely abiotic processes can produced out of equilibrium products, e.g. OH and H as the result of H2O photodissociation. Accurate planetary models has to be worked out in order to qualify or falsify this criterion as any other.

4. Ozone, a better tracer than Oxygen

Considering the massive presence of 02 and H2O as a good criterion to evidence exo-life, T. Owen [2] has proposed to search for the spectroscopic signatures of the former gas in the visible (A and B-band at 760 and 720 nm, respectively). In practice, this proposal is not very operational because it faces a huge difficulty: in the visible, the star is much brighter than the planet (5 109 for Sun / Earth) and the spectrum of the latter would be extremely difficult to obtain.

R.Bracewell [15] has pointed out that this contrast is more favourable in the IR (7 107 at 10 mm for Sun/Earth) than in the visible and R. Angel, A. Cheng and N. Woolf [16] showed that the mid IR region is of special interest because it contains many valuable

spectroscopic signatures: 15 mm (CO2) whose presence or absence would indicate a major similarity or difference with the atmospheres of solar terrestrial planets, 6-8 mm (H2O) telling whether it is habitable or not and 9.6 mm (O3) whether 02 is present, as the former species is made from the latter. The IR spectrum of the Earth illustrates the informative power of this part of the spectrum (Fig. 1). Observing the Solar terrestrial planet spectra would indicate that Venus, the Earth and Mars contain CO2 but the Earth alone is habitable (H2O) and inhabited (O3).

In addition, the ozone 9.6 mm band intensity is a very sensitive tracer

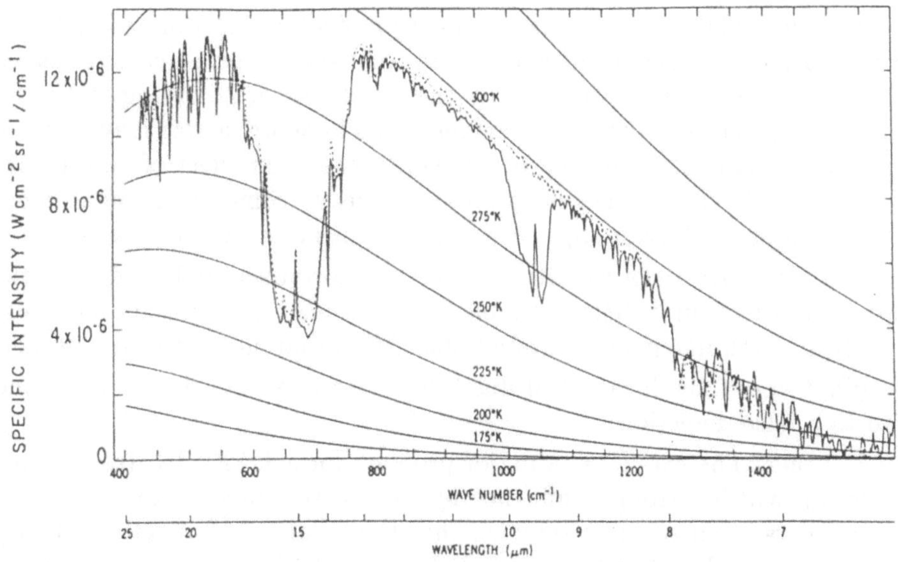

Figure 1. Earth emission as measured from Space over the Pacific Ocean near Guam by Nimbus 4 [13]. The dotted curve is a model calculation that does not include the ozone component. Note the strong CO2 band at 15 mm, which is present in the spectra of all Solar terrestrial planets that have an atmosphere, the O3 band at 9.6 mm and the structure shorter than 8 mm due to H2O . The latter two spectral features are specific to the Earth and reveal that the planet is habitable (H2O) and that a photosynthetic activity takes place on it at a large scale (O3). (Courtesy of Hanel [14])

of O2. The dependence of the former gas abundance upon the latter one is not linear but logarithmic [17] [18] [19]. On Earth, a small amount of O2 (10 mbar) would lead to a O3 abundance only 1.7 times smaller than the present one [19]. The O2 bands would be much more reduced (by a factor 20), making the detection of the latter species more difficult than that of the former. As an illustration, an outer observer of the Earth with the capability of measuring these bands at their present level with a signal to noise ratio of 12, would have detected life on Earth (at 7 sigma) for 2 Gyr if observing the O3 band but only for 0.5 Gyr if observing the O2 ones (20).

Space missions dedicated to this remote sensing of life indices have been described [21] [22] and are presently under study by ESA (DARWIN mission) and NASA (Planet Finder).

5. Discussion

Richard Brockie: In the Solar System, most of the mass is in the Sun, whilst most of the angular momentum is in the orbit of Jupiter. Would the absence of a "Jupiter" containing the angular momentum of an extra-solar system have consequences on an "Earth" in the same system?

A. Léger :The excess angular momentum of a forming star can be transferred to its surrounding disk that can, or cannot, form giant (gaseous) planets, before been swept away by the stellar wind, according to the respective time scales of the two phenomena. Therefore, a star may have, or not, giants planets. According to Wetherill, the main effect of Jupiter like planets. on Earth like ones is to present a heavy cometary bombardment after the first 0.5 Gyr or so, a necessary condition for life to evolve.

F. Malbet :How can we estimate the of exo-zodiacal light? Most of the dust is seen in young stellar systems. Does it mean that it would be easier to detect Earth-like planets around older stars than around young ones?

What about the location of dust? Is it possible that clumps exist and lead to a false detection?

A. Léger :The exo-ZL is a major problem for a DARWIN like mission. Its design will be strongly influence by the fraction of stars that have a Z cloud orders of magnitudes stronger that the Solar one. Indications should be provided by ISO but likely, we shall need either a Ground based measurement at 10 mm with the new generation large telescope interferometers (VLTI, KECK, Magellan, LBT...) or possibly a Space precursor mission (a nulling interferometer, as suggested by Bracewell).

Concerning false planet detection due to Z dust clumping: in the Solar System, such clumps are observed to be fainter at 10 mm, than the Earth. Larger Z clouds may lead to brighter clumps but there are likely to be build and maintained by planets. If alone, they are unstable, due to differential rotation with distance to the star. Anyhow, their spectra should be clearly distinguishable from those from a planet with an atmosphere (possibly silicate 9.7 mm feature but no CO2 15 mm one).

I. Trifalenkov: I think that there are two advantages in considering microlensing search for planets around background stars, not lensing stars. This advantages are :

1) Events are more probable

2) Light curve, in such cases are easier to interpret.

What is your opinion?

A. Léger :Your suggestion is interesting, however I do not think it is operational because source stars have to be very bright, e.g. giants, to be observable 8 Kpc away and therefore are not likely to have planets.

S. Loiseau :Listening to you, it seemed to me that we know a great deal about life and, in particular, about information storage and propagation : I thought we were in the early stages of that kind of knowledge, could you comment on that ?

A. Léger :I cannot tell you if we know much about all life forms in the Universe but, as I have tried to show, our present knowledge of physical chemistry and planetary atmospheres points to Carbon and liquid Water based lives. Perhaps this knowledge is misleading and we are missing a whole domain, but presently we have no indication in that direction. This is one of the subjects that should be worked out by young scientists...

X You have talked about a temporary run away episode in an atmosphere because of the Greenhouse Effect, and mentioned Venus as an example. Could Venus atmosphere change (back) into the atmosphere of a habitable planet somehow? And could Earth's atmosphere under go such a run away and come back to today's state?

A. Léger: Venus cannot come back to a habitable planet because the very thick CO2 atmosphere provides a high T due to its strong Greenhouse Effect that prevents CO2 to be fixed in the rocks (at high T, carbonates dissociate and release CO2). Therefore, its runaway looks irreversible unless the Sun flux decreases a lot, which is not expected. The runaway process has be trigged by the high energy flux that impacts a planet at 0.7 AU from a Sun like star.

For planets in the habitable zone (at least 0.9 - 1.1 AU) a regulation mechanism exists. Rock weathering by liquid water fixes CO2 as carbonates and reduces the Greenhouse Effect whenever T increases, as a result of additional H2O evaporation. This situation looks stable as far as the stellar flux input is not too high and carbonates dissociate. Such planets seem to be protected from a runaway process.

X :Way to achieve the achromaticity of the nulling interferometer?

A. Léger: Several possibilities are presently under study. A concept is by Mike Shao as explained in Shao and Colavita (An. Rev. of Astr. & Astroph., 30, 457; 1992), another by Roger Angel and Nike Woolf (N. Woolf in the proceedings of the IR Space Interferometry Workshop, Toledo, Spain March 1996, to be published in Experimental Astronomy).

In both cases, the incident fluxes will have to be divided into several spectral ranges, narrow enough for the achromaticity be achieved at a high level.

References

Brack A., 1993: Origin of Life and Evol. of the Biosphere , 23,3.

Owen, T., 1980: in Strategies for the Search for Life in the Universe, Papagiannis (ed), Reidel, p177.

Kasting J.F., Whitmire, D.P. and Reynolds R.T., 1993: Habitable Zones around Main Sequence Stars, Icarus 101, 108.
Williams D.M. and Kasting J.F., 1996: Habitable Planets with High Obliquities,LPS 27, 1437.
Gautier D., 1992: Primitive planetary atmospheres: Origin and evolution in Frontiers of Life , Trn Thanh Vn J.&K. et al (eds) Editions Frontires, 91192 Gif, Fr.
Broecker W.F. and Teng T.H., 1982: Tracers in the Sea, Lamont-Boherty Geol. Observ., Columbia U. Press, N.Y..
Rosenqvist J. and Chassefire E., 1995: Planet. Space Sci., 43,3
Walker J.C. 1977, Evolution of the Atmosphere, Macmillan, N.Y..
Chassefire E., 1996: a/ Hydrodynamic Escape of Oxygen from Primitive Atmospheres: application to the cases of Venus and Mars, Icarus (in press)
b/ Hydrodynamic Escape of Hydrogen from a Primitive Water Rich Atmospheres: the cases of Venus, Icarus (submitted).
Kasting J.F., 1995: Planet. Space Sci., 43, 11.
Lovelock J.E., 1975: Proc. R. Soc. Lond. B 189, 167
Sagan C., Thompson W.R., Carlson R., Gurnett D. & Hord C. 1993: Nature 365, 715.
Kunde V.G. et al, 1974: J. of Geoph. Research 79, 777-84.
Hanel R.A., Conrath B.J., Jennings D.E. & Samuelson R.E.,1992: Exploration of the Solar System by IR remote sensing , Cambridge Univ. Press, USA.
Bracewell R.N., 1978. Detecting nonsolar planets by spinning IR interferometer, Nature 274, 780-781
Angel J.R., Cheng A.Y. and Woolf N.J., 1984: Nature 322,341.
Paetzold H.K., 1962: "La Physique des Plantes" vol. 7, Mem. Soc. Roy. Sci., Lige, p 452.
Kasting et al., Oxidant abundances in rainwater..., J. Geophys. Res. 90, 1049, 7-10510, 1985.
Léger A., Pirre M. & Marceau F., 1993: Astr. Astroph. 277, 309 - 313.
Kasting J.F. 1993, Earth's Early Atmosphere, Science, 259, 920- 926.
Léger A., Mariotti J.M., Mennesson B., Ollivier M., Puget J.L., Rouan D. and J. Schneider, 1996: Could we search for Primitive Life on Extra-Solar Planets in the near Future? - The DARWIN project -, Icarus (in press).
Angel J.R. and Woolf N.J., 1996: An Imaging Nulling Interferometer to study Extra-Solar Planets, Ap. J. (in press)

HIGH ANGULAR RESOLUTION STUDIES OF STARBURST GALAXIES

DANIEL ROUAN
Observatoire de Paris-Meudon - Département Spatial - CNRS - F92195 Meudon, France - Email: rouan@obspm.fr

1. Introduction: importance of the starburst phenomenon

Episodes of violent star formation in the nuclear regions of spiral galaxies are currently observed. Because they would exhaust the gas in a time of a few 10 millions years, and thus, must be of short duration, those episodes are generally called burst of star formation or, more concisely, "starburst", using the term first proposed by Feldman et al. (1982). This is an important process in extragalactic astrophysics because it strongly affects the evolution of the galaxy and its environment and because it should have played a major role in the evolution of the early universe, the starburst phenomenon being most likely the result of interaction and/or merging between galaxies. A non-exhaustive list of topics where understanding starburst physics is essential, follows:

- *Galaxy evolution*: the starburst strongly interacts with the host galaxy since it will convert a major fraction of gas in stars, will contribute largely to the enrichment in heavy elements and, through Supernovae explosions and ionization fronts, will produce shocks and fast winds in the ISM; in addition it can even affect the large-scale structure of the galaxy since it may be responsible of the formation of the bulge (Rich, 1992).
- *Intergalactic medium:* In many cases, starbursts are at the origin of hot bubbles of fast expanding gas resulting from the merging of SN remnants, (sometimes called "superwind"), and that can interact energetically with the intergalactic medium at a large distances.
- *Star formation*: the study of the star formation mechnism is often based on the observation of galactic regions, but obviously, the fact that initial conditions and intensities of the starburst phenomenon can

A-M. Lagrange et al. (eds.), High Angular Resolution in Astrophysics, 293–316.
© 1997 *Kluwer Academic Publishers.*

span several orders of magnitude, represents a unvaluable data base for modelization.

- *Extragalactic population:* the most extreme SBs, the UltraLuminous IRAS galaxies (ULGs), represent in fact the dominant population, in term of space density, of objects with $L > 10^{11}$ $L_\odot$ (Sanders & Mirabel, 1996): this population deserves thus a peculiar attention.
- *AGNs and Quasars:* Several years ago, it has convincingly been proposed (Sanders et al., 1988) that the huge starburst in the ULGs could be the progenitors of the most luminous objects in the universe: quasars and radio-galaxies. This very important prediction must be fully confirmed by observations.
- Cosmology: distant starburst galaxies trace the history of early universe when interactions between galaxies was much more frequent than now, with the likely consequence that most of the ellipticals formed during those merging processes.

The origin of the nuclear starburst phenomenon and the reason of its frequent association to an Active Galactic Nucleus are not yet clearly understood: although it is generally accepted that several basic conditions are required, such as a large concentration of gas, the detailed process is still to be established. Because of the distance of the objects, observations at high angular resolution (HAR) represent a key tool in this field, as I propose to show in the following, illustrating the different topics with recent results obtained either with Adaptive Optics, speckle camera or HST instruments.

2. Characterizing the starburst phenomenon

Review papers on the starburst activity in central regions of galaxies can be found in the following references: Rieke & Lebofsky (1979), Soifer et al. (1987), Telesco (1988), Sanders & Mirabel (1996). A large part of the information of this section derives from this bibliography.

2.1. A SHORT EVENT

Let's first give a justification, with raw numbers, of the term of "burst". Because the luminosity is dominated by massive, luminous, *young* stars with a short lifetime, the bolometric luminosity of the burst is proportional to the star formation rate: $L_{bol} \propto (dM_*/dt)$. On the other hand, the maximum duration of the burst is set, by supposing that all the gas is used for star formation :
$t_{burst} = M_{gas}/(dM_*/dt) = M_{gas}/L_{bol} \times (L/M)_{O,B} \times t_{O,B}$.
Using $(L/M)_{O,B} \approx 450\ L_\odot M_\odot^{-1}$ and $t_{O,B} \approx 5\ 10^6\ yr$, and considering the case of M82 – the archetype of starburst galaxies – characterized by:

	L_{IR} ($L_\odot$)	I_{IR} ($L_\odot$ pc^{-2})	IR/Tot
M31 (800pc)	3 10^7	60	.03
Milky Way(800 pc)	7 10^8	1.4 10^3	.3
M83	4.3 10^9	6 10^4	
M82	3 10^{10}	1.4 10^5	.84
NGC1068	3 10^{11}	3. 10^4	.76
Arp 220	1 10^{12}	2.5 10^6 ?	98

TABLE 1. A few global infrared properties of selected star-forming galaxies. First column: IR luminosity, second column: surface luminosity, third column: IR to visible ratio. Partly adapted from Telesco (1988).

$L_{bol} \approx 3\ 10^{10}\ L_\odot$ and $M_{gas} \approx 2\ 10^8\ M_\odot$, we finally find $t_{burst} = 2\ 10^7 yr$ which is indeed much shorter than the Hubble time: any important star formation activity one detects in a galaxy must be just a short event during the lifetime of the galaxy.

2.2. A WIDE RANGE OF BURST INTENSITY

Table 1 (adapted for a part from Telesco, 1989) gives a few global quantities for different starburst galaxies, from enhanced star formation in the central region of the Milky Way to ultraluminous galaxies: it illustrates the fact that the starburst activity covers indeed an extremely large range of intensity. For instance, bolometric luminosities span 3 orders of magnitude (typically $3\ 10^9 - 4\ 10^{12}$ $L_\odot$), the dust temperature can vary by up to a factor of two (25 – 50 K) and the extension of the burst region is between 50 pc and 3 kpc.

Despite this dispersion of absolute values, all those objects clearly belong to a unique class, characterized by several common features: a) a high surface luminosity compared to normal galaxies, with peculiar morphological structures, so that those objects are often refered to as "Hot spot" or "HII" galaxies (e.g. Serzic & Pastoriza, 1967); b) a strong concentration of the activity in central parts (< 1 kpc) of the galaxy, often under the form of a ring; c) all the distinctive signs that this activity results from active star formation.

2.3. TRACERS OF THE STARBURST

All those galaxies exhibit the classical tracers of the star formation, i.e. of the presence of young, massive, luminous stars :

- *Large IR luminosity:* it is due to the emission from dust heated by the young OB clusters; a huge concentration of dust and molecular gas generally lead to a high conversion efficiency from visible to IR radiation. All those galaxies are detected by IRAS which first revealed how large was the fraction of infrared galaxies in the universe (Soifer & al., 1987). One can add that the IR luminosity is shown to be proportional to the number of ionizing photons measured by infrared or radio means (see Fig. 2 of Telesco, 1988): an evidence that young ionizing stars dominates the energy injection finally degraded to infrared radiation.
- *Strong UV emission:* originating in the photosphere of the hottest stars, it explains why, most of those objects belong to the Markarian catalog (Balzano, 1983).
- *Emission lines:* numerous recombination lines and fine-structure lines, formed in the ionized gas (hydrogen and other ions), are emitted at visible and infrared wavelengths. Because they allow a deeper probing of the – generally – opaque region, infrared lines are especially of interest and $Br\alpha/Br\gamma$ observation (Ho & Turner, 1990) is a powerful tool to evaluate the rate of Lyc photons.
- *Shock excited H_2:* fluorescent emission lines of excited molecular hydrogen is frequently observed in starburst galaxies. Line ratio indicate, in most of the case, a shock excitation rather than UV absorption. The relation with the starburst activity is either indirect if the shock reflects rather the conversion into heat of kinetic energy at the interface between colliding galaxies, or direct if it is rather due to young stellar outflows or even turbulent gas within heavily ionized region.
- *Dust features:* when detected in absorption, such as the 10 μm silicate fature, those bands characterize the large opacity of the medium; when in emission (PAH bands at 3.3, 6.2, 7.3, 11.3 μm) they generally indicate a large UV density that excite PAH molecules or Very Small Grains to a high temperature. Taken from Roche et al. (1991), the average spectrum of 12 starburst galaxies shown on Fig. 1 illustrates the main features observed in the mid-IR: a deep silicate absorption, and two lines in emission, one from a carboneous component (likely PAH molecules) at 11.3 μm and one from NeII at 12.7 μm.
- *Evolved stars features:* a deep CO absorption band at 2.35 μm can be considered as a rather good tracer of supergiants, a late phase in the evolution of massive stars (e.g. Doyon et al., 1994)
- *Strong non-thermal radio emission:* it is always associated to starburst activity (Condon et al., 1982) and extremely well spatially correlated with the 10 μm sources (Telesco, 1988, Condon et al., 1992); it is generally attributed to the synchrotron radiation from high-energy electrons

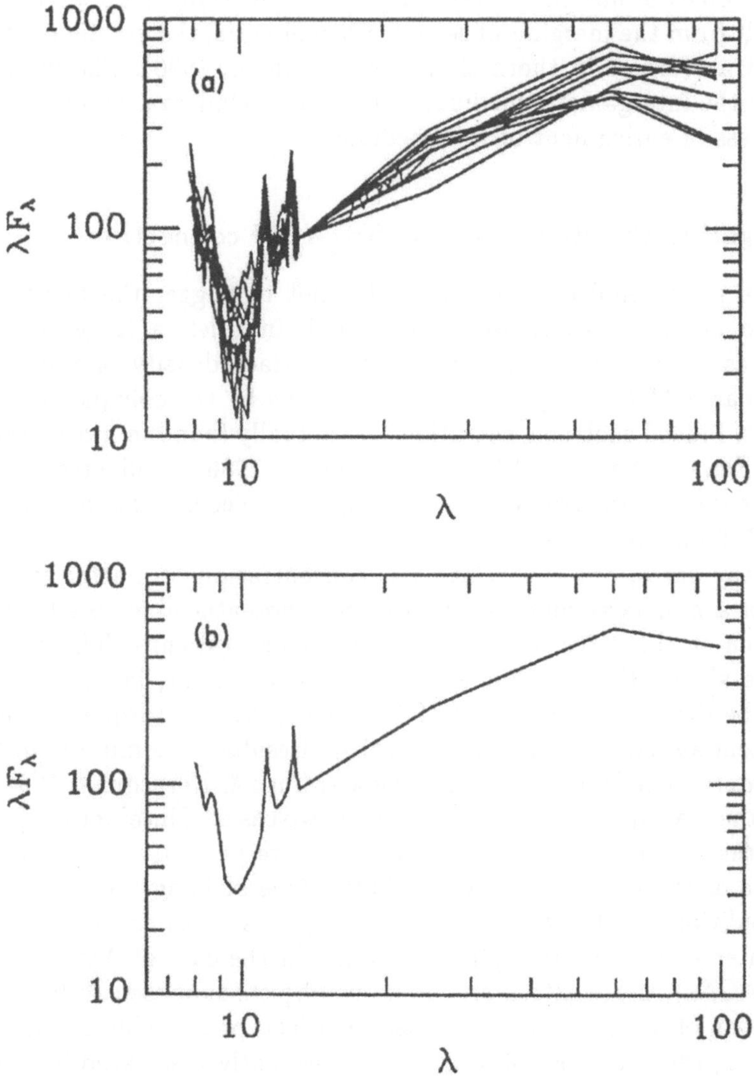

Figure 1. a) mid-IR spectra of 12 nuclear HII region galaxies; b) the average spectrum. Taken from Roche et al. (1991)

produced in type II SNs explosions. The high rate of SNs (typically one every 4 years) is generally consistent with the stellar population producing the IR luminosity.

– *Superwind* : recently put in evidence in several SBGs (e.g., Veilleux et al., 1996), this wind corresponds to fast (100 km/s) expanding superbubbles of hot gas at very large scale (typically 1 kpc), that could result from the merging of several SN remnants, the kinetic energy being transformed in thermal energy (Heckman, 1990). The interaction with the intergalactic medium of this superwind can be an important process of enrichment of this medium.

3. Triggering the starburst: the SB/AGN connection

The large scale conditions that must be met to trigger the burst of star formation are now rather well understood: in brief, a large concentration of gas is required, with a very large surface density of the order of $1000\,M_{\odot}\mathrm{parsec}^{-2}$ (Larson, 1988), followed then by the collapse of molecular clouds. Indeed such concentrations are actually found in most situations (see e.g. Fig. 4 of Sanders, 1992; Sanders et al., 1986), as illustrated by the extreme case of Arp 220, where $10^{10}\ M_{\odot}$ of molecular gas is found in a region of 300 pc diameter.

In many cases, one can relate this concentration of matter to the existence of a non-axisymmetric gravitational potential due to either a bar (Hawarden et al., 1986) or an interacting galaxy (Wynn-Williams, 1987): it can be shown that, within a few galactic rotations, large amounts of gas can be driven to the center of the galaxy by the torques induced in such potentials (Larson & Tinsley, 1978). Results of computer model of galaxy interaction fully support this view (Mihos & Hernsquist, 1994): Fig. 2 shows the star formation rate vs time for two cases of interacting galaxies: the shortness and the intensity of the burst are perfectly clear, especially when one of the galaxies has a rich bulge. One striking point is that the same conditions that favour a star forming burst (gas concentration and bar or interaction) are also generally found in the case of Active Galactic Nuclei (AGNs), those extremely luminous objects supposed to be powered by accretion of matter on a supermassive object, presumably a Black Hole. In addition, the two types of activity are frequently associated in the same galaxy, as examplified by NGC1068 or NGC 7469. There is, thus, good reasons to look for a direct link between the two types of activity.

Sanders et al. (1988) were first to propose that, indeed, the most powerful starburst are the progenitors of QSO and radio-galaxies. They based their assumption that Ultra Luminous Galaxies (ULGs) probably host dust-enshrouded quasars, on several observations: *i)* the increased frequency of the association at high IR luminosity between an AGN and a starburst (see table II of Sanders, 1992); *ii)* the similarity of near-IR colors and of line ratios for high excitation transitions; *iii)* the luminosity function of ULIRAS and QSO which become very similar at luminosity $> 10^{11}\ L_{\odot}$. In this scenario, no mechanism is actually proposed to explain how the AGN is

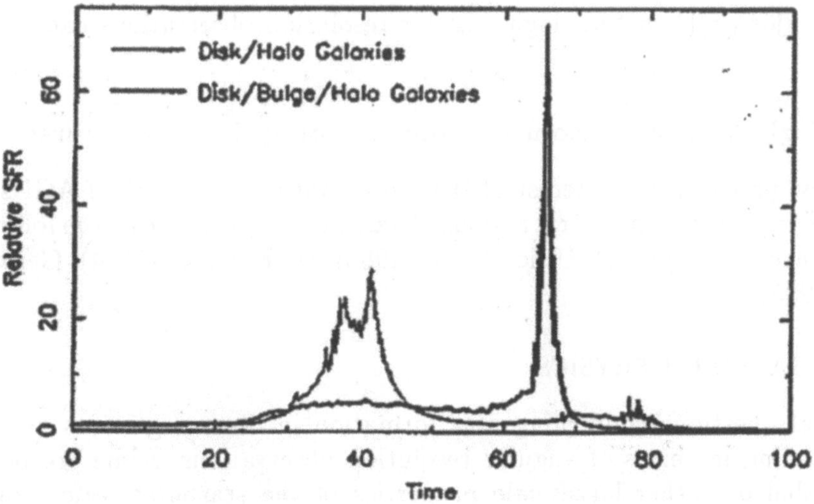

Figure 2. Evolution of the star formation rate with time during a direct encounter of two galaxies that ends in a complete merging. When one galaxy has a developed bulge, the burst is stronger and shorter. Taken from Mihos & Hernquist (1994)

formed as a result of the starburst activity, but it is assumed that the AGN is a "classical" accreting supermassive object which is the more easily fed with fresh matter that molecular gas is overabundant.

A quite different approach, but that also directly links the starburst activity to the presence of an AGN is the one by Terlevitch and co-workers: here, it is proposed that the AGN itself is nothing but a monstruous and extremelely compact starburst. According to those authors (Terlevich & Melnick 1985; Terlevich, 1992) all the traces of activity could be then explained by the evolution of a massive young cluster of coeval high metallicity stars: *i)* extremely massive stars called *Warmers* with temperature up to 100 000 K would be responsible of Seyfert 2 or LINERS characteristics, i.e. those of the Narrow Line Region (NLR); *ii)* SNs explosions of type I (very massive stars) produce superbubbles of hot material explaining the observed radio jets; *iii)* type II SN, i.e. those following Red supergiants ejecting their envelope, would form high-velocity ejecta that would be detected as the Broad Line Region (BLR); in addition, the very high rate of SNs would explain the variability.

Whatever the scenario to explain the AGN is (accretion on supermassive object or superstarburst), the main problem remains the way the AGN can be fueled with fresh matter, at a rate of typically 1 $M_\odot$ yr^{-1}: no simple mechanism is able to provide the loss of angular momentum that will bring matter from a 100 pc radius to 0.1 pc (Schlosman, 1992). This is certainly one major problem that high angular resolution observations can help to solve.

4. High Angular Resolution and Astrophysics of starburst

Review paper on the interest of HAR observations for study of AGNs and Starburst activity in central regions of galaxies can be found in the following references: Rees (1981), Ulrich (1981, 1988), De Robertis (1994), Genzel et al. (1995).

4.1. STARBURST PHYSICS

Several questions concerning star formation in distant galaxies are very demanding in terms of angular resolution observations. Some are related to global or rather large-scale properties of the starburst region, others requiring a detailed study of the young stellar clusters themselves.

A first role of HAR imaging is simply to identify the different components in the starburst region, not an easy task when a patchy and high obscuration – usually the rule – makes the morphology confusing. Infrared is well suited to do so, as examplified by Sams et al. (1994), who, using HAR color map obtained with a speckle camera, were able to identify the weak, but true, nucleus of NGC 253 – another starburst template – within a complex region of hot-spots.

How does vary the star formation rate with gas density, in particular, is there a universal Schmidt law that relates the star formation rate to the local gas density ($SFR = a\rho_g^n$) ? This is an important question that has to be explored on orders of magnitude of burst intensity; however, because they are far away (typically 200 Mpc), ULGs require imaging with an angular resolution better than 0.1" in order to reach scales below 100 pc where the starburst activity is generally concentrated. Fig. 3 shows an image at 2.2 μm of NGC3690, one component of the advanced merger Arp 299 (= NGC3690/IC694), an ULG at a distance of 42 Mpc; this image was obtained with the *Pueo* Adaptive Optics System at the CFHT telescope. The excellent resolution allows to separate clearly the nucleus and several complexes corresponding to clusters of very young stars.

Related to the previous one, another important topic on the large scale properties of starburst is the degree of clustering of stars versus the burst intensity: for instance, O'connel et al. (1995) have been able to resolve, thanks to the sharpness of the HST images, a large number of individual OB clusters in M82 – the prototype of nearby starburst galaxies – each of them has a size of $\approx 3.5pc$ and is brighter than any cluster in the Local Group ! An equivalent work, but for a more distant galaxy, was recently conducted by O. Lai and co-workers (1996) and is illustrated on Fig. 6.

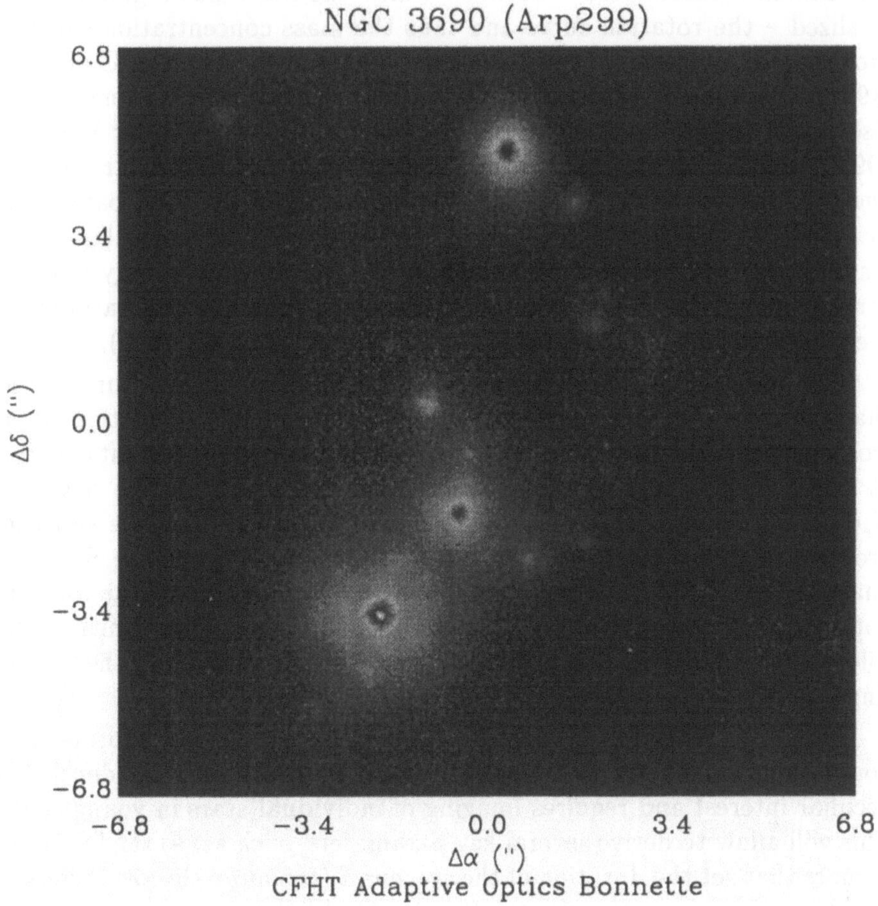

Figure 3. K-band image of the NGC3690, one component of the Ultra Luminous system of merging galaxies Arp 299 (=NGC3690/IC694). Imaging was performed with *Pueo*, the Adaptive Optics system at CFHT (Lai et al., 1996), using the *Monica* camera. The scale reached on this distant galaxy (D= 42 $(H_o/75kms^{-1}\ Mpc^{-1})^{-1}$ Mpc) is typically 40 pc; the nucleus and several compact sources are resolved: they likely are extremely young OB clusters. Several regions appear extremely reddened and should be the currently most active star forming sites.

To check the consistency between the burst intensity and the molecular gas surface density, it is important to trace the gas and dust concentration with a good angular resolution in millimetric CO (interferometry), and at 10 μm (adaptive optics). Alternatively, one can evaluate this mass as a byproduct of the visible/near-IR imaging: if a starburst ring is put in evidence, then from dynamical considerations on the position of the inner Lindblad resonances – between which the starburst region should be localized – the rotation curve and thus the mass concentration can be approximated; an example of such a reasoning can be found in Gallais et al. (1991). Mapping fluorescent H_2 at a good resolution in the near-IR, can also reveal the shock of molecular clouds in mergers (van der Werf et al., 1993; Angonin et al. 1995) or concentration of hot molecular gas in the nucleus (Tacconi et al., 1994). HAR line-imaging in the PAH bands (3.3 to 11.3 μm) can provide, in addition to the location of the molecular material, a supplementary information on the local UV – or high energy particles – density, by tracing the destruction of those molecules in the starburst or in the superwind (e.g., Rouan et al., 1994; Normand et al., 1995).

The question of a possible propagation of the starburst through some chain reaction has been envisioned (Elmegreen, 1987): detection of spectroscopic features (e.g. CO line at 2.35 μm tracing supergiants, FeII line at 1.69 μm revealing SNs) or of color effect characterizing the relative age of different regions in the starburst should be a way to make substantial progress in this matter. To illustrate the method, we show in Fig. 4-b an image of NGC 863, a Seyfert galaxy with circumnuclear star formation, obtained with adaptive optics (Lai et al., 1996): the coding of near-infrared color-color indexes reveals a possible evolution of ages along the starburst ring.

The study of the Initial Mass Function (IMF) of the newly formed stellar population, versus the burst intensity and/or the initial gas density is of peculiar interest and requires imaging of individual stars in young clusters. This will allow to derive several key parameters, such as: *a)* the lower cutoff in mass that set the duration of the starburst (the more the gas is converted into low mass stars, the shorter the burst); *b)* the maximum stellar mass which will determine the evolution of the cluster and of its environment (ionization of the gas, disruption of the clouds, SN-induced shocks etc.); *c)* the slope of the IMF, i.e. the exponent of the power-law $\zeta(m) = m^{\Gamma}$ (its value is -1.5 in Milky Way) which is a fundamental parameter in the models of fragmentation of parental clouds. Using adaptive optics, such a study has been conducted by Brandl et al. (1995) for R136 in 30 Doradus (LMC), the most massive HII region in the Local Group: from a census of 1000 stars, those authors were able to measure an evolution of the IMF from center to outside, probably due to some dynamical evolution. Fig. 5 shows the K-band exposure of R136 taken with the camera Sharp installed on the Comeon AO system of the ESO 3m60 telescope.

In nearby starburst, the evolution of young stellar clusters with very high stellar densities is a new and important field that instruments combining spectroscopy and high angular resolution will certainly open in the future: it will be possible to study the role of collisions, tidal disruptions,

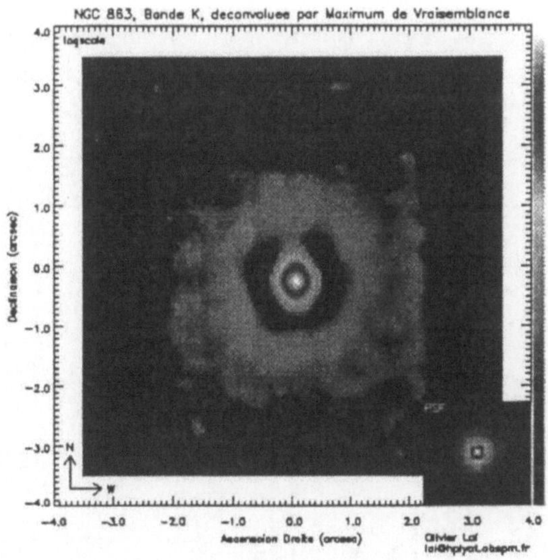

(a)

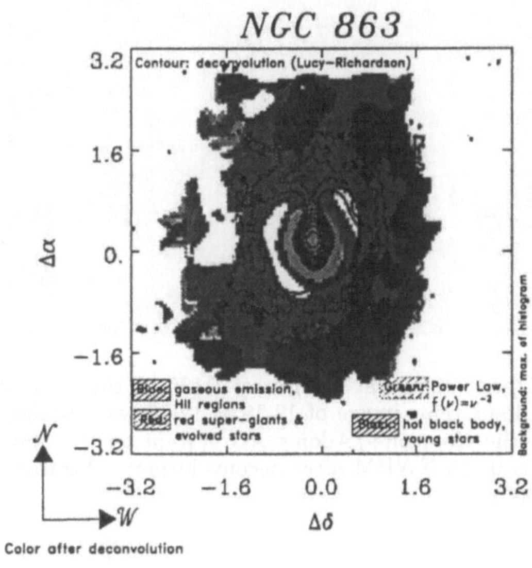

(b)

Figure 4. Near-IR Adaptive Optics Image of the starburst/AGN galaxy NGC 863 obtained with Adonis at the 3m60 ESO telescope. Left panel: K-band image; the Seyfert nucleus is well separated from the starburst ring; one noteworthy feature is the "bridge" of emission that links the ring to the AGN: is it a counterpart of some inner bar that could manage to drive gas from the molecular ring down to the innermost region ? Right panel: "color-color" image of the same object: the grey scale is coded so as to represent the dominant source of radiation, according to position in the J-H/H-K color-color dagram; such coding may be helpful to look for some gradient of age along a ring of star formation (Lai et al., 1996)

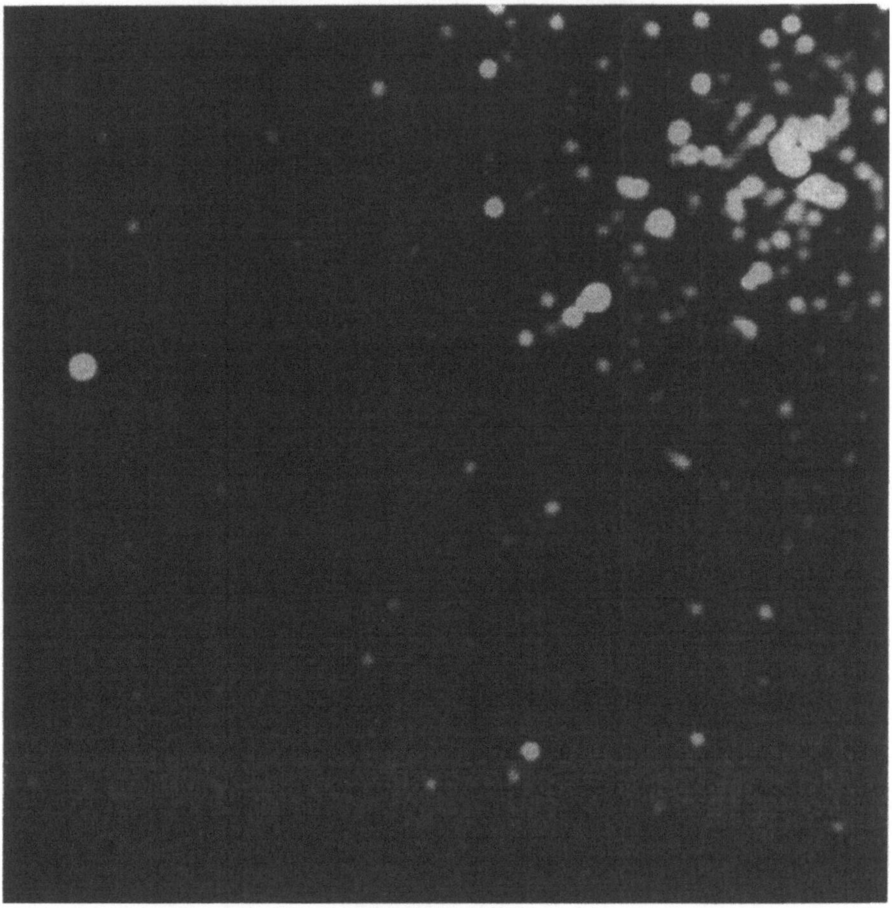

Figure 5. Adaptive Optics Image at 2.2 μm of R136 in the 30 Doradus complex of the Large Magellanic Cloud. This image of 12."8 × 12."8 was obtained with the Sharp camera at the focus of the Comeon+/Adonis AO system on the 3m60 ESO telescope. The spatial resolution is 0."15 FWHM after deconvolution; a logarithmic scale is used. (From Brandl et al., 1995)

stripping of enveloppes etc.

4.2. THE STARBURST/AGN CONNECTION

The proper separation of the AGN from the starburst is a first necessary step in order to assess the respective contributions to the bolometric liminosity, to delineate the line emitting regions etc. As an example, Fig. 6 shows, in the case of the active galaxy NGC 7469, that separation of the Seyfert nucleus from the starburst ring can be clearly done; in addition, many details on the inner spiral structure and on the location of young clusters can be distinghuised.

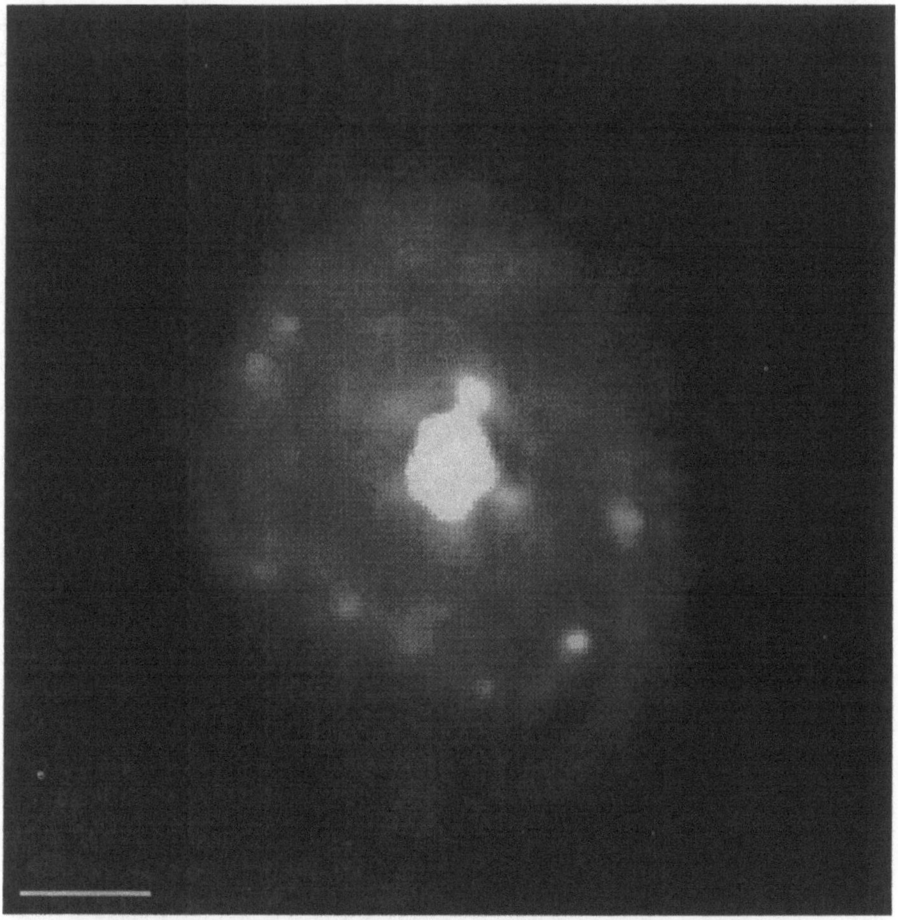

Figure 6. Adaptive Optics H-band Image of the SB/AGN galaxy NGC 7469 obtained at the 3m60 CFHT telescope with the IR camera Monica. The white line at bottom left corresponds to 1 arcsec. The seyfert nucleus appears as a pointlike source, well separated from the starburst ring. The scale reached on this distant galaxy (D= 68 Mpc) is typically 40 pc and the young clusters of star forming regions in the ring are clearly distinguishable within the inner spiral structure (Lai et al., 1996).

A fundamental question is to know if a phase of interaction/merging is always required to obtain an AGN; in this respect, one must look for clues indicating that such an event actually took place, for instance double nuclei, wisp, tails. Fig. 7 shows a K-band image of the central region of NGC 1365, a close face-on barred spiral galaxy which is hosting an AGN and a starburst active site at its center. The HAR image, taken with Comeon+/Adonis at the 3m60 ESO telescope, reveals that the nucleus is complex and probably double, a possible indication that NGC 1365 is a merger. The same image reveals two "voids" along a SE-NW direction which is in fact the direction along which a radio jet (Sandqvist et al., 1995) and fan-shaped cone of [OIII] emission (Jörsater & Linblad, 1989) have been put in evidence; this suggests that the voids are the counterparts of those structures: the energetic flow that produces the cone and the jet, may be responsible of low density cavities on each side, that appear as a lack of continuum emission.

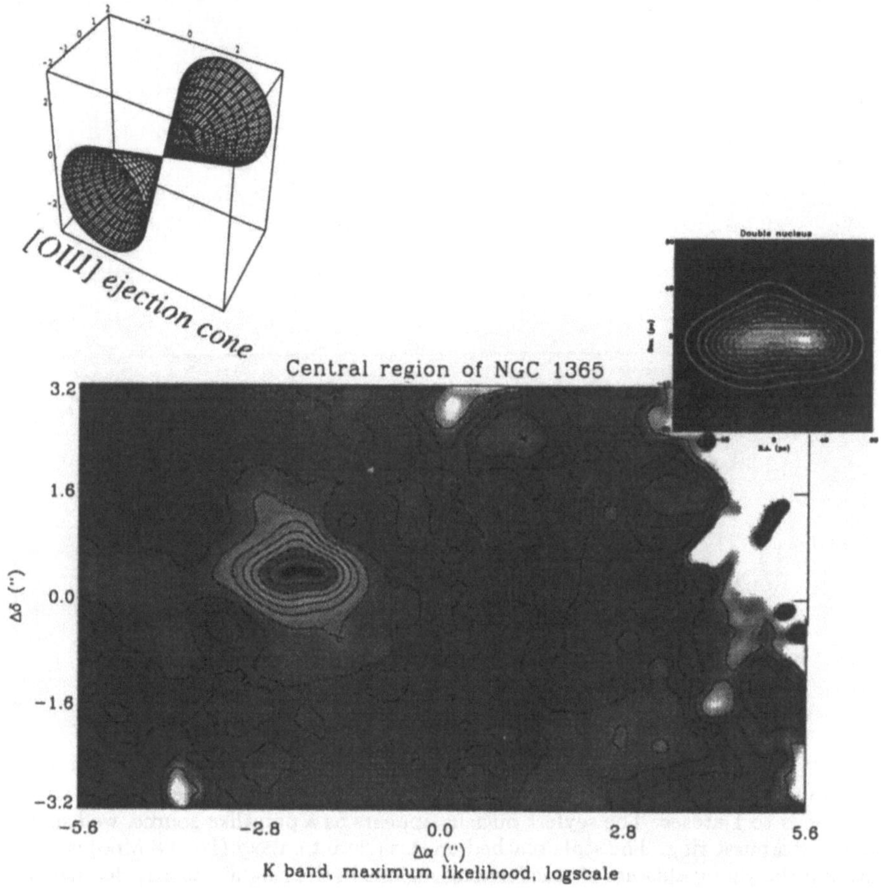

Figure 7. Adaptive Optics K-band Image of the SB/AGN galaxy NGC 1365 obtained with the AO system COMEON+/ADONIS at the 3m60 ESO telescope (Lai et al., 1996). The cartoon in the upper left panel shows the geometry and direction of the [OIII] fan-shaped cone. The upper right insert shows the detail of the nucleus which appears as a double one. The main image reveals two voids, aligned on the [OIII] cone, bracketting the nucleus: they could be the counterpart of the high energy flow that creates the [OIII] cone and the radio jet.

Spectroscopy with a good resolving power, allowing dynamical studies of nuclei must also become an irreplaceable tool, since dynamically decloupled nuclei, and even counter-rotating ones, such as in IC 1459(Franx

& Illingworth, 1988), are strong indications of the occurence of a major interaction in the past.

As important would be the proof that hidden AGN are deeply embedded in the dust of ULGs; this quest is particularly difficult because it is believed that opacities can be in excess of 100 in those objects: 10 μm HAR imaging is probably the most powerful tool in this respect.

Understanding how the monster can be fueled with gas is, by itself, a critical problem that probably only detailed observations at HAR will allow to elucidate. The rate of conversion of mass in radiated energy required to supply typically 1. $M_{\odot} yr^{-1}$, a very large number if we compare it to the star formation rate which is rather 2-12 $M_{\odot} yr^{-1}$ in a standard burst.

Among the proposed mechanisms (see a review by Schoslman, 1992), there is the hypothesis of a superdense stellar cluster associated to the AGN, that would feed the latter through mass-loss, tidal disruption of stars, star collisions (Norman & Scoville, 1988). Another proposition (Schlossman et al., 1989) is that an internal gaseous bar could produce the required torque. Observation of one of those structures, cluster, that could be of the parsec size with a high rate of supernovae, or gaseous bar, would be extremely important. Fig. 4 shows, in the case of the SBG/AGN NGC 863, that a "bridge" of emission links the Seyfert nucleus to the starburst ring. Several other galaxies exhibit such a bridge, that may indicate the existence of this inner bar.

4.3. AGN STUDIES

This paper is not intended to deal thoroughly with the question of AGN, and excellent reviews on the subject exists (see beginning of this section); however, given the importance of this issue, and the lack of a dedicated lecture on the subject during the School, I give briefly in the following indications on the importance of HAR observations related to AGN understanding.

The structure of the central 100 pc of a SB/AGN nucleus is schematically described on Fig. 9 in the frame of the largely accepted unified model of AGN (Antonucci, 1993). This model invokes the presence of a thick torus around the central engine, whose inclination on the line of sight would be the key parameter fixing the respective importance of the Narrow Line Region and of the Broad Line Region when observing a given object; in particular, this model explains convincingly that Seyfert 1 and 2 are actually a same object seen under a different viewing angle and conducts to a pretty good fit of the observed IR spectrum (Pier & Krolik, 1993).

HAR visible imaging will probably be xconcentrated on the mapping of cones of photo-ionized gas in [OIII] or other high excitation lines associated to the NLR. Fig. 8 shows a juxtaposition of an Adaptive Optics K-band image of NGC 1068, the well-studied closest Seyfert2 nucleus, and a HST [OIII] image of the same region. An active continuum source must lie at the apex of the fan-shaped [OIII] emission.

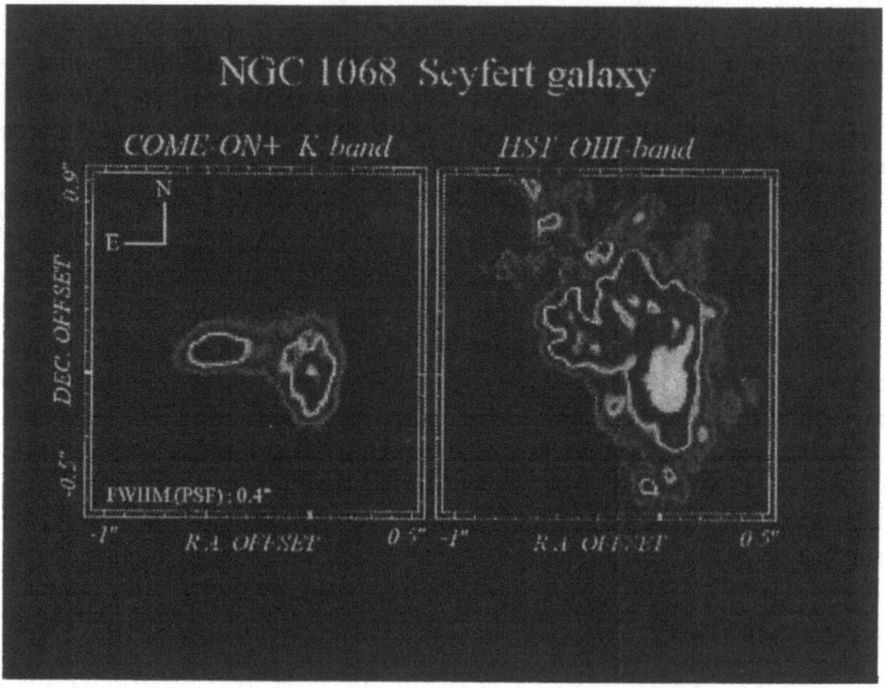

Figure 8. a) Adaptive Optics (Comeon/Adonis at ESO 3m60) K-band Image of the SB/Seyfert2 galactic nuclei NGC 1068 Comeon/Adonis (Marco et al., 1996). b) [OIII] HST image of the same region (Evans et al., 1991). A massive star formation event is inferred from the tight spatial correspondence ofsource at East, while the brightest K component is probably a tracer of the hot dust directly heated by the active continuum source revealed by the [OIII] emission.

The previous result introduces the important role of the infrared range which has the strong advantage of a deeper probing capability. Clearly, an imaging infrared capability which gives access to the 100 mas domain will allow, in a first step, to tackle the question of the nature of the NLR on a large sample of objects, and possibly of the thick torus of dust. Establishing the 2 μm compactness and resolving the 10 μm region, in order to find the dust temperature distribution $T_d(r, \theta)$, are certainly two first important goals. Evidencing details on the torus morphology and on a possible warp of the dust disk that could explain a lack of UV (Phinney, 1989) are among the following steps.

Several recent papers argued that indeed the torus was detected (Simpson et al., 1996; Ford & Ferrarese, 1995): Fig. 10 shows the HST image of NGC4261, taken by Ford & Ferrarese (1995), that reveals a structure interpreted as a spiral-shaped disk of dust around a supermassive object; however, the size of this structure (240 pc) is much larger than the dust torus predicted by the unified model. It is probably only with an interferometric technique that it will be possible to explore the 0.1–1pc scale and assess the validity of this prediction. In the meantime, evidence of a super massive object would have probably been demonstrated on a large sample of objects, by measurement of proper motions of stars and gas thanks to integral field spectroscopy and/or measurements at different epochs.

Detection of optical counterpart to the extremely collimated radio jets currently observed in active nuclei or radio-galaxies is certainly a goal that HAR will be able to reach routinely in the near future. Fig. 11 shows an HST image of the nucleus of the giant elliptical galaxy M87, where such a jet is put in evidence: it corresponds to relativistic electrons emitted by the putative Black Hole engine revealed by the rapid rotation of the inner disk. (Ford et al., 1996)

5. A few practical considerations

Let's share the domain of HAR between interferometry and Adaptive Optics / HST at the angular scale of 50 mas. What should be practically possible in the near future ?

5.1. ADAPTIVE OPTICS & HST

The advantage of a high core/halo contrast given by AO is essential, since in many cases the presence of an AGN can dazzle the field. Strehl ratio larger than 60 % are probably required to be able to detect the direct environment of a conventional AGN. The scale of 0.1" translates to d = 10 pc at D = 20 Mpc this means that the NLR is accessible with AO technique for a significant sample of relatively nearby objects.

Using a nearby star to servo the AO loop is certainly possible, but in a very limited number of cases. Using the nucleus itself as the reference object for wavefront sensing appears more promising, at least in a first step. The number of targets can be evaluated using available catalogs, provided that the listed magnitude corresponds to a small diaphragm. It can be checked that this is generally the case for the AGN catalog by Véron-Cetty & Véron (1996), by cross-checking the magnitude of a dozen nuclei with measurements at a good angular resolution (Kotilainen, Ward & Willinger, 1993). Based on this catalog, one finds that 205 objects brighter than m_V

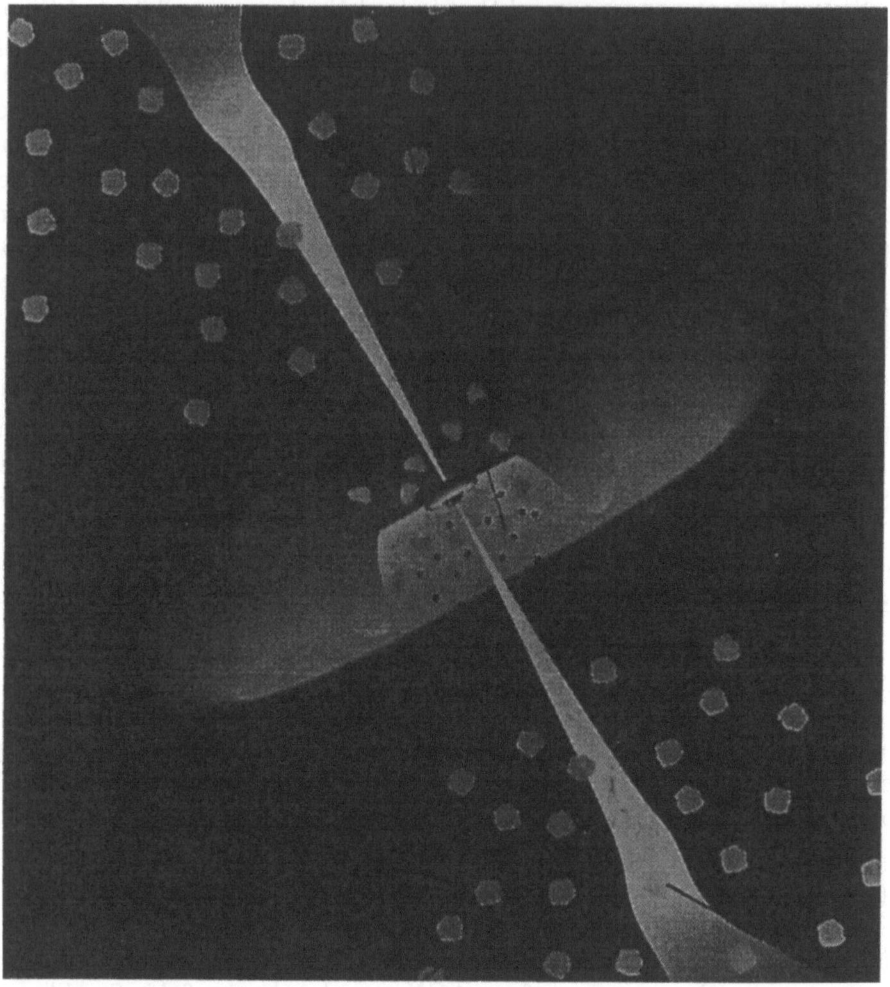

Figure 9. Cartoon depicting the supposed structure of an AGN in the frame of the Unified Model (Antonucci, 1993). Taken from Urry & Padovani (1995)

= 14 would be good candidates, this number becoming 765 if the wavefront sensor limit is rather $m_V = 16$.

Fig. 12 gives the number of objects brighter than respectively 16 (upper histogram) and 14 (lower histogram) according to their distances: one can see that at a limit of $m_V = 16$, 180 objects could be observable, where a resolution of 25 pc or better, is achievable (134 for $m_V = 14$).

In the case of more distant objects, especially of the ULGs, laser guide star is required; in this case, tip-tilt correction on the nucleus itself should be probably possible and should allow a fairly large sample of nuclei to be

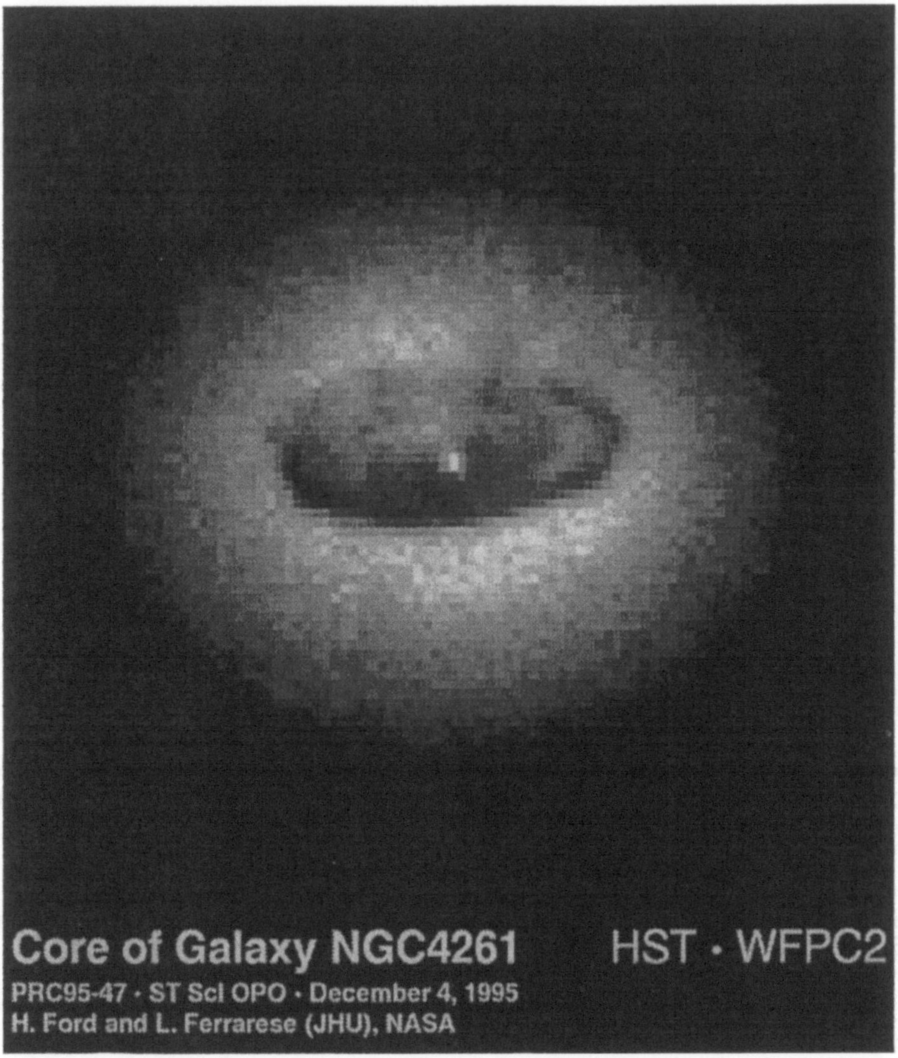

Figure 10. HST image of the core of the galaxy NGC 4261. The dark structure at the center is interpreted as a spiral-shaped disk of dust around a supermassive object, which is inferred from velocity measurements (Ford & Ferrarese, 1995). Taken from HST web server

observed.

Imaging is the basic starting tool, but spectroscopy will rapidly become necessary and should have to be associate to AO in integral field spectrograph with typical resolution ≈ 3000 (100 km/s).

Because a large fraction of the light from an AGN is in fact reflected, a strong polarization is often measured and a polarimetric capability of an AO system is certainly desirable. This is also true for a coronographic

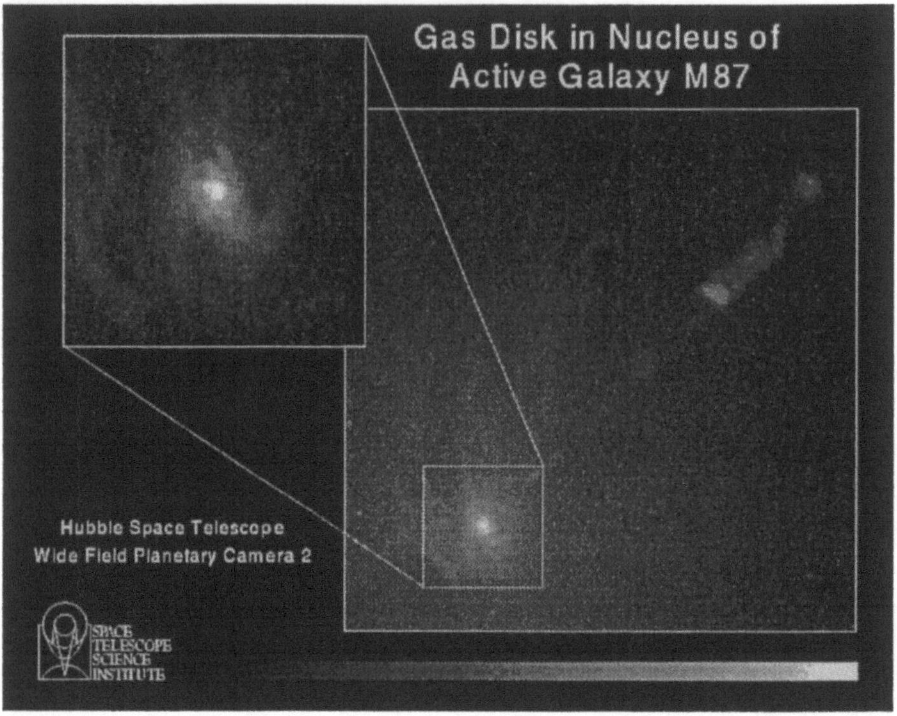

Figure 11. HST image of the nucleus of M87. One can see clearly a jet in the NW direction, tracing relativistic electrons ejected by the central Black Hole engine (Holland et al, 1996). Taken from HST web server

mode that should be extremey useful when tackling the question of the direct environment of an AGN.

5.2. INTERFEROMETRY

The sensitivity issue is a major one and it is generally accepted that measurement of a significant sample of objects requires fully phased pupils 8 m telescopes (Ulrich, 1988). Fig. 13, taken from Ulrich (1988) gives the theoretical sensitivity limit that an interferometer such as the ESO-VLTI should reach in the infrared domain. Note that fully phasing an interferometer requires Adaptive Optics servoed on a bright reference source, a condition that may severely limit the number of actually available objects.

An imaging mode is required, with a field of 0.1 to 1 arcsec for most programs. Note that, to separate individual stars in a dense young cluster at a distance of 10 Mpc, it is required to reach an angular resolution of $\approx$ 1 mas, that only interferometry will be able to provide. However, the large number of sources in such a cluster will require a dense coverage of

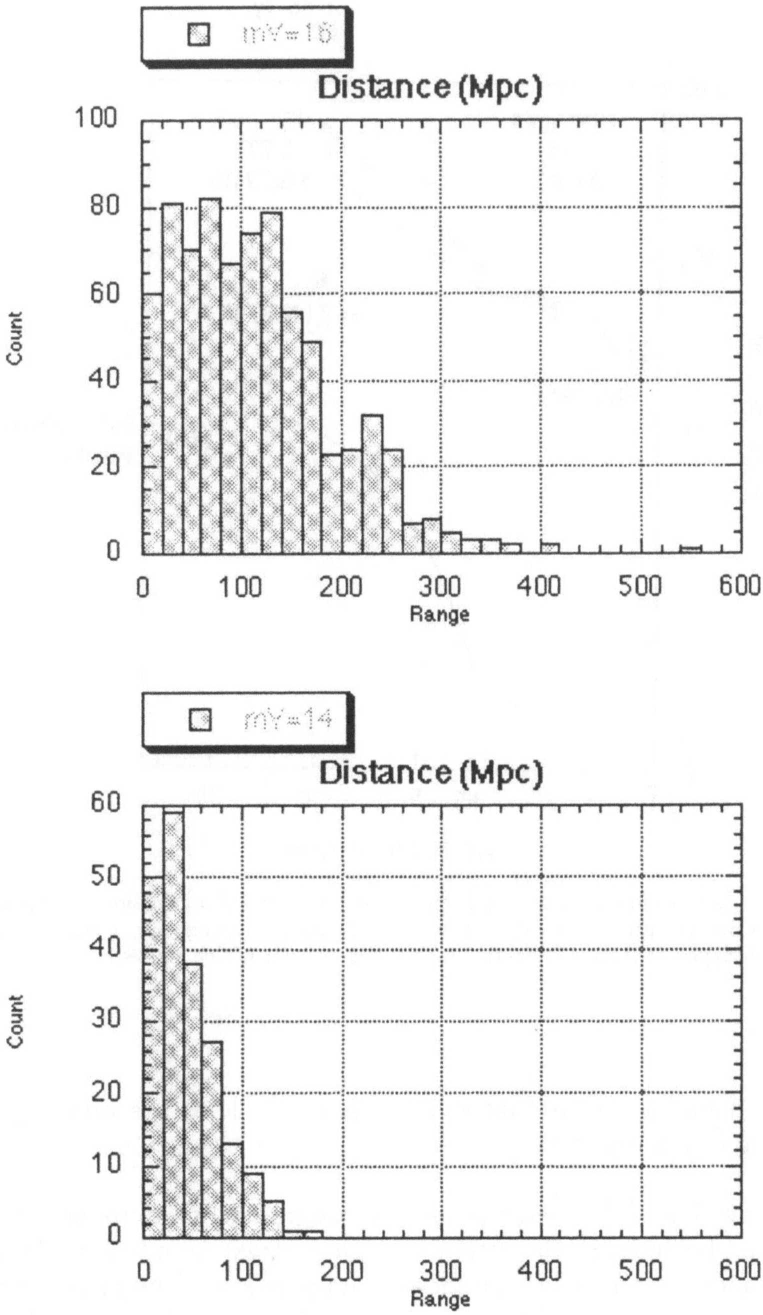

Figure 12. Histogram of the distances ($H_o = 75kms^{-1}\ Mpc^{-1}$) of all the nuclei in Véron-Cetty-Véron (1996) that are brighter than (upper) $m_V = 16$ and (lower) $m_V = 14$.

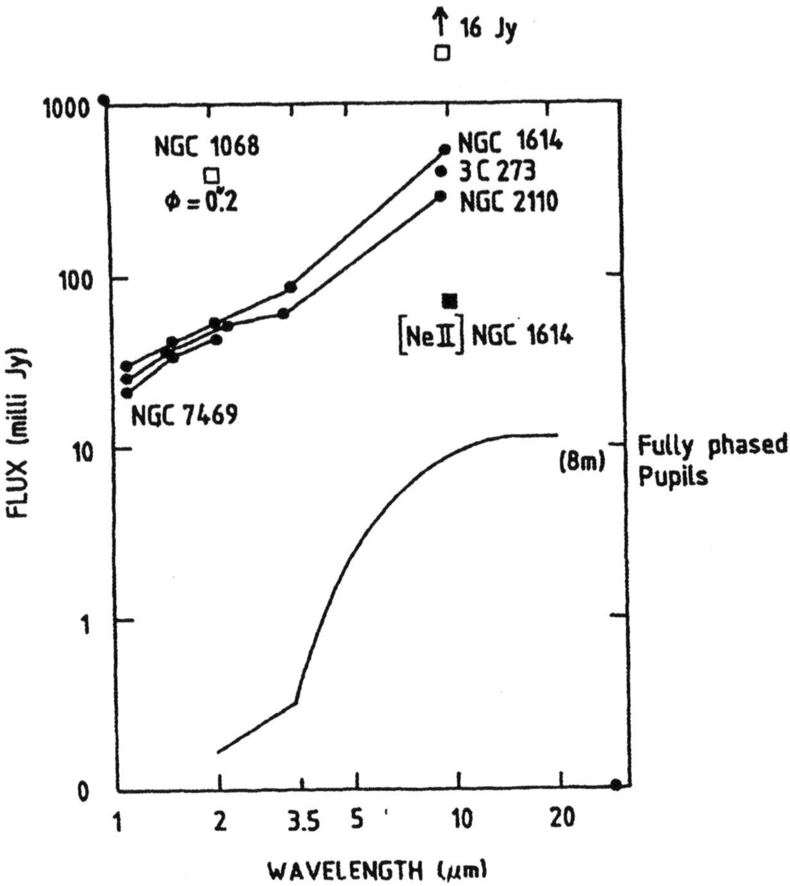

Figure 13. Lower curve: sensitivity limit, in the infrared, of a two 8m telescopes interferometer when the pupils are fully phased with Adaptive Optics; upper curves: fluxes of several archetypes of Active Galactic Nuclei. Taken from Ulrich (1988).

the (u,v) plane, a feature that will not be available at the very beginning of imaging interferometry.

The question of the usefulness of a coronographic interferometer (Angel, 1989) could be debated: the extremely high contrast close to a bright object it would allow, is certainly a major advantage to study the surroundings of the central engine in AGNs; however, maintaining a high rejection factor is may be not an easy task.

Acknowledgements. I warmly thank Olivier Lai for providing me with recent (and fully processed...) results from his thesis work.

6. References

References

1. Angel R., 1989, *Use of a 16m Telescope to detect Earth-like planets*, in Bély P., Burrows C., Illingworth G. (eds), *The Next Generation Space Telescope*, pp. 81–88
2. Angonin-Willaime M.-C., Casoli F., Gerin, M., Rouan D., 1996, BCFHT 34, 8
3. Antonucci R., 1993, ARAA 31, 473–521
4. Balzano V.A., 1983, ApJ 268, 602
5. Brandl B., Sams B.J., Bertoldi F., Eckart A., Genzel R., Drapatz S., Hofmann R., Loewe M., Quirenbach A., 1996, ApJ, in press
6. Condon J.J., Condon M.A., Gisler G., Puschell J.J., 1982, APJ 252, 102
7. Condon J.J., Huang Z.P., Yin Q.F., Thuan T.X., 1992, *Co,pact Starburst in Ultraluminous Infrared Galaxies*, in A.V. Filippenko (ed.), *Relationship between Active Galactic Nuclei and Starburst Galaxies* ASP conference Series, Vol. 31, pp. 79–82
8. De Robertis, M.M., 1994, *AGNs with Adaptive Optics*, in R. Arsenault (ed), *Astronomy with the CFHT Adaptive Optics Bonnette*, Proc. of workshop in Quebec, May 1994
9. Doyon R., Joseph R.D., Wright G.S., 1994, ApJ 421, 101–114
10. Elmegreen B.G., 1987, in M.Peimbert, J. Jugaku, (eds), *Star Forming Region*, IAU Symp. No 115, p.457
11. Evans, I.N., Ford H.C., Kinney A.L., Antonucci R.R.J., Armus L., Caganoff S., 1991, ApJ 369, L27.
12. Feldman F.R. & al., 1982, ApJ 256, 427
13. Ford H., Ferrarese L., 1995, BAAS 187, 560
14. Ford H., Tsvetanov Z., 1996, Sky & Telescope 91, 28
15. Franx M., Illingworth G., 1988, ApJ 327, L55
16. Gallais P., Rouan D., Lacombe F., Tiphene D., Vauglin I., 1991, A&A 243, 309
17. Genzel R., Eckart A., Drapatz S., Krabbe A., 1995, High Angular Resolution Infrared Imaging and Spectroscopy with the VLT: Galactic Nuclei, in J.R. Walsh, I.J. Danziger (eds), *Science with the VLT*, pp. 281–292
18. Hawarden T.G., Mountain C.M., Leggett S.K., Puxley P.J., 1986, MNRAS , 221, 41
19. Heckman T.M., Armus L., Miley G.K., 1990, ApJ Sup 74, 833
20. Ho P.T.P., Turner J.L., 1990, ApJ 349, 57
21. Jörsater S., Linblad P.O., 1989, *The [OIII] plume in NGC 1365*, in Meurs E.J.A., Fosbury R.A.E. (eds), ESO conf and workshop *Extranuclear Activity in Galaxies*, pp. 39
22. Kotilainen J.K., Ward M.J., Willinger G.M., 1993, MNRAS, 263, 655
23. Lai O., Rouan D., Gendron E., Rigaut F., 1996, in preparation
24. Larson R.B., Tinsley B.M., 1978, ApJ 219, 46
25. Larson R.B., 1988, in *Starburst and Galaxy Evolution*, 22nd Rencontre de Moriond, eds. T. Montmerle, Thuan, Tran Thanh Van, Editions Frontières, p. 467.
26. Marco O., Alloin D., Beuzit J.L., 1996, A&A, submitted
27. Mihos J.C., Hernsquist L., 1994, ApJ 431, L9
28. Normand, P., Rouan, D., Lacombe, F., Tiphène, D., 1995, A&A 297, 311
29. Norman C., Scoville N., 1988, ApJ 332, 124
30. Pier E.A., Krolik J.H., 1993, ApJ 418, 673
31. Rich R.M., 1992, *The Bulge/Starburst connection*, in A.V. Filippenko (ed.), *Relationship between Active Galactic Nuclei and Starburst Galaxies* ASP conference Series, Vol. 31, pp. 93–104
32. Rieke G.H., Lebofsky M.J., 1979, ARAA 17, 511
33. Roche P.F., Aitken D.K., Smith C.H., Ward M.J., 1991, MNRAS 248, 606-629
34. Rouan D., Tiphène D., Lacombe F., Normand P., 1994, *3.28 μm Feature Mapping in Starburst Galaxies: destruction of PAHs by UV and superwind*, in I.S. Mc Lean (ed),

Infrared Astronomy with Arrays: the next generation, Kluwer Academic Publishers, pp. 81-84
35. Sams B.J., Genzel R., Eckart A., Tacconi-Garman L., Hofmann R., 1994, ApJ 430, L33
36. Sanders D.B., Soifer B.T., Elias J.H., Madore B.F., Matthews K., Neugeubauer G., Scoville N.Z., 1988, ApJ, 325, 74
37. Sanders D.B., Scoville N.Z., Young J.S., Soifer B.T., Schloerb F.P., Rice W.L., Danielson G.E., 1986, ApJ, 305, L45
38. Sanders D.B., Mirabel F., 1996, ARAA 34, in press
39. Sanders D.B., 1992, *Luminous infrared galaxies: interactions and environment*, in A.V. Filippenko (ed.), *Relationship between Active Galactic Nuclei and Starburst Galaxies* ASP conference Series, Vol. 31, pp. 303–316
40. Sandqvist Aa., Jörsater S., Lindblad P.O., 1995, A&A 295, 585
41. Schlosman I., 1992, *Mass Supply to Active Galactic Nuclei and Starburst Galaxies*, in A.V. Filippenko (ed.), *Relationship between Active Galactic Nuclei and Starburst Galaxies* ASP conference Series, Vol. 31, pp. 335–346
42. Schlosman I., 1989, Frank J., Begelman M.C., Nature 338, 45
43. Sérsic J.L., Pastoriza M., 1967, PASP 79, 152
44. Simpson C., Mulchaey J.S., Wilson A.S., Ward M.J., Alonso-Herrero A., 1996 ApJ 457, L19
45. Soifer B.T., Houck J.R., Neugebauer G., 1987, ARAA 25, 187
46. Tacconi L.J., Genzel R., Blietz M., Cameron M., Harris A.I., Madden S., 1994, ApJ 426 L77
47. Telesco C.M. , 1988, ARAA 26, 343
48. Terlevich R., Melnick J., 1985, MNRAS 213, 841
49. Terlevich R., *The Starburst Model for AGN*, in A.V. Filippenko (ed.), *Relationship between Active Galactic Nuclei and Starburst Galaxies*, ASP conference Series, Vol. 31, pp. 133–146
50. Ulrich M.-H., 1981, *Optical Observations of Active Galaxies and Quasars at High Angular Resolution*, in Proceedings of the *ESO conference on the scientific Importance of High Angular Resolutionat Infrared and Optical Wavelengths*, Garching 24-27 March 1981
51. Ulrich M.-H., 1988, in *ESO conference on Very Large Telescopes and their instrumentation*, p.33
52. van de Werf P.P., Genzel R., Krabbe A. Blietz M., Lutz D., Drapatz S., Ward M.J.; Forbes, D.A., 1993, ApJ 405, 522
53. Urry C.M., Padovani P., 1995, PASP 107, 803
54. Veilleux S., Cecil G., Bland-Hawthorn J., 1996, Sci Am 274, 86
55. Véron-Cetty M.-P., Véron P., *A catalogue of Quasars and Active Nuclei* (6th edition), 1993, ESO Science Report 13
56. Wynn-Williams C.G., 1987, in M. Peimbert, J. Jugaku (eds) *Star Forming Region*, IAU Symp. No 115, , pp. 587.

ORAL CONTRIBUTIONS

A NEW DETECTOR FOR WAVEFRONT SENSING IN ADAPTIVE OPTICS APPLICATIONS

A. W. KLUTTIG AND J. L. A. FORDHAM
Department of Physics and Astronomy
University College London
Gower Street, London, WC1E 6BT.

1. Introduction

All of the new large telescope projects, and a number of existing telescopes are planning to incorporate an adaptive optics (AO) system to minimize the affect of atmospheric turbulence and thus increase the spatial resolution obtainable.

A typical AO system comprises four elements :

1. The optical input. Base-lined for most AO systems is the Shack Hartmann approach where the incoming wavefront is divided into a number of sub-apertures over the pupil plane of the telescope primary mirror. The image in each sub-aperture is focussed onto a wavefront sensor using a lens array.
2. A wavefront sensor that captures the Hartmann images
3. A processor that centroids each Hartmann image and provides the wavefront slope in each sub-aperture. From a map of the distortions associated with each image, coefficients for correction (Zernike coefficients) can be then derived.
4. An actuator system for de-forming the telescope mirror (primary or secondary) and thus compensating for the wavefront perturbations.

Within these systems, the demand is to provide accurate corrections in as fast a timescale as is reasonably achievable. The speed restriction is primarily governed by the frame rate of the wavefront sensor and the processing time required to derive the corrections that need to be applied to the actuators on the mirror. The accuracy of the corrections is dependent

A-M. Lagrange et al. (eds.), High Angular Resolution in Astrophysics, 319–323.
© 1997 *Kluwer Academic Publishers.*

upon the signal to noise ratio (SNR) on the data acquired by the wavefront sensor.

Two approaches are available for acquiring the wavefront error vectors: (1) laser beacons and (2) natural guide stars where, with the laser beacon, short exposures on natural guide stars are required to obtain tip/tilt information. For the natural guide star approach, the highest SNR is demanded within the detector frame period to increase the probability of having a bright enough guide star in the field of view – i.e. be able to guide on a fainter star. For the laser beacon approach, the same demand is required to minimize the power requirements on the laser and increase the star selection range for obtaining tip/tilt information.

Here we propose a new wavefront sensor based on advances in photon counting detector technology that are implemented in the MIC photon counting detector developed at University College (Ref.1), that will:

1. provide increased signal to noise when compared against the currently base-lined fast scan CCD cameras
2. minimize the overheads associated with capturing and processing of the images.

In addition the detector area will inherently be larger than that available from a fast scan CCD. This will allow lower de-magnification in the optical train or increased separation between the Hartmann images.

2. Application of the MIC Photon Counting Detector to Wavefront Sensing

The MIC photon counting system is an intensified CCD detector developed for high resolution and sensitivity astronomical applications, particularly in the blue and UV regions of the spectrum. The system is based around a micro-channel plate image intensifier developed by Imperial College (Ref.2) specifically for photon counting.

Normally the detector is operated in a true photon counting mode, where individual events are amplified by the intensifier, captured by the CCD, centroided to provide high resolution, and then accumulated in computer memory. The centroiding accuracy is $\sim 2.5\mu m$ FWHM.

For the wavefront sensing application the detector would not *operate in a photon counting mode but, within each frame period, capture the Hartmann images on the intensified CCD.* Then, by applying in hardware the algorithms derived for centroiding individual events, each Hartmann image is centroided to very high resolution. Finally, the X and Y centroid positions are transmitted to the processor for determination of the correction coefficients.

3. Comparison between Photon Counting and Fast Scan CCD Technologies

At present, direct fast frame rate CCD cameras with, typically, a format of 64 x 64 pixels are being base-lined as the wavefront sensor. The frame period is 1 millisecond and a maximum readout noise of 10 electrons rms is aimed for. Each of the frames must then:

1. be read into a data processing system requiring, in the base-lined direct CCD case, (64x64x1000) = 4 x 10^6 transfers/sec.
2. have the data on the separate Hartmann images extracted,
3. have centroiding of the Hartmann images carried out,
4. have the correction coefficients deduced prior to operation of the mirror actuators.

A simplistic comparison between this CCD approach and a photon counting approach is shown in Fig 1.

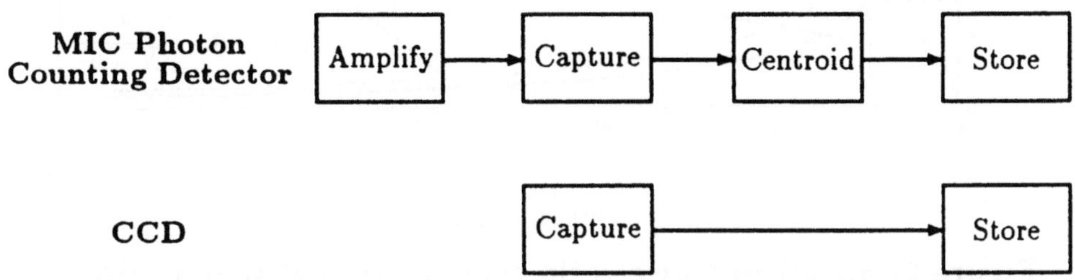

Fig.1 Comparison between Photon Counting and CCD technologies

As can be seen, the MIC detector contains two additional components, the image intensifier and the Centroiding, both of which provide advantages for AO applications. The image intensifier increases the signal to noise in the acquired data and the Centroiding minimizes data transfer overheads and Hartmann image processing.

4. Signal to Noise

The amplification stage provided by an image intensifier gives an essentially noise free system. The potential advantage of this can be seen in Figure 2, where a SNR comparison is carried out. For a direct CCD:

$$\mathrm{SNR} = \frac{\mathrm{N}}{\sqrt{\mathrm{N} + \mathrm{P}^2 \mathrm{r}^2}}$$

where P is the number of pixels that each Hartmann image subtends (two curves are plotted, one for P=4 (minimum) and one for P=9), N is the number of photons accumulated (taken as 100, this being the nominal minimum photon flux that allows adequate data processing) and r is the CCD readout noise.

It can be seen that for $10e^-$ readout noise an SNR of 4.47 is obtained with P=4 and an SNR of 3.16 obtained with the more reasonable P=9.

For the photon counter, as no noise is present, the equation simplifies to SNR $= \sqrt{N'}$. For the current MIC detector the peak detective quantum efficiency is ~14% and that of a thinned CCD ~80%. Thus N=100 for the CCD is equivalent to N′=20 for the current MIC detector. For N′=20, the current MIC SNR is already identical (4.47) to that of the base-lined AO CCD with P=4.

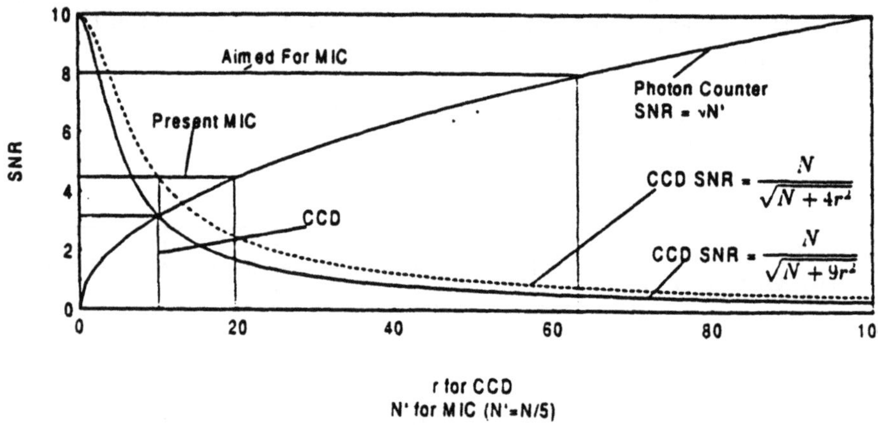

Figure 2. Signal to Noise ratio comparison between a direct CCD and the MIC photon counting detector.

However, by including a commercially available vacuum, proximity focussed, photo-diode with a photocathode efficiency peaking at ~50% in the wavelength range 500nm to 600nm in front of the image intensifier an SNR of 8 can be obtained.

5. Centroiding

In the MIC system event centroiding is used to find the centre position of each detected photon event thus overcoming the resolution loss in the image intensifier (Ref.3). The centroiding error has been measured as 2.5μm FWHM. The circuit associated could also be used for providing the centroid position of a Hartmann image to as high an accuracy. This is shown in Figure 3 where a pseudo star image was moved in 1μm steps along an X-Y table and the centroid position at each step then found.

The requirement for AO applications is, typically, to provide a centroid position within $\pm 2\mu$m and this will be easily achievable providing the following advantages over a direct CCD system:

- lowering the transfer requirements to the data processor. If, for example, a 10 x 10 array of Hartmann images is utilized then the number of transfers per second at 1KHz frame rate will be reduced by a factor 10 to 400K transfers/sec assuming that two bytes of data for each X and Y centroid position are required;
- reducing the processing requirements to just deriving the correction coefficients.

It is believed that the processing time is dominated by reading in of the data, locating of the Hartmann images, and the centroiding. Thus by implementing a hardware centroiding approach the data processing time can be minimized.

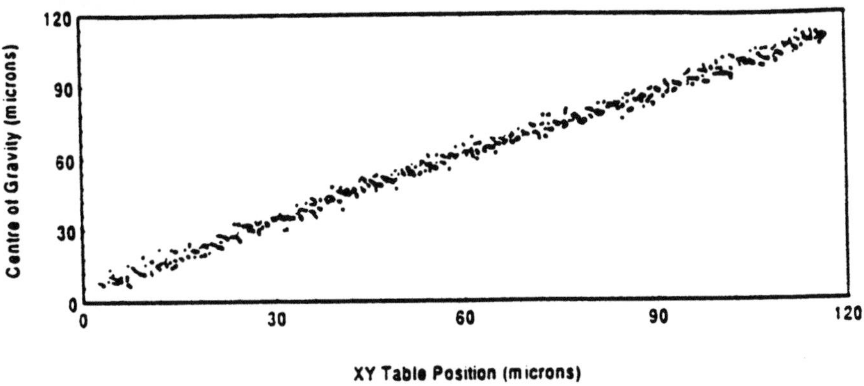

Figure 3. Measurement of Centroid Accuracy on a Star Image.

6. Summary

Here we have shown that by utilizing current advances in photon counting detector techology a wavefront sensor can be manufactured that (1) provides higher SNR when compared against the currently baselined CCD detectors – in particular, for baselined minimum signal (100 photons/image) and maximum CCD noise (10 electrons) an SNR of 8 can be achieved with a photon counting detector compared against 3.6 for a CCD – and (2) minimizes data processing overheads which allows corrections to be applied at a faster rate

References

Fordham,J.L., Bellis,J.G., Bone,D.A., Norton,T.J. (1991) The MIC photon counting detector *SPIE, Electron Image Tubes and Image Intensifiers II*, Vol.1449.

Airey,R.W., Norton,T.J., Morgan,B.L., Fordham,J.L., Bone,D.A., Powell,J.R. (1990) DQE enhancement of MCP intensifiers: results of the MIC II programme *SPIE Instrumentation in Astronomy VII*, Vol.1235.

Bellis,J.G., Bone,D.A., Fordham,J.L.A. (1990) A new real-time centroiding technique for photon counting detectors *PASP*, Vol 103.

FACTORS AFFECTING THE PERFORMANCE OF STELLAR INTERFEROMETERS

Application to the VLTI Error Budgets.

B. KOEHLER
European Southern Observatory
Karl-Schwarzschild Strasse,2, D-85748 Garching, Germany

Abstract

This presentation intends to provide a basic and didactic introduction to the various effects affecting the performance of stellar interferometers in terms of fringe visibility. For each effect, we first briefly describe the physical phenomenon involved and give the basic formulae useful to assess the associated visibility loss. Then, we illustrate their respective magnitude for the case of the Very Large Telescope Interferometer (VLTI) using the error budget developed for observation in the K band with fringe tracking.

1. Introduction

The formation of interference fringes in the optical domain is affected by many different physical effects which tend to spoil the quality of the fringe pattern. Optical stellar interferometers present some specific complication with respect to laboratory experiments. First, they tend to be complex systems with a large number of optical elements spread over large distances; second, several of these optical elements need to be moved continuously during the observation to track the stellar object and compensate for the diurnal motion; third the stellar light propagates through the atmosphere and is affected by the atmospheric turbulence. It is therefore of paramount importance to identify, understand, and analyse all potential sources of performance degradation. This is necessary to verify the feasibility of the scientific objectives but also to identify the areas where calibration of systematic errors is the most critically required. This is best achieved by the generation of error budgets which consist in distributing the top-level performance requirements into individual error sources and sub-system requirements as illustrated in figure 1.

We will review, in this paper, the various contributions to the error budget for the instrumental fringe visibility loss shown in figure 2. This error budget has been

A-M. Lagrange et al. (eds.), High Angular Resolution in Astrophysics, 325–335.
© 1997 *Kluwer Academic Publishers.*

developed for the VLTI for observation in the K band with the VLT 8-m telescopes equipped with adaptive optics and with fringe tracking.

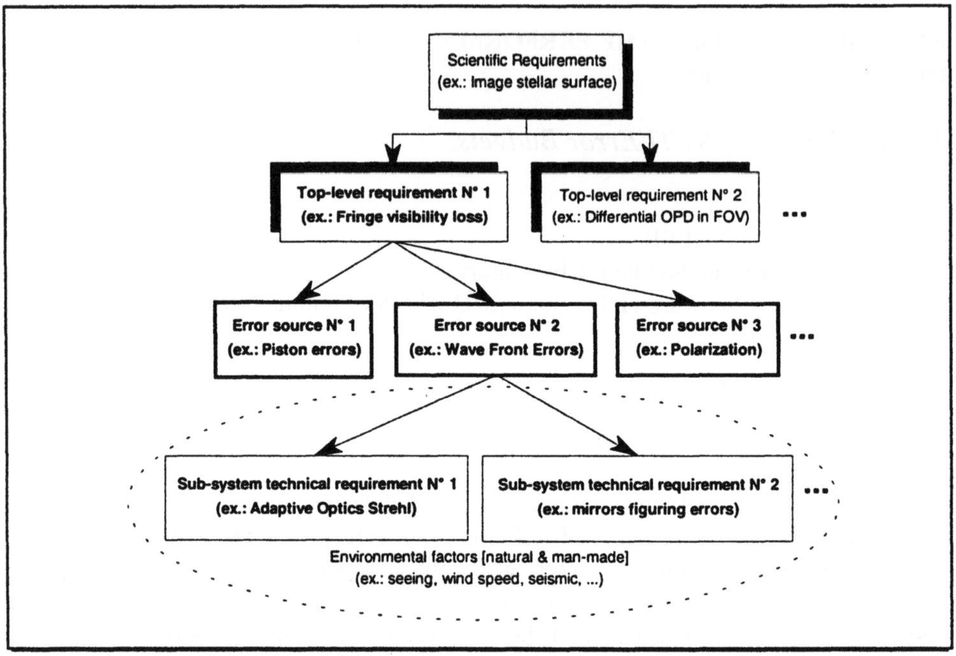

Figure 1: An error budget is a break-down of the top-level requirements into individual error sources and subsystem requirements. It is an essential tool for complex systems like stellar interferometers in order to identify and track the various factors affecting their performance.

The generic equation describing the fringe pattern obtained with a perfect instrument by combining two monochromatic beams of equal intensity is of the form:

$$I(\delta) = 2.I_0\left(1 + V_0.\cos\left(\frac{2.\pi.\delta}{\lambda} + \phi\right)\right) \qquad (1)$$

where
- I_0: Intensity in each beam
- V_0: Degree of mutual coherence of the incoming beams
- δ: Modulation delay between the two beams
- λ: The observing wavelength
- ϕ: Phase difference between the two beams

We recall that the fringe visibility (or fringe contrast), characterises the amplitude of the intensity modulation in the fringe pattern. It is defined by:

$$V = \frac{I_{max} - I_{min}}{I_{max} + I_{min}}$$

where I_{min} and I_{max} are respectively the minimum and maximum intensities.

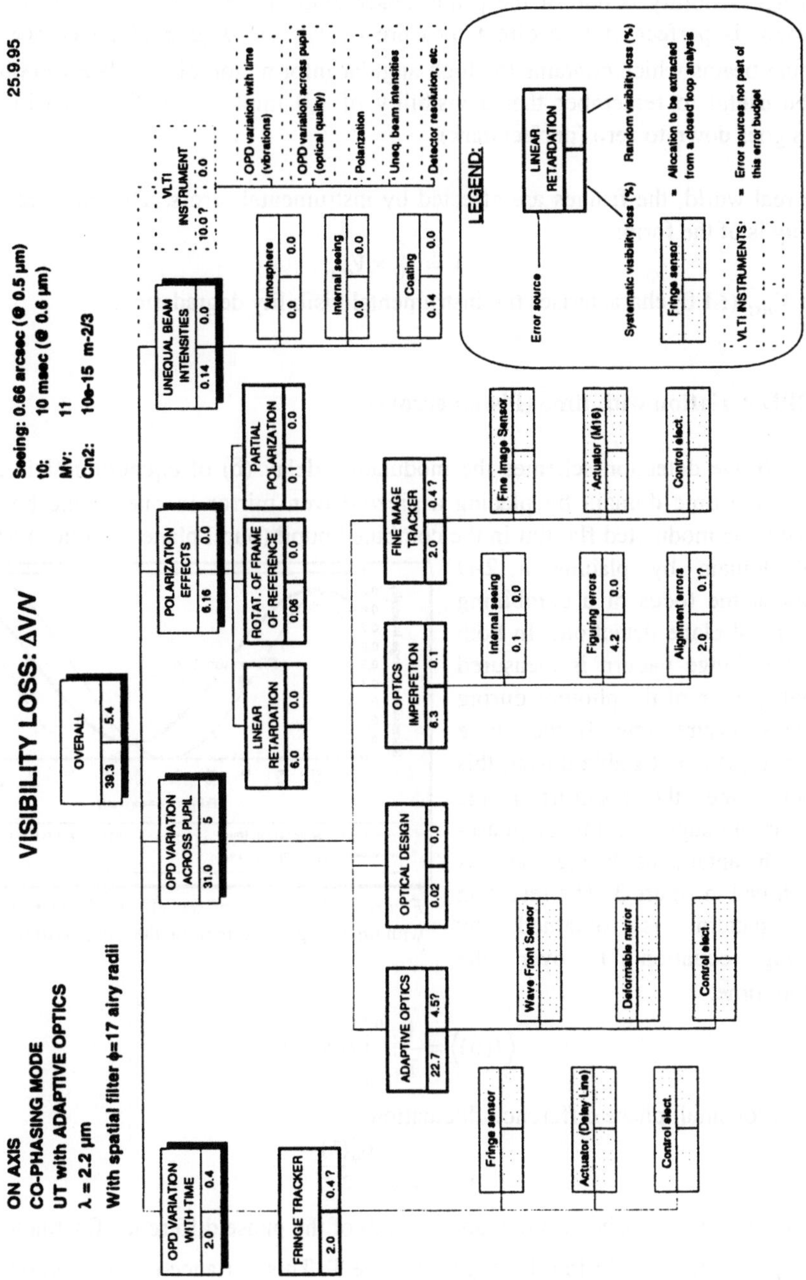

Figure 2: The VLTI Fringe Visibility Error Budget for observation in the K band using the 8-meter telescopes with adaptive optics and fringe tracking.

The fringe visibility is one of the prime observable of a stellar interferometer. If the instrument is perfect, it is a direct measure of V_0, the degree of coherence of the incoming beams which contains the high angular information of the observed object. It may be useful to remember that a visibility of 1.0 implies that the intensity in the fringes goes down to zero (perfect dark).

In the real world, the fringes are affected by instrumental errors. The measured fringe visibility is of the form:

$$V = V_0 \times V_{inst}$$

where V_{inst} (≤ 1.0) characterises the instrumental visibility degradation.

2. OPD variation with time (Piston errors)

In most fringe detection schemes the modulation delay (δ) of equation n°1 is created either in the time domain by placing a piezo-driven mirror in one of the beam and observing the modulated flat tint in the combined pupil (pupil plane detection) or in the spatial domain by placing a 2-D detector at the focus of a combining optics (focal plane detection). In both cases, the fringe pattern is measured after integration of the photons during a finite exposure time. If the phase difference (ϕ) is not stable during this exposure time, the resulting fringe pattern is the superposition of phase-shifted instantaneous fringe patterns as illustrated in figure 3. The resulting fringe pattern is obtained by averaging equation n°1 over the exposure time:

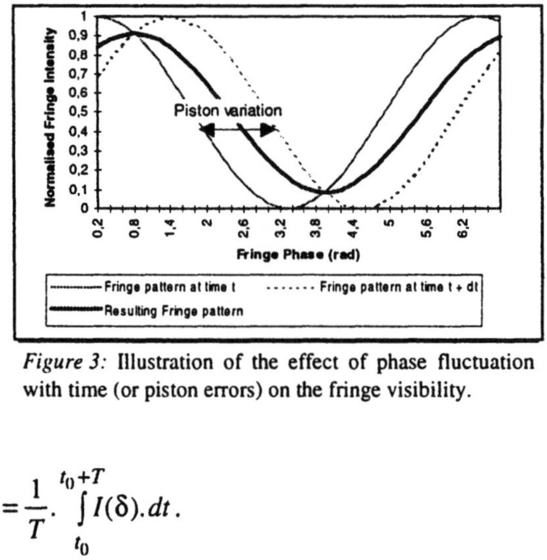

Figure 3: Illustration of the effect of phase fluctuation with time (or piston errors) on the fringe visibility.

$$\langle I(\delta)\rangle = \frac{1}{T} \cdot \int_{t_0}^{t_0+T} I(\delta).dt .$$

It gives, for small phase difference fluctuation,

$$V_{inst} = e^{-\frac{\sigma_\phi^2(T)}{2}}$$

where $\sigma_\phi^2(T)$ is the variance (in radians squared) of the phase difference fluctuation (also called "piston errors") during the exposure time T. This variance can be computed, for any exposure time (T), from the Power Spectral Density of the phase fluctuation (S_ϕ) by:

$$\sigma_\phi^2(T) = \int_0^\infty S_\phi(f).\left(1-\left(\frac{\sin(\pi.T.f)}{\pi.T.f}\right)^2\right).df .$$

The various sources of piston errors in a stellar interferometer are:

- the atmospheric turbulence
- the air turbulence inside the interferometer (internal seeing)
- motion of the optical element due to vibration (wind shaking, natural & man-made seismic motion, acoustics, vibrating equipment installed on or close to the telescope like pumps, etc.)

When the interferometer is equipped with fringe tracking capability, the piston errors are partially compensated and long integration time can be performed. In that case, the fringe contrast is computed as above with the variance of the residual phase fluctuation as delivered by the fringe tracking loop over the integration time.

In the case of the VLTI, a detailed control loop model including the fringe sensor dynamic and noise characteristic, the delay line mechanical model, the control law, and the various piston error sources as listed above has shown that a 2% visibility loss at 2.2 μm can be guaranteed.

3. OPD variation across the pupil (Wave Front Errors)

The effect of the Wave Front Errors (WFE) can best be understood when considering a pupil plane detection scheme. In the absence of WFE, the pupils after combination generate a perfect flat tint alternatively black (zero intensity) and white (full intensity) when the OPD is modulated (see figure 4, *top*). In the presence of WFE, the combined pupil contains spatial interference fringes as indicated schematically in figure 4, *bottom*, the intensity integrated over the pupil area by the monopixel detector never goes to zero and never reaches the full intensity. This means that the fringe contrast is lower that unity. By integration of equation n°1 over the pupil area, it can be easily shown that the instrumental visibility factor is

$$V_{inst} = e^{-\frac{\sigma^2_{(\phi_2-\phi_1)}}{2}}$$

where $\sigma^2_{(\phi_2-\phi_1)}$ is the variance over the pupil area of the differential WFE (in radians squared). When the WFE in each interferometer arm can be considered as statistically uncorrelated and of the same order of magnitude (case of the atmospheric corrugation, and random figuring & alignment errors) the above expression simplifies in

$$V_{inst} = e^{-\sigma^2_{\phi}},$$

where σ^2_{ϕ} is the variance of the WFE in *each* individual pupil.

It is worth to note here that the effect of WFE depends very much on the actual detection scheme. Indeed, if we place an array detector in the pupil plane to sample properly the pupil interference pattern shown on figure 4, each pixel of the detector will "see" a perfect fringe pattern of visibility $V = V_0$ when the path is modulated. Similarly, in the image plane, placing a spatial filter in the focal plane (e.g. a monomode fiber) will remove the high spatial frequency WFE present in the pupil and will greatly improve the instrumental visibility factor. This example shows the importance of considering the actual beam combining instrument when describing the performance of the interferometer. For the VLTI which is designed to feed several instruments, an assumption has therefore to be taken in the error budget. The example shown in figure 2 assumes a spatial filter with a diameter of 17 airy radii. This means that the high spatial frequency WFE which in any case cannot be corrected by the Adaptive Optics system (with 17 actuators across the pupil diameter) are filtered out. A smaller visibility loss, and therefore better calibration accuracy, could be obtained with a smaller filter (e.g. monomode fiber) at the price of some energy loss and more stringent alignment requirements.

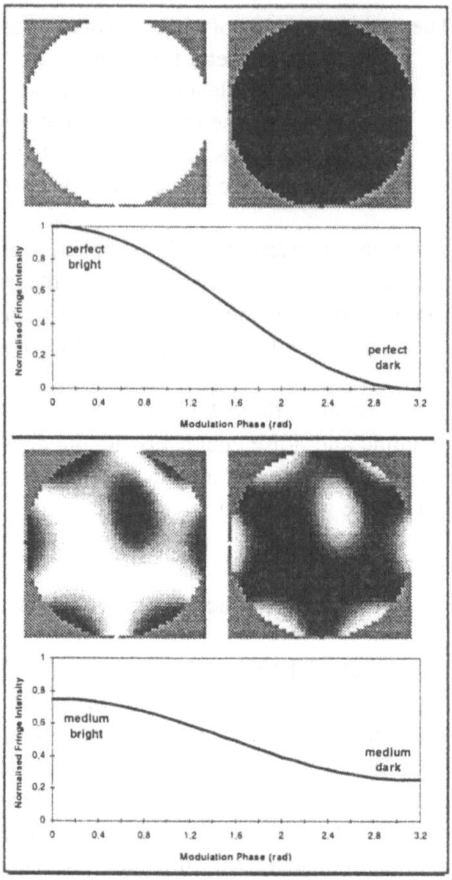

Figure 4: Illustration of the effect of Wave Front Errors on the fringe visibility when using a monopixel detector in the pupil plane.

The sources of WFE are:

- the atmospheric turbulence
- the air turbulence inside the interferometer (internal seeing)
- the optical design
- the optics misfigure & misalignment

WFE can be partially compensated by an Adaptive Optics (AO) system. The degree of compensation depends on the performance of the AO system and on its location in the optical train.

For the VLTI, an important part of the WFE will be compensated by the AO system located at the coudé focus of the 8-m telescope. The residual atmospheric WFE after compensation remains nevertheless the dominating source of error in the whole error budget with a corresponding visibility loss of ≈25% including residual image tracking errors. The next most important WFE source is the optics misfigure which, inspite of state-of-the-art optical quality, amounts to about 4% because of the large number of optical elements.

4. Unequal beam intensities

The effect of unequal beam intensities is easily computed considering that the complex amplitude of the combined electro-magnetic field is the sum of the complex amplitudes in each individual aperture: $A = \sqrt{I_1}.e^{j.\varphi_1} + \sqrt{I_2}.e^{j.\varphi 2}$. The combined intensity is therefore $I = A.A^* = (I_1 + I_2).\left[1 + 2.\frac{\sqrt{I_1}.\sqrt{I_2}}{I_1 + I_2}.\cos(\varphi_2 - \varphi_1)\right]$. This evidences the visibility loss factor which, in the general case of partially coherent incoming beams, leads to the formula :

$$V_{inst} = 2.\frac{\sqrt{I_1}.\sqrt{I_2}}{I_1 + I_2}.$$

This expression shows that the visibility is only weakly affected by "reasonable" unequal intensity. As an example, a drastic 50% intensity difference generates a visibility loss of only 6%. Therefore, in most cases, the visibility loss due to unequal intensity remains negligible with respect to the other sources of visibility degradation.

Sources of unequal beam intensities are:

- the atmospheric scintillation
- the scintillation due to internal seeing
- unequal throughput in the interferometer arms (e.g. differential coating efficiency).

Atmospheric scintillation, and internal seeing effects, are generally completely negligible. Scintillation rate, defined as the variance of the intensity fluctuation (normalised by the mean intensity), is at Paranal of the order of 0.1 in average with maximum instantaneous values reaching 0.6. These values, measured on small aperture (e.g. 3 cm) are drastically attenuated by the spatial averaging effect of the large VLT apertures. The resulting visibility loss is virtually zero.

Unequal beam intensity in the VLTI error budget is dominated by the assumption that up to 10% differential throughput can result from differences in the coating efficiency (e.g. ageing).

5. Polarisation effects

5.1. PARTIAL POLARISATION

Partial polarisation is the relative attenuation of one of the polarisation with respect to the other. If partial polarisation is not equal in the two arms, it affects the fringe visibility in a way similar to unequal beam intensities but with a dependence on the polarisation state. For unpolarised incoming light the visibility factor is:

$$V_{inst} = \frac{2(1+p)}{(3+p^2)}$$

where p is the partial polarisation affecting one of the two arms. As for unequal beam intensities, this effect is usually very small. As an example, a 10% polarisation (i.e. $1-p^2 = 0.1$) generates only a 0.07% visibility loss.

5.2. LINEAR RETARDATION

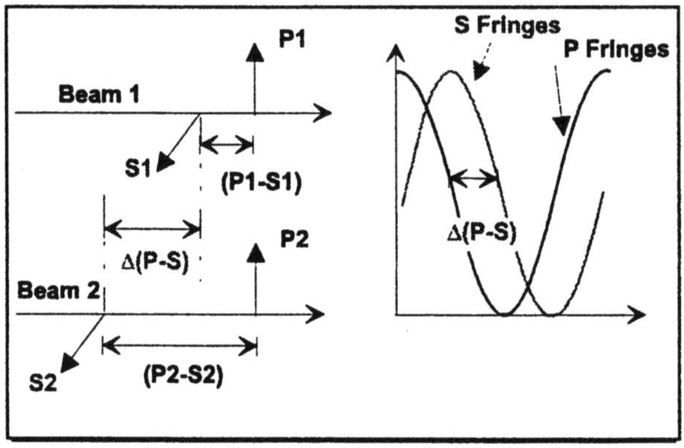

Figure 5: Illustration of the effect of polarisation linear retardation on the fringe visibility.

Figure 5 illustrates how linear retardation affects the fringe contrast. After passing through an optics, the two orthogonal polarisations of an incident light beam are phase shifted one with respect to the other by a quantity (P-S), the linear retardation. If the two beams in each arm of the interferometer are affected by exactly the same linear retardation, the fringes produced by the S and P polarisations will be superimposed. If a differential retardation Δ(P-S)=(P2-S2)-(P1-S1) exists, the two fringe patterns will be shifted one with respect to the other. For unpolarised light, these two patterns add incoherently on the detector to produce a fringe pattern with a visibility affected by the factor:

$$V_{inst} = \left|\cos\left(\frac{\Delta(P-S)}{2}\right)\right|.$$

Differential linear retardation are related to the coatings and occurs in the following cases:

- different coatings are used in the two arms
- different incidence angles are seen by the beam in the two arms
- differential polarisation characteristics due to coating manufacturing errors and ageing

In most interferometers including VLTI, the first two cases are eliminated by a perfectly symmetric design (i.e. same number of P and S reflections and same coating in the various arms). The third error source can be very significant for complex coatings. For the VLTI, it is minimised by the use of simple metallic coatings with a minimum number of protecting layers. As an example, a 4-layers boosted silver coating with less than 2% error in the layer thickness ensures visibility losses below 1% and 5% respectively in the K band and in the visible. Due to the highly coating-dependent nature of this effect and because the exact coating design is not yet frozen, a 6% visibility loss is allocated in the current VLTI error budget shown on figure 2.

5.3. ROTATION OF POLARISATION FRAME OF REFERENCE

The incident polarisation frame of reference experiences rotation when propagating through the optics of each interferometer arm. If the rotation in each arm is not exactly the same, the situation illustrated in figure 6 occurs after recombination. Each polarisation can be projected on the two reference axes (e.g. P1, S1). If the incoming light is unpolarised (or linearly polarised), the only components generating fringes are P1 with P2.cos(θ) and S1 with S2.cos(θ), where θ is the frame rotation angle. The other two components simply add constant intensities IP2*sin(θ)^2 and IS2*sin(θ)^2. The instrumental visibility factor is

$$V_{inst} = |\cos(\theta)|.$$

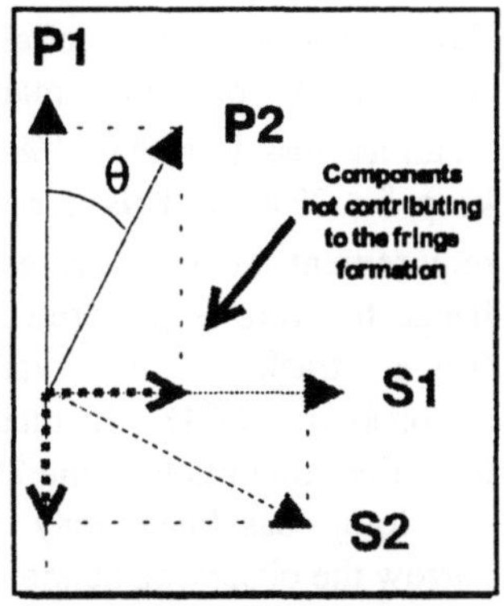

Figure 6: Effect of the rotation of the polarisation frame of reference for unpolarised light.

Rotation of the frame of reference can originate from:
- the optical design (i.e. non symmetric sequence of reflection in the two arms)
- non identical field/pupil rotation in the two arms when the star is tracked
- misalignments.

The first two error sources are avoided by design in the VLTI and the third one remains negligible even with a conservative assumption of a 2° pupil alignment error.

6. OPD variation with wavelength

6.1. SPECTRAL BANDPATH EFFECTS

It can be seen from equation n°1 that, as soon as the delay (δ) is not zero, the intensity in the fringe pattern at a given delay depends on the wavelength. When the observing spectral bandpath is not infinitely small, the resulting fringe pattern is the sum of each individual monochromatic pattern and presents therefore a reduced contrast. This is illustrated in figure 7. With the assumption that I_0 is independent of λ, integration of equation n°1 over the spectral bandpath $[\lambda_1 ... \lambda_2]$ shows that the instrumental visibility factor is:

$$V_{inst} = \frac{\sin(\pi.\delta / \Delta l)}{\pi.\delta / \Delta l}$$

where: $\Delta l = \lambda_0^2 / \Delta\lambda$ is the coherence length with $\Delta\lambda = \lambda_2 - \lambda_1$ and $\lambda_0^2 = \lambda_1.\lambda_2$.

In the general case, it can be shown that the function $V_{inst}(\delta)$ is the Fourier transform of the spectral distribution of the light being combined (taking into account the source distribution and the instrumental transmission characteristics).

As an example, the above equation shows that with a spectral resolution $R = \lambda_0 / \Delta\lambda$ of 300, maintaining a visibility loss less than 1% requires to keep $\delta \leq 23 \times \lambda_0$. This is not a difficult requirement as soon as even a poor fringe tracking is performed. When no fringe tracking is available, the atmospheric OPD fluctuation which typically amounts to some 20 μm r.m.s. for long baselines may require to narrow the observing bandpath.

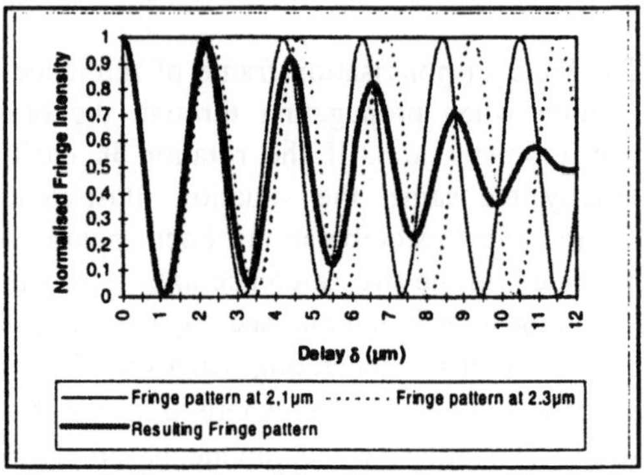

Figure 7: Illustration of the effect of spectral bandpath on the fringe visibility as a function of delay.

The VLTI error budget shown in figure 2 does not include this effect since the fringe tracking loop design ensures tracking around the so-called "white-light fringe" (i.e.δ =0). This effect can nevertheless exist if the OPD seen by the instrument differs from that measured by the fringe sensor because of misalignment. It is considered part of the instrument error budget and will be limited by an internal calibration system enabling to measure and cancel this misalignment.

6.2. LONGITUDINAL DISPERSION

The longitudinal dispersion effect occurs when the external delay due to the zenith angle of the observed star is compensated by delay lines placed in air instead of vaccum. In that case, the external delay (D) which has occured in vaccum above the atmosphere is compensated by placing the delay line at a physical position (X) so that, at a given wavelength (λ_0) the OPD is zero. This means that $OPD(\lambda_0) = n(\lambda_0).X - D = 0$ and therefore $X = D / n(\lambda_0)$ where n is the index of refraction of the air. Because n varies with the wavelength, the OPD at an other wavelength will be different from zero and equal: $OPD(\lambda) = n(\lambda).X - D = \left(n(\lambda)/n(\lambda_0) - 1\right).D$. If the observing spectral bandpath is not infinitely small, the resulting fringe pattern is the sum of each individual monochromatic pattern as described by equation n°1 where ϕ is replaced by $2.\pi.OPD(\lambda)/\lambda$. The resulting fringe visibility can be computed numerically from the dispersion characteristics of the air.

The magnitude of the dispersion effect depends on the observing wavelength and bandpath. It is usually negligible in the infrared but can be very significant in the visible as soon as the spectral bandpath exceeds some Angströms. In that case, partial compensation can be achieved by inserting in one arm an appropriate glass of adjustable thickness.

This effect does not appear in the error budget shown in figure 2 because it depends essentially on the instrument characteristics and is therefore considered part of the instrument-related error budget.

7. Conclusion

We have described the numerous physical phenomena affecting the fringe visibility in stellar interferometers and have illustrated their respective importance for a specific case of VLTI observation.

Being aware of these sources of performance degradation is certainly a must for engineers building interferometers but we believe it is also important for scientists involved in the definition and/or the use of interferometric facilities. Indeed, they drive the technical feasibility of the scientific objectives and their relative importance can define the calibration needs required to achieve a given measurement accuracy.

It shall also be kept in mind that the exact definition of the combining instrument is required to define the final performance of the interferometer.

OPTICAL INTERFEROMETRY AND GROUND-BASED ASTROMETRY AFTER THE HIPPARCOS MISSION

a niche for intermediate angle optical astrometry

G. DAIGNE
Observatoire de Bordeaux
BP 89, 33270-Floirac, France

Abstract. Magnitude extension of the Hipparcos catalogue is supposed to be realized through the Tycho Reference Catalogue. The deteriorating precision of this catalogue is already a limiting factor. A significant improvement can be realized with new ground-based observations of the Tycho stars, a dual-object astrometric interferometer in the K band being a very promissing instrument for such an objective.

1. Introduction

A clear hierarchy between its different instruments has long been prevailing in ground-based optical astrometry : the wider the angle, the more fundamental was the measurement. The success of the Hipparcos mission and the achievements of VLBI techniques in the fields of geodynamics and astrometry have greatly modified this situation. With its link to the celestial (radio) reference frame, the Hipparcos catalogue has been adopted as the Stellar Reference System and will be available as such from 1997 on. In the realization of reference systems, the contribution of the astrometric space mission is twofold : a global reduction, on the whole sphere with a single instrument, and a much better precision in the measurements.

Is there a need for secondary catalogues and the next ? The answer is yes certainly, as most of the optical observations will not have a direct access to objects of the Hipparcos catalogue, the fields observed with modern instruments being usually too small and/or the reference stars too bright to be measured. The Hipparcos catalogue is nearly complete up to magnitude $V = 9$ with a density of about 3 stars per square degree ; a secondary catalogue will be a *magnitude extension* of the primary one.

A-M. Lagrange et al. (eds.), High Angular Resolution in Astrophysics, 337–341.
© 1997 *Kluwer Academic Publishers.*

In this contribution, I shall first recall the precisions and densities expected with the planned extensions, and stress some inadequacy between the different levels in this new astrometric hierarchy in formation. Then I shall show that a dedicated IR interferometer is a way towards a more direct access to the astrometric accuracy of the Stellar Reference System.

2. Magnitude extensions of the Hipparcos catalogue

Small transit telescopes used in the drift-scan mode (Flagstaff, Tokyo, Bordeaux, Sao Paulo) are supposed to greatly contribute to the extension of astrometric measurements into the magnitude range 15 – 16, at least for the half sky about the equatorial zone. Development and validation of new reduction methods are in progress [1] and it has been shown that a reference catalogue with a density of at least 10 objects per square degree is required to determine instrument parameters and correct for some random refraction effects.

In the present situation, the extension of the Hipparcos catalogue to the magnitude range $V = 15 - 16$ is hampered by the absence of a secondary catalogue with adequate stellar density and accuracy. The Tycho experiment, which made use of the star tracker aboard the astrometric satellite, is supposed to greatly contribute to fill this gap [2]. The magnitude limit of the Tycho experiment is $V = 11$, that is a density of about 25 stars per square degree. Its precision is expected to be 50 mas in the magnitude range $10 \leq V \leq 11$. Due to the uncertainty on proper motions, the Tycho catalogue could not be used as such at other epoques and data from the Astrographic Catalogue and from the POSS-1 are being reprocessed at Heidelberg. With proper motion estimates, the position measurements of the Tycho experiment will form the Tycho Reference Catalogue or TRC [3].

The TRC will be published slightly after the Hipparcos catalogue, and its density seems to be quite adequate for further magnitude extensions of the astrometric catalogues. Will its precision be adequate also for the next decade ? A simple relation gives the precision σ_t of a catalogue in terms of the epoque t :

$$\sigma_t^2 = \sigma_0^2 + (t - t_0)^2 \sigma_\mu^2 \tag{1}$$

where σ_0 is the precision at the catalogue mean epoque t_0 and σ_μ the precision in proper motion. The catalogue mean epoque is a weighted average of the different contributions. With a precision of 250 mas for the AC (epoque 1900), 150 mas for the POSS-1 (epoque 1954) and 50 mas for the Tycho experiment (epoque 1991.25), the three parameters of the TRC have been estimated as well as the expected precision in year 2005 (Table 1).

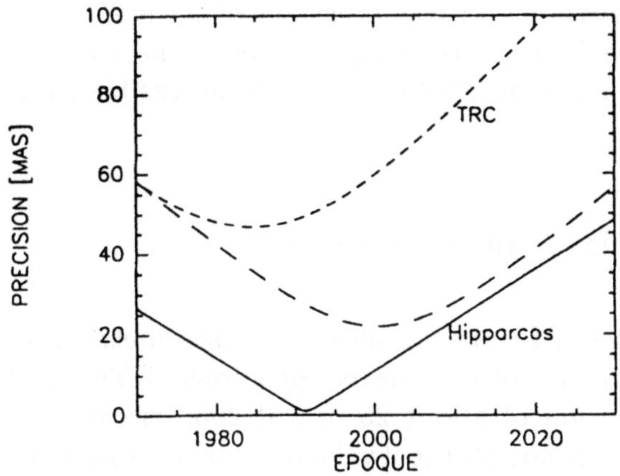

Figure 1. Precision of the astrometric catalogues in terms of epoque. The precision of a suggested upgrade, with new ground-based observations in year 2005, is shown by a long-dashed line.

The Table shows also the parameters and expected value for the Hipparcos catalogue, and Fig.1 represents the precisions of the two catalogues along the period 1970-2030.

TABLE 1. Parameters of the two astrometric catalogues and of the suggested upgrade (TRC+). σ_{2005} is the expected precision in year 2005.

	Mean epoque	σ_0[mas]	σ_μ[mas/y]	σ_{2005}[mas]
Hipparcos	1991.25	1	1.25	17
TRC	1984.6	47	2.4	68
TRC+	2000.4	22	1.8	23

The Tycho Reference Catalogue seems to be mainly useful for the first half of the period. For the next decade, it will degrade appreciably, 68 mas in year 2005 compared to 17 mas for Hipparcos. A factor of merit of a catalogue can be roughly defined as the ratio of its density to its variance. In year 2005, the factor of merit of the TRC will be about half that of the Hipparcos catalogue , which means that further extensions of catalogues will not have a clear access to the first level and its quality. A way to correct such a situation is to carry new observations of the Tycho objects. With observations in year 2005 and a precision of 25 mas per object, we have estimated the 3 parameters of a new catalogue 'TRC +'. The catalogue parameters are shown in Table 1, and the precision is plotted on Fig.1. New ground-based observations with a precision of 25 mas definitely improve the situation for the decades to come, to a point where the precision of the Hipparcos catalogue in year 2010 or so will be accessible to a large number of stars.

3. An astrometric IR interferometer

One of the objective of future ground-based astrometric measurements is clearly defined : to measure the relative positions of Tycho and Hipparcos objects with a precision of about 25 mas if 2005 is the mean epoque for these measurements. If not all the Tycho objects are to be observed, nearly half should be in order to have more than 10 objects per square degree. The distance between the reference and program objects is typically less than a degree, that is it lies in an *intermediate angular range* as compared to wide angles on the one hand and narrow angles (a few 10 arc minutes) on the other. To our knowledge, a relative precision of 25 mas, one degree apart, cannot be reached with mono-pupil instruments, it's the reason why an optical interferometer has been considered.

The magnitude limit is a problem for an interferometer operating in the visible, as the size of the (atmospheric) coherent cells does not allow to observe beyond $V = 8 - 9$ at a good site, that is nothing else than the reference stars in our astrometric problem. The situation improves on longer wavelength, the size of the coherence cells increasing as the wavelength to the power 2.4. The detection threshold of an IR interferometer in the K band can be estimated with the best instrument parameters.

A magnitude $K = 10 - 11$ should be reached with an interferometer observing at 2.2μm and an entrance pupil with about 1 meter in diameter. If so, nearly all the Tycho objects could be observed, the K magnitude of stars with a positive colour index being smaller than the V one. As a matter of fact, the Besançon model for the Galaxy shows that the average density of stars with $K \leq 11$, outside the galactic disk itself, is about 4 times its counterpart in the V band. In order to preserve some degree of freedom in the choice of the objects to be observed, the detection threshold of the interferometer should certainly lie beyond $K = 10$, if not $K = 11$.

A demonstration experiment on the achievable detection threshold is a prerequisite to more detailed investigation of the instrument. One or the other optical interferometer at Calern could be a candidate for a sensitivity test, although a demonstration experiment for the VLT interferometer appears to be more promissing.

4. Precision and output

Due to the large number of stars to be observed with an IR astrometric interferometer, the observing time required to reach a nominal precision of 25 mas is a crucial parameter. This parameter is strongly dependent on the atmospheric turbulence, that is on the observing site as well as on the specific night considered. The variations in the observing conditions are usually averaged with several independent observations of the same targets, let's say 4 or 5, which means a precision of 50 mas per measurement. The precision allowed by atmospheric turbulence is dependent on the angle to be measured. For intermediate angles, Shao and Colavita [4] have shown that it is only weakly dependent on the interferometer baseline and varies as $\sqrt{T}$ where T is the observing duration. The figure given for a 1 degree angle at Mauna Kea is 5 mas/$\sqrt{\text{hour}}$, which means that a duration of 36 s might be enough for individual measurements, or a total duration of about 3 mn for each Tycho star to be observed with a dual-object interferometer.

The previous evaluation is just a rough estimate. The outer scale of the turbulence has not been considered, nor the average wind speed, nor the interferometer baseline and its optimisation. Further investigations on the achievable precision are needed. For the time being, a figure of 5 mn per object, including pointing and star captation, can be retained. It means that 2.10^5 stars could be observed in less than 7 years. This object number is about half the Tycho stars observable from a site at mid-latitude. It means that trying to improve the secondary astrometric catalogue through ground-based observations is a sustainable objective.

References

1. Stone, R.C., Monet, D.G., Monet, A.K.B., Walker, R.L., Ables, H.D.,Bird, A.R. and Harris, F.H. (1996) The Flagstaff Astrometric Scanning Transit Telescope (FASTT) and star positions determined in the extragalactic reference frame, *Astron. J.* **111** (4), 1721-1742.
2. Høg, E., Bastian, U., Egret, D., Grewing, M., Halbwachs, J.L., Bässgen, G., Bernacca, P.L., Donati, F., Kovalevsky, J., van Leeuwen, F., Lindegren, L., Pedersen, H., Perryman, M.A.C., Petersen, C., Scales, D., Snijders, M.A.J. and Wesselius, P.R. (1992) Tycho data analysis : overview of the adopted reduction software and first results, *Astron. Astrophys.* **262**, 177-185.
3. Röser, S. and Høg, E. (1993) Tycho Reference Catalogue (TRC) : a catalogue of positions and proper motions of one million stars, in A.G. Davis Philip, B. Hauch and A.R. Upgren (eds.) *Workshop on Databases for Galactic Structure*, L. Davis Press, N.Y., pp. 137-143.
4. Shao, M. and Colavita, M.M. (1992) Potential of long-baseline infrared interferometry for narrow-angle astrometry, *Astron. Astrophys.* **262**, 353-358.

IMPROVING THE NATURAL GUIDE STAR MAGNITUDE LIMIT FOR WAVEFRONT SENSING IN AN OPEN LOOP SYSTEM

O. ESSLINGER AND M.G. EDMUNDS
Dept. of Physics and Astronomy
University of Wales College of Cardiff
CF2 3YB Cardiff
United Kingdom

Abstract. We present a way of improving the limited sky coverage of natural guide star adaptive optics by increasing the limiting magnitude of the wavefront sensor. We take advantage of the occurrence of a special turbulence situation where a single phase screen slides with time in front of the telescope pupil. This enables us to add predictive input to the normal temporal integration performed by an adaptive optics system, therefore to improve the quality of the correction, hence the limiting magnitude. This paper presents some preliminary results obtained in the simulation of an open loop system working with a Shack-Hartmann wavefront sensor.

1. Introduction

The largest limitation to the application of adaptive optics to astronomy is the very limited sky coverage when using natural guide stars, mainly due to the limitation in magnitude for the reference star. The expectation for the visible limiting magnitude for the Very Large Telescope adaptive optics system is 16 [1]. This value falls too short, for example, for the study of a large sample of QSOs [2]. An improvement in limiting magnitude by a factor 1, though small, would lead to a large increase in the sky coverage and the number of possible targets (see e.g. [3]).

We discuss here the possibility of improving adaptive optics correction, and therefore the limiting magnitude of a system, by using the fact that wavefronts can be predicted on the basis of their past measurements. Earlier studies have mostly applied prediction to solve the problem of the time delay in the adaptive optics servo-loop (see e.g. [4], [5], [6], [7]). The aim of

A-M. Lagrange et al. (eds.), High Angular Resolution in Astrophysics, 343–347.
© 1997 *Kluwer Academic Publishers.*

this article is slightly different. We intend to use several wavefront measurements and to combine them, after correction for time evolution, in order to obtain a better estimation of the correction to be applied by the deformable mirror. The result is a better adaptive optics correction for a given reference star magnitude, or an improved limiting magnitude for a given level of correction.

The goal of our work is not a comprehensive analysis of such a procedure, but a first evaluation of the improvement it can yield. We take advantage of the presence under certain circumstances of a single turbulent layer. Gendron [8] has observed several occurrences of a special turbulence situation, where the major source of time dependence of the refractive index variations is the effect of a global wind carrying the perturbations. The random pattern of perturbations can be considered as frozen during the time it crosses the telescope pupil. The occurrence of this phenomenon has been roughly estimated for the site of the ESO observatory in La Silla (Chile) by processing data corresponding to arbitrarily chosen nights over a period of 2 years. This situation seems to occur about 10 % of the nights and, when present, it remains usually stable during the whole night (Gendron, private communication). Of course, turbulence is never entirely due to a single layer but, in this work, we will assume it is. With this hypothesis, the wavefront prediction simply consists of a translation, which enables us to make a first evaluation of the method without going into complex considerations about prediction in the general case.

2. The method

This paper deals with an open loop system and with a Shack-Hartmann wavefront sensor. The prediction is carried out by shifting a set of wavefront slopes and transforming it to take into account the different geometry of the subaperture array in regard to the wavefront.

The preliminary step in the method is the determination of the wind parameters by application of the same method as Gendron [8]. Cross-correlations between sets of wavefront slopes are calculated and show a peak of correlation whose motion can be used to determine the direction and the velocity of the wind. We denote $\mathbf{v}$ the corresponding vector.

With this information we can shift a set of wavefront slopes to make a prediction. We start with a grid of 16×16 values and we want to shift it by the vector $\Delta t \cdot \mathbf{v}$ where Δt is a time gap. The method is illustrated in Fig. 1. The grid G (filled circles) represents the position of the 16×16 values $\mathbf{S}(i,j)$. The grid G' is that grid shifted with the vector $\Delta t \cdot \mathbf{v}$. We intend to compute, from the slopes $\mathbf{S}(i,j,t)$ at the positions on G', new values for the slopes $\mathbf{S}(i,j,t+\Delta t)$ at the positions on G, which correspond to the

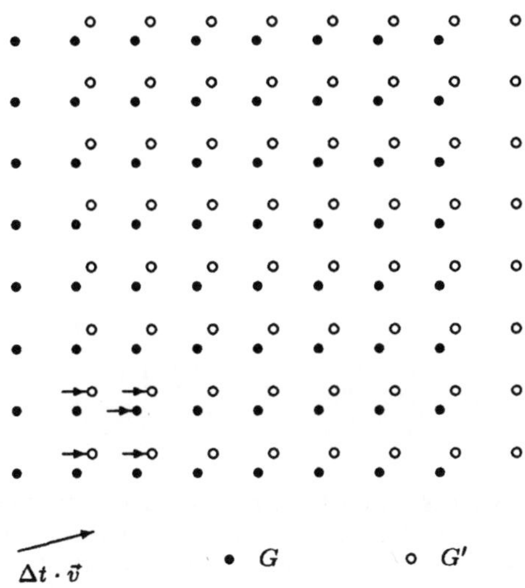

Figure 1. Schematic representation of two sets of slopes. The filled circles represent the position of the subapertures. The circles represent the new equivalent positions of a previous set of slopes shift by $\Delta t \cdot \mathbf{v}$. The arrows show 4 circles corresponding to 4 old slopes at 4 new positions, and 1 filled circle (a subaperture) which will be assigned a new predicted value for the slope.

subapertures. For example, from the values $\mathbf{S}(1,1,t)$, $\mathbf{S}(1,2,t)$, $\mathbf{S}(2,1,t)$, $\mathbf{S}(2,2,t)$ on the grid G' (the circles with arrows), we can estimate the value $\mathbf{S}(3,2,t+\Delta t)$ on G (the filled circle with an arrow) by computing a weighted average of the 4 old slopes.

We apply this procedure to different previously measured frames (with the adequate shift) to get different sets of predicted slopes. We then combine these to the current measured set (for example by a weighted average) to obtain a new set of improved slopes which can be used by the system.

3. Results

In order to evaluate the gain provided by our method, we created a code simulating the whole system (atmosphere, wavefront sensor, noise, phase screen reconstruction) in the open loop case. This enables us to study the improvement in the Point Spread Function. Fig. 2 is an example of result. On the left, we have the PSF calculated from the noisy slopes improved by our method, and on the right, the one computed from the usual noisy slopes. There is clearly a significant improvement between the PSF obtained in the

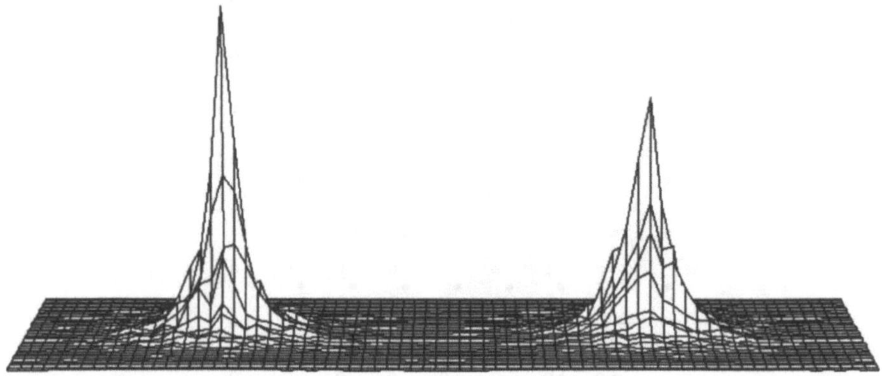

Figure 2. The PSF in two cases : on the right it is reconstructed using the raw data, affected by the noise corresponding to a 16th magnitude star (about 8 photons per subaperture per sampling period with our parameters), on the left, the same thing when the noise has been reduced by our method. The improvement is carried out by using the current frame plus the 12 previous ones.

normal way and the one computed from the improved slopes, of about 40 % in height.

We show if Fig. 3 the evolution of the heights of the PSFs obtained in the normal way and with our method as a function of the guide star magnitude. It can be seen that if the limiting magnitude is defined by a given height of the PSF, for example 40 % of the height in the case without photon or readout noise, our method yields an improvement of about 1 magnitude.

4. Conclusion

We have developed a new method of improving the measurements of the most common type of wavefront sensor. This means a better point spread function for a given guide star magnitude, but also a better limiting magnitude for a given correction level, therefore an significant increase in sky coverage.

This paper deals with results of a preliminary work in the open loop case. The improvement is shown to be very good. The increase in the height of this PSF is about 40 % in the interesting region and can even reach 70 % for different conditions. For a given level of correction, the algorithm leads to an improvement in the guide star limiting magnitude of about 1, yielding a large increase in the sky coverage.

We are currently working on the application of this method to an adap-

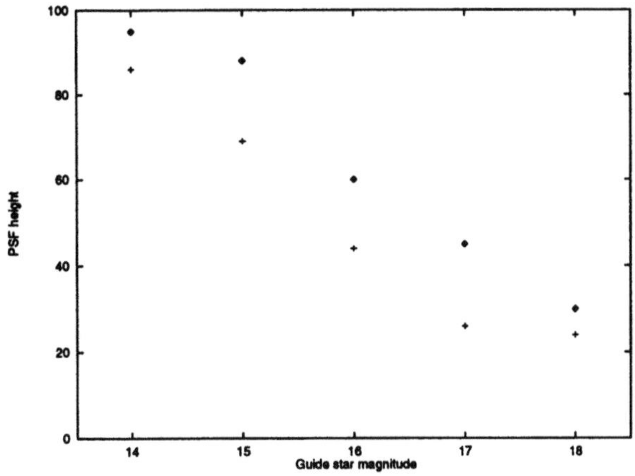

Figure 3. Improvement as a function of the reference star magnitude. 100 % corresponds to the height of the PSF when the slopes are unaffected by photon or readout noise. The crosses are the results when the raw data are used, the diamonds correspond to slopes improved by our method. The shift is a quarter of a subaperture between successive frames. It can be seen that if the limiting magnitude is defined by a given height of the PSF, for example 40 % of the height in the case without photon or readout noise, our method yields an improvement of about 1 magnitude.

tive optics system and on the generalisation of the algorithm to any kind of wavefront sensor [9].

References

1. Hubin, N., Théodore, P., Petitjean, P. & Delabre, B., 1994. In: *Astronomical Telescopes and Instrumentation for the 21st Century*, Proc. SPIE 2201.
2. Théodore, B., Petitjean, P. & Hubin, N., 1994. *The Messenger*, **77**, p.20.
3. Bienaymé, O., Robin, A.C. & Crézé, M., 1987. *Astr. Astrophys.*, **180**, p.94.
4. Zuev, V.E., Lukin, V.P., 1987, Applied Optics 26, 139
5. Jorgenson, M.B., Aitken, G.J.M., 1993, In Active and Adaptive Optics, Garching
6. Paschall, R.N., Anderson, D.J., 1993, Applied Optics 32, 6347
7. Schwartz, C., Baum, G., Ribak, E.N., 1994, J. Opt. Soc. Am. A 11, 444
8. Gendron, E., 1995, Optimisation de la Commande Modale en Optique Adaptative: Applications à l'Astronomie. *PhD thesis*, Université Paris VII.
9. Esslinger, O., Edmunds, M.G., 1996, *Astr. Astrophys.*, in preparation.

DISPERSION COMPENSATION AND FRINGE TRACKING

P.R. LAWSON
MRAO, Cavendish Laboratory
Madingley Road, Cambridge CB3 OHE, UK

Abstract. Stellar interferometers use delay lines to stabilize the location of white-light fringes, which would otherwise drift as a star moves across the sky. In principle a vacuum delay line should be used, but in practice many interferometers have used air delay lines. We examine the effects of dispersion and illustrate how it degrades our ability to track fringes.

1. Introduction

The delay line in a stellar interferometer is positioned to match a vacuum path-difference that arises in space. The vacuum difference arises due to the orientation of the baseline with respect to the source and changes as the source moves across the sky. For economic reasons many interferometers use an air delay line and match an air path to the vacuum path. Consequently the light is dispersed longitudinally *within* the interferometer and light of different wavelengths arrives at the beam combiners with different relative delays.

The fringe visibility can be approximated by the modulus of the Fourier component at the fringe frequency. With channeled spectra, where the fringe frequency is determined by the group optical path-difference, the fringe visibility is highest for an optical path-difference that is independent of wavelength. When using an air delay line this is only true over suitably narrow bandwidths. The effects of longitudinal dispersion therefore limit the bandwidth of observations. These limitations are straightforward to determine: if one uses an air delay-line and tolerates no more than a 5% reduction in the visibility amplitude, the fractional bandwidth for measurements at 950 nm must be less than 10% when the path-difference of air is 10 m, 6% for 30 m, and 3% for 100 m. At shorter wavelengths the restrictions are yet more severe. For interferometers with only modest-sized baselines,

A-M. Lagrange et al. (eds.), High Angular Resolution in Astrophysics, 349–353.
© 1997 *Kluwer Academic Publishers.*

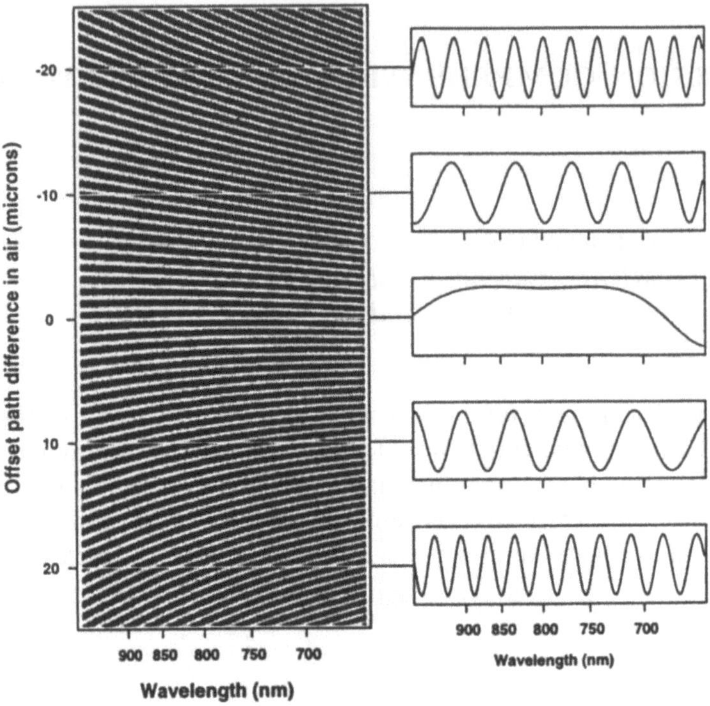

Figure 1. Channeled spectra for an interferometer with an air delay line compensating for 1.0 m of vacuum path. The image on the left shows the fringes as a function of the offset in air delay. The origin of the vertical axis is the path-difference in air that yields a zero group optical path-difference at 800 nm. The examples on the right show the channel fringes at offsets of -20, -10, 0, 10, and 20 μm. Changes in fringe frequency at a given offset are due to dispersion. The largest allowable bandwidth for observations at wavelength of 800 nm is ~200 nm.

fringe visibilities may be degraded by an amount that varies according to the hour angle of the source. Simultaneous observations across the entire visible spectrum are normally impossible, as the path-difference in air may be several tens of meters.

2. Dispersion and Channeled Spectra

Figures 1 and 2 illustrate the effect of dispersion. In both cases a prism spectrometer is simulated that uses a BK7 prism whose prism angle is 60°. Over the band 650-950 nm this is an almost ideal spectrometer; wavenumber is mapped as a quasi-linear function of detector coordinate. Each figure shows a simulated channeled spectrum as a function of a changing offset in an air delay line. The cross section of channeled fringes are shown on

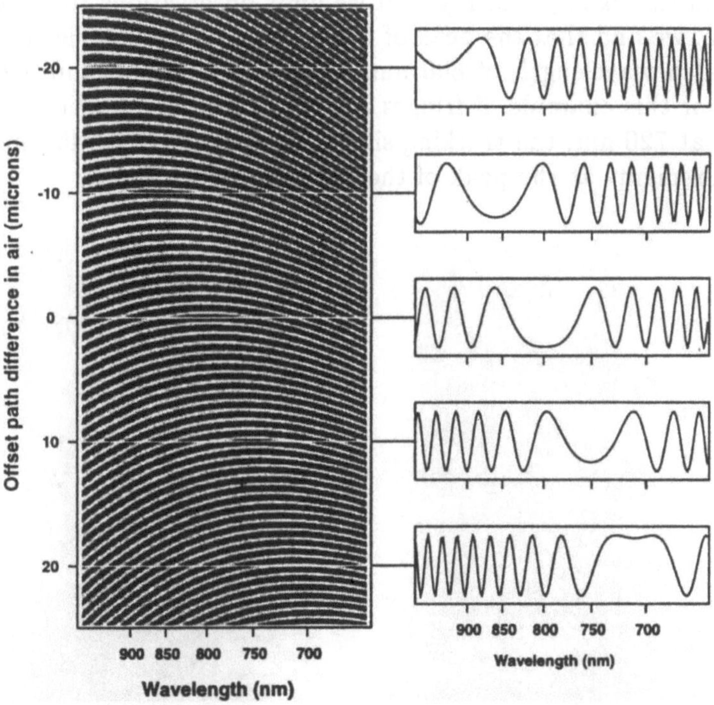

Figure 2. Channeled spectra for an interferometer with an air delay line compensating for 10.0 m of vacuum path. The origin of the vertical axis has shifted to a new delay but still corresponds to a zero group optical path-difference at 800 nm. The allowable bandwidths for visibility measurements are now reduced to ~70 nm for measurements at a wavelength of 800 nm.

the right for increasing offsets at 10 μm intervals. The origin of the vertical axis corresponds to a group optical path-difference of 0 μm at a wavelength of 800 nm. In Fig. 1 the delay line corrects for 1.0 m of vacuum path-difference; although the fringes should be regularly spaced in each example on the right, and be constant at the origin, some distortions are apparent. By inspection we find that the optical path-difference is constant over bandwidths of ~200 nm. Figure 2 shows a more extreme example, where the delay line now corrects for a vacuum path-difference of 10.0 m, and the optical path-difference is constant only for bandwidths less than ~70 nm.

A group-delay fringe tracker operating with an air delay line must abide by the same bandwidth limitations but also reference its detected delay to the wavelength of the visibility measurements. The fringe tracker seeks to locate the peak of the coherence envelope—where the number of fringes across a given bandwidth is minimized. This location is wavelength dependent. In the example of Fig. 2 if we limit observations to a bandwidth of ~70 nm, we find that the peak of the coherence envelope lies at an offset of 0 μm at a wavelength of 800 nm, -20 μm at 950 nm, and +20 μm at 720 nm. For this example, if fringes are tracked at 950 nm but visibility is measured at 720 nm, the tracking signal must be offset by 40 μm to place the measurements at the peak of the coherence envelope.

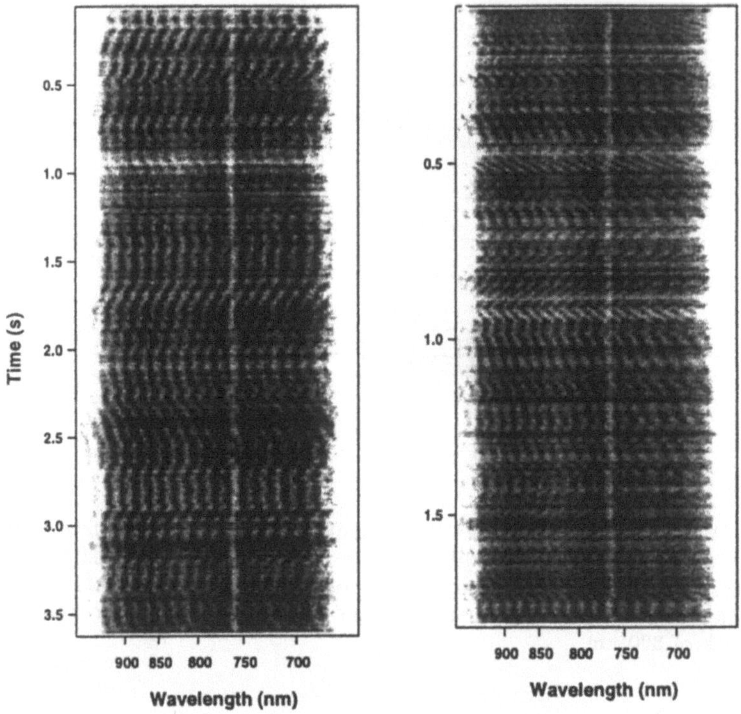

Figure 3. Channeled spectra observed at COAST with the star α Boo. On the left the air path-difference is 1.7 m but shows little evidence of dispersion. However, on the right the air path is now 3.1 m and the fringe spacing visibly increases towards shorter wavelengths.

3. Channeled Spectra at COAST

The Cambridge Optical Aperture Synthesis Telescope (COAST) is a long baseline stellar interferometer designed for synthesis imaging using visibility amplitudes and closure phases [1]. Within it a spectrometer has been installed consisting of a 60° prism of BK7 glass, a 400 mm achromatic doublet, and a fast CCD camera. Fringes from any combination of available baselines can be recorded, either separately or simultaneously. The

CCD camera was used with vertical binning to achieve an exposure time of 5 ms per frame. In each strip of binned data, channeled spectra were recorded spanning a wavelength range of approximately 650–950 nm. In these wideband spectra the effects of dispersion can be seen.

Figure 3 shows (negative) images of channeled spectra obtained with the star α Boo, Arcturus. The effects of scintillation, visibility fluctuations, and atmospheric path variations are all present in these time series. An absorption line of atmospheric O_2 is also present as a constant feature at 760 nm. The image on the left shows very little evidence of dispersion and the fringe spacing is approximately constant. However, on the right the dispersion is obvious with the fringe frequency increasing towards shorter wavelengths.

4. Comments and Conclusion

COAST currently operates using baselines up to 10 m and observes at 840 nm with a bandwidth of only 40 nm. Under these conditions the longitudinal dispersion has a negligible effect on the measured visibilities, and only when the baselines exceed 30 m will it be necessary to either compensate the dispersion or reduce the bandwidth.

Three basic approaches have been proposed for dispersion compensation: matching the optical path-difference at two wavelengths [2]; setting the lower derivatives of the phase with respect to wavenumber to be zero at a central wavenumber [3,4]; or performing a least-squares fit to the fringe phase across a band defined by two wavelengths [5]. For each of these methods, the vacuum path would be given and a system of equations then solves for the air delay and the path differences of the compensating glasses. Note that the phase of the fringes is a function of these parameters and should not be restricted to zero by the equations! The least-squares approach [5] is no doubt optimum, although the differences between these methods are probably unimportant in practice. The indices of the glasses at hand will differ from the catalog values, and the predicted thicknesses can only serve as a first guess in the calibration of a compensator.

References

1. Baldwin, J.E. et al. (1996) The first images from an optical aperture synthesis array: mapping of Capella with COAST at two epochs, *Astron. Astrophys.* **306**, L13–L16.
2. ten Brummelaar, T.A. (1995) Differential path considerations in optical stellar interferometry, *Appl. Opt.* **34**, 2214–2219.
3. Tango, W.J. (1990) Dispersion in stellar interferometry, *Appl. Opt.* **29**, 516–521.
4. Lawson, P.R. and Davis, J. (1996) Dispersion compensation in stellar interferometry, *Appl. Opt.* **35**, 612-620.
5. Burnett, J.G. et al. (1995) Balancing optical path lengths in broadband fiber interferometers, *Appl. Opt.* **34**, 2194–2201.

LOCAL PISTON DETECTION OF A SEGMENTED MIRROR TELESCOPE WITH CURVATURE SENSING OF WAVEFRONTS AFFECTED BY ATMOSPHERIC TURBULENCE. NUMERICAL SIMULATIONS.

J.M. RODRIGUEZ-RAMOS and J.J. FUENSALIDA
Instituto de Astrofísica de Canarias

1. Introduction.

Segment alignment is one of the fundamental factors affecting the shape of a segmented mirror telescope PSF. Tilt and piston of each segment must be well adjusted with the rest of the segments. The Hartmann-Shack Sensor is very efficient in detecting the local wavefront tilt, but it is unsensitive to detect local piston errors, except if lenslets at the edges of the segments are used. This task is however complicated enough especially when the wavefronts are affected by atmospheric turbulence. On the other hand, it would be very convenient to be able to do alignment tests during observing time with natural stars.

The Curvature Sensor is based on the irradiance transport equation (Roddier,F. 1988) [1]. Through the intensity difference between two defocused images, the local wavefront laplacian and the radial gradient at the edge of the pupil are obtained. Phase can be recovered using this information. We show, with numerical simulations, that curvature sensing is sensitive enough in order to detect piston errors in the segmented mirrors.

2. Procedure of the simulations.

The simulations have been made for a segmented mirror like that of the Keck Telescope (D=10m. and 36 segments), and $r_0 = 10\ cm.$ $\lambda = 500\ nm$. The complete procedure of the simulations has three parts:

a) Simulation of the segmented mirror with piston errors between the segments. We generate the pistons following a gaussian distribution with zero mean.

b) Simulation of a curvature sensor using the Fourier Transform to carry the electromagnetic field from the fore-space respect the telescope to the rear space (Roddier,C. and Roddier,F. 1993) [2]. This amount of defocus fixes the distance between the focus of the system and the planes where the intensities are measured.

c) Recovering the wavefront phase using complex exponential expansion. Then we can use the Fourier Transform with iterations applying the Gerchsberg algorithm to avoid edge efects (Gerchsberg,1974) [3].

In Figure 1, we show an scheme of the complete procedure. We start from different piston distributions (in the first row, we present two extremes cases: $\sigma =$

A-M. Lagrange et al. (eds.), High Angular Resolution in Astrophysics, 355–358.
© 1997 *Kluwer Academic Publishers.*

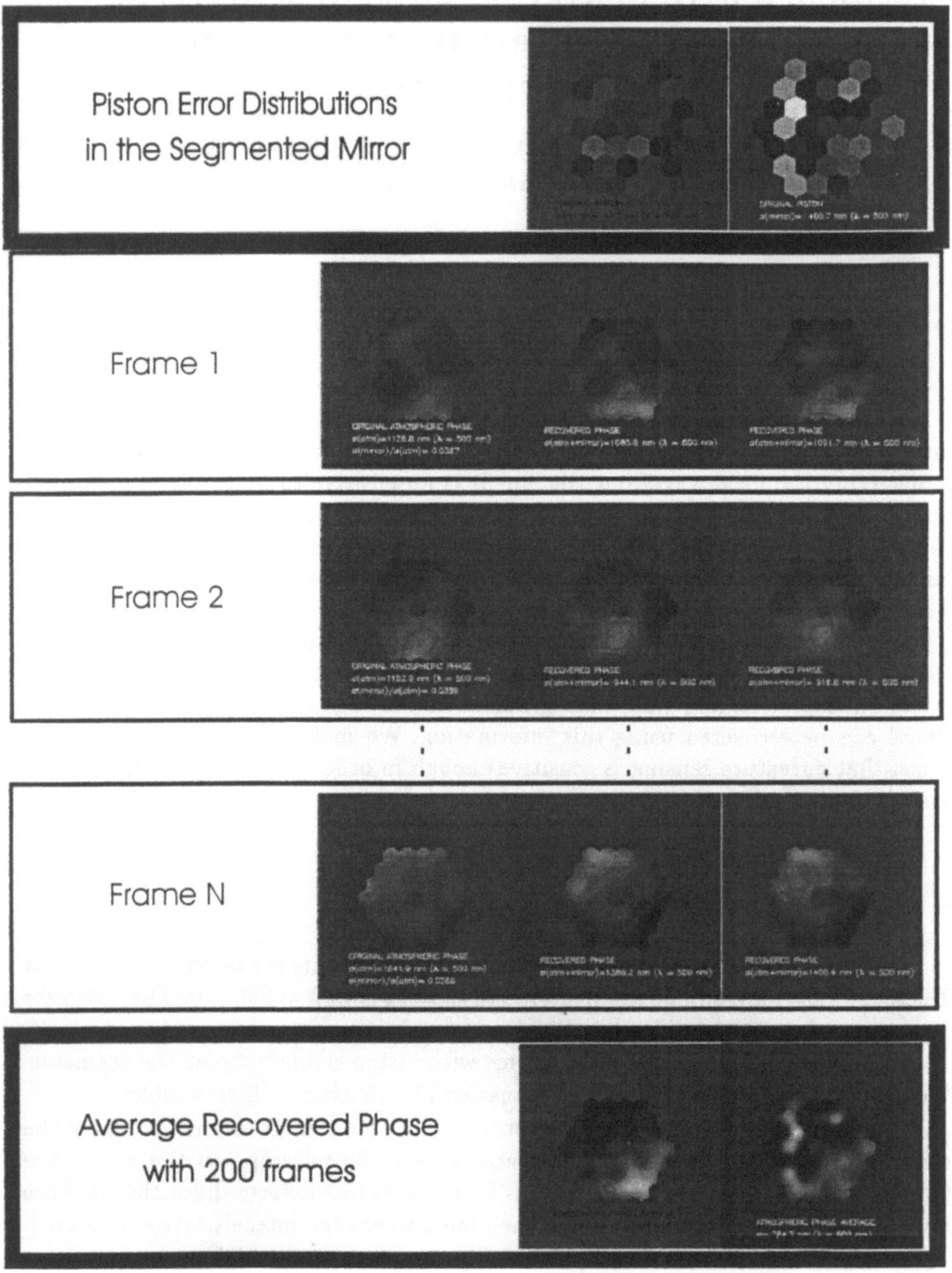

Fig. 1. Procedure of the simulations. In the first row, two different piston errors distributions are shown ($\sigma = 44\ nm.,\ and 481\ nm.$). The following rows are examples of independent frames: the first column is the representation of the atmospheric phase, and the second and third one are the recovered phase by curvature sensor. The last row is the average recovered phase for each piston distribution (first row).

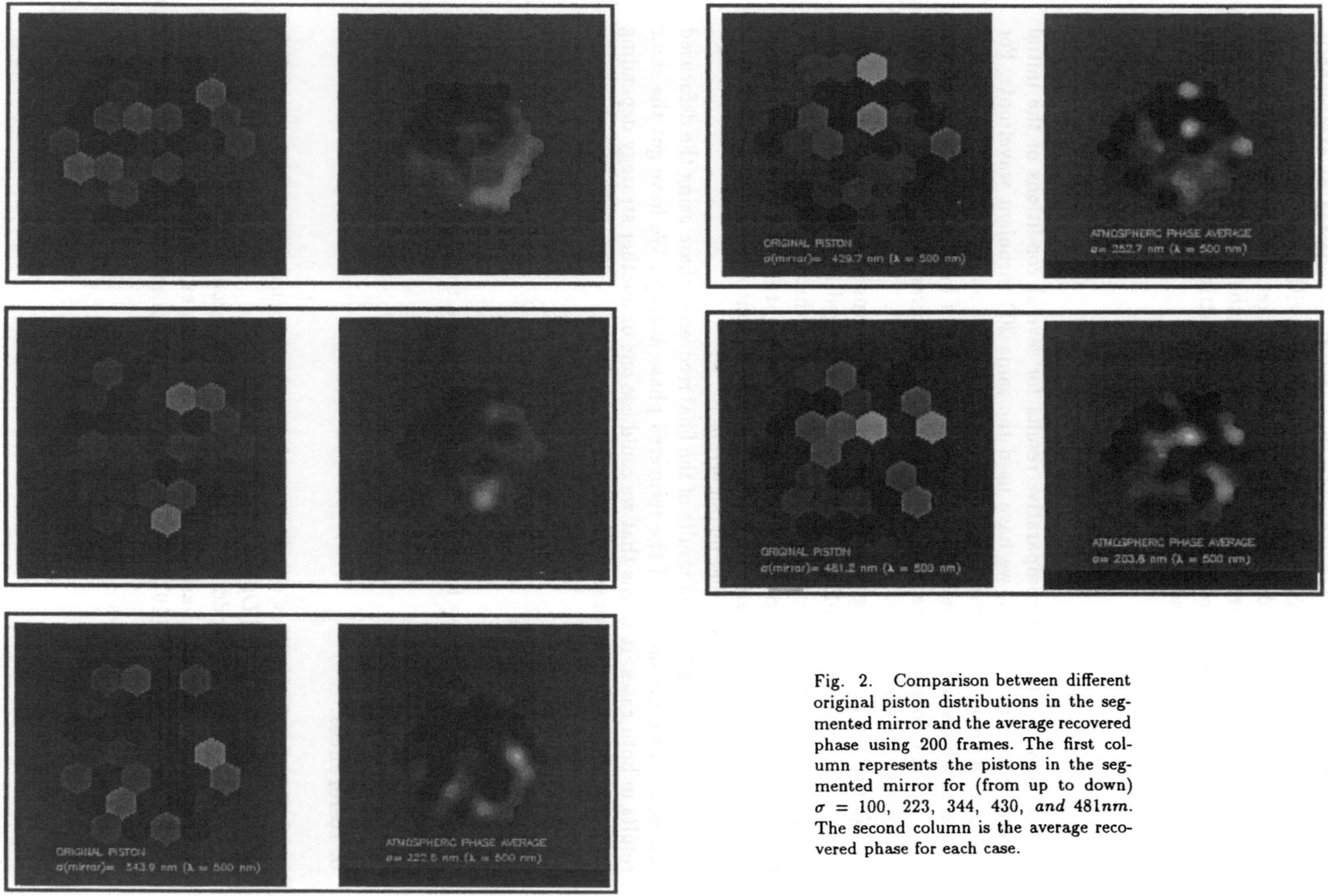

Fig. 2. Comparison between different original piston distributions in the segmented mirror and the average recovered phase using 200 frames. The first column represents the pistons in the segmented mirror for (from up to down) σ = 100, 223, 344, 430, *and* 481*nm*. The second column is the average recovered phase for each case.

44 *nm.* and 481 *nm.*). We proceed generating wavefronts affected by atmospheric turbulence (first column) and add them to the mirror with piston errors. Then we apply the second and third part of the simulation to get the recovered phase for each frame (second and third columns). The last row is the average recovered phase of 200 independent recovered phases for the piston distribution $\sigma = 44$ *nm.* (first image) and $\sigma = 481$ *nm.* (second image).

3. Results.

We show in Figure 2 comparative results for several conditions of the initial piston errors. For all cases, we have used the same 200 turbulent wavefronts. For each pair of images, the representation of the mirrors are on the left, and the average recovered phases are on the right. Also for each pair of images, we have fixed the scale of the colors, and the zero, so that each color corresponds to the same level of intensity for each image.

We can see in these figures that the smaller the rms of the initial piston distribution the larger the difference between the averaged recovered phase and the piston error pattern. If we suppose the time scale of the atmosphere is 10ms., so that frames of 10 ms. could be considered independent, then the average recovered phase would be obtained after only 2 s. It is evident that a longer integration time would be necessary for smaller piston errors.

We have also checked the results of the final recovered phase using the defocused intensities averages instead of the recovered phase average. We have got the same results in both cases. It seems that we could use one or another strategy depending on the conditions of the observations.

4. Acknowledgements.

We thank N. Devaney for providing us the routine to generate the hexagonal segmented mirror.

5. References.

1. Roddier,F. (1988) Curvature sensing and compensation: a new concept in adaptive optics, *Applied Optics* **27** , 1223-1225.

2. Roddier, C. and Roddier, F. (1993) Wave-front reconstruction from defocused images and the testing of ground-based optical telescopes,*JOSA A* **10** , 2277-2287.

3. Gerchsberg, R.W. (1974) Super-resolution through error energy reduction,*Optica Acta* **21** , 709-720.

THE FIRST CALTECH–JODRELL BANK VLBI SURVEY:

A morphological classification scheme of powerful radio sources

ANTONIOS G. POLATIDIS
Onsala Space Observatory
Onsala, S-43992, Sweden

1. Introduction

The Pearson & Readhead (1988) [2] (PR) survey is the first VLBI imaging survey and consists of a sample 65 objects of which 47 were observed with VLBI. This sample provided the first well-defined classification scheme for the milli-arcsecond scale morphologies of strong radio sources. However the size of the sample is too small for some of the most interesting statistical studies.

The first Caltech–Jodrell Bank VLBI survey (CJ1 survey) is an extension of the PR survey whose major aim is to treble the number of objects studied. It contains 135 objects selected from the NRAO-MPI **S4** and **S5** surveys with flux density 1.3 Jy> S_{5GHz} >0.7 Jy, declination $\delta > 35°$ and galactic latitude $|b^{II}| > 10°$.

In combination with the PR sample, the CJ1 survey forms a *complete flux–limited sample* of 200 objects for statistical studies, of which 131 objects can be mapped with Mark 2 VLBI; the remaining objects are large, lobe-dominated sources. Complementary observing programs to obtain arcsecond scale radio maps, optical identifications and redshifts for all of the objects has also been carried out ([9], [10])

Our principal astronomical aims are:

1. To study the morphologies of the parsec scale radio sources in a large sample of active galacti nuclei, in order to test and if necessary extend the classification scheme introduced by PR.

2. To provide a sample large enough to make interesting statistical studies, including the distribution of the misalignments between the milliarcsecond scale and the arcsecond scale, the cosmological evolution of different classes and the applicability of "Unified Schemes" relating radio galaxies, quasars and BL Lac objects.

A-M. Lagrange et al. (eds.), High Angular Resolution in Astrophysics, 359–364.
© 1997 *Kluwer Academic Publishers.*

3. Provide first epoch observations of candidate sources for superluminal motion to enable statistical studies of superluminal motion with redshift to be made.

2. Observations

We made both $\lambda 6$ cm and $\lambda 18$ cm images of all objects that are accessible to the Mark 2 VLBI system (ie. those with compact cores stronger than about 0.2 Jy), excluding those for which images had already been published. The dual wavelength observations improve the reliability of the source classification - the $\lambda 18$ cm maps have superior surface brightness sensitivity for extended emission compared with the higher resolution $\lambda 6$ cm maps. During the course of the survey, from 1990 to 1991, 81 CJ1 sources and 31 PR sources were observed at $\lambda 18$ cm and 82 CJ1 sources at $\lambda 6$ cm.

In order to complete the CJ1 survey as fast as possible and make efficient use of the VLBI networks time we used a "snapshot" observing technique, using the Mark 2 VLBI recording system with a bandwidth of 1.8 MHz on an array of 12-16 telescopes. Three 30 minutes scans were scheduled for each source, separated in hour angle to maximise the *uv* coverage. The noise levels achieved are close to the theoretical thermal noise (rms noise $\simeq$ 1 mJy/beam in uniformly weighted maps) yielding images with dynamic ranges $\geq$ 100:1 and of comparable quality with those of PR.

3. Results

The collection of the images is presented in three papers: Polatidis *et al* (1995) [4] (18 cm), Thakkar *et al* (1995) [7] (18 cm) and Xu *et al* (1995) [10] (6 cm). A selection of images are shown in Figure 1. The main result is that the classification of PR is still appropriate though it can be simplified. In broad terms the new classification is:

- $\sim$ 80% asymmetric core-jets, with wide range of core/jet strengths, jet lengths and jet bends.
- $\sim$ 10% "Compact Symmetric Objects" (CSOs)
- $\sim$ 10% of the objects are not classified yet.

The most interesting morphological result is the discovery of the class of "Compact Symmetric Objects" which encompasses the "Compact S Double" class of PR and consists of a significant fraction of the whole sample.

Another significant result is the confirmation from the CJ1 data of the bimodality of the distribution of misalignment angle between the milliarcsecond and the arcsecond scale radio structure that was first noted in the PR sample.

4. The Compact Symmetric Objects (CSOs)

The Compact Symmetric Objects (CSOs) have high luminosity structure (in the form of "mini lobes", hotspots and/or symmetric jets) on each side of the central engine on scales of a few hundred parsecs and no or very weak extended emission at scales larger than 1000 parsecs. They also show weak variability and low polarisation at GHz frequencies ([8]). An examples CSO is 1943+546 in Figure 1.

Both the morphology and the relative abundance of CSOs in complete samples show that they are not seriously affected by relativistic beaming and that they are intrinsically small and highly luminous. Arguments based on kinematics, jet power and lobe energy indicate that CSOs are young ($\sim 10^3 - 10^4$ yrs) objects ([5]) and that they probably evolve in luminosity to form the 1000 times larger FR I and FR II objects ([6]).

5. The bimodal distribution of the misalignment angle

Pearson and Readhead first noted the relatively large fraction of sources showing large bend angles ($\sim 90^o$) between the dominant axes of the milliarcsecond and arcsecond scales (ie. $\theta_{VLBI} - \theta_{arcsec}$ or ΔPA). The distribution they found had one peak near 0° (aligned sources) and another near $\sim 90^o$ (misaligned sources). The combination of the PR with the CJ1 sample yields 86 sources for which we have measured ΔPA (compared with the 18 sources in PR). The position angle of the milliarcsecond scale structure is determined by connecting the core to the closest gaussian component, while the arcsecond scale position angle was determined between the core and the brightest component in an arcsecond scale (VLA or MERLIN) map.

The ΔPA distribution of the 86 sources in the combined PR and CJ1 sample is shown in the top panel of Figure 2. The distribution is still bimodal but shows a broader peak near 90°, compared with that found by PR. Most interestingly by dividing the sources according to indicators or relativistic beaming e.g. flux variability, fraction of flux density in the mas core (F_c) (Figure 2), degree of optical polarisation, we found that only the sources that are beamed towards us had a strong tendency to exhibit the $\sim 90^o$.

A possible explanation of this peculiar observational result has been given by Conway and Murphy (1993) [1] in the form of "saturated helical jets". A possible cause for the creation of the helical motion is precession due to binary black holes.

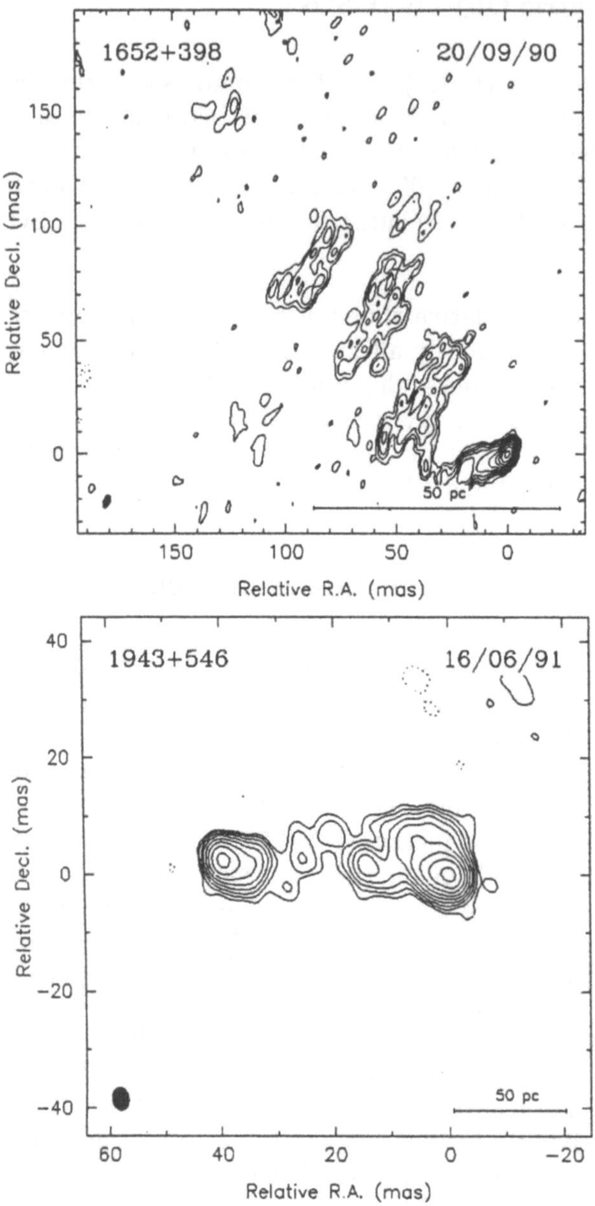

Figure 1. Two λ18 cm images from the CJ survey; the quasi-helical core-jet source 1652+398 and the Compact Symmetric Object 1943+546. The contours are spaced by factors of two in brightness, with the lowest at three times the rms noise level

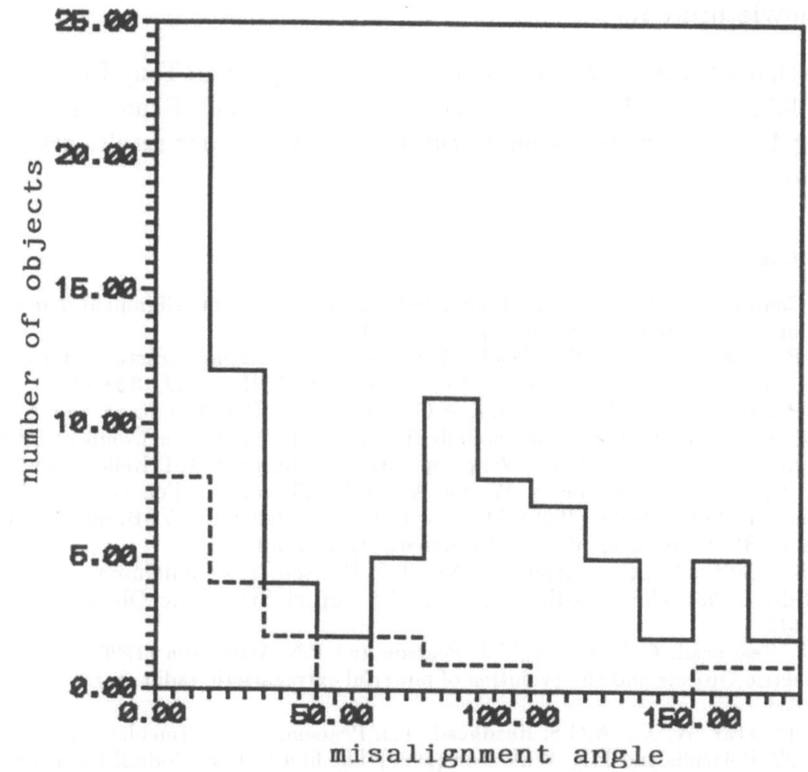

Figure 2. Histograms of the misalignment angle distribution: *(bold line)* 86 objects from the PR+CJ1 sample; *(dashed line)* objects with less misalignhan 30% of the flux density from the sub-mas core (ie. $F_c < 0.3$);

6. Conclusions

The first Caltech Jodrell Bank VLBI survey has been a great success. We have shown that it is possible to make images of high quality using the snapshot technique. This technique will be particularly important with the advent of the VLBA. We have mapped a large sample of extragalactic radio sources at two wavelengths and combined these results with those from the Pearson and Readhead survey to make a systematic study of the morphology of compact radio sources. Further to the two significant results reported above, second epoch observations of all sources are in progress to obtain values of superluminal motion. The study of other statistical problems like the angular size to redshift relation for compact extragalactic objects ([3]) and tests of unified schemes are also in progress.

7. Aknowledgments

I wish to thank the other members of the CJ1 survey team: Tony Readhead, Peter Wilkinson, Tim Pearson, Wenge Xu, Diraj Thakkar, Rene Vermeulen and Greg Taylor for allowing me to summarise some of our results prior to publication.

References

1. J.E. Conway and D.W. Murphy (1993) Helical jets and the misalignment disrtibution for core dominated sources, *Ap.J.*,**411**, 89.
2. T.J. Pearson and A.C.S. Readhead (1988) The milliarcsecond structure of a complete sample of radio sources. II First epoch maps at 5GHz, *Ap.J.*, **328**,114.
3. T.J. Pearson,W. Xu, D.D. Thakkar, A.C.S. Readhead, P.N. Wilkinson and A. G. Polatidis (1994) The First Caltech-Jodrell Bank VLBI Survey, in *Compact Extragalactic Radio Sources*, NRAO Workshop, eds. J.A. Zensus & K.I. Kellermann, 1.
4. A.G. Polatidis, P.N. Wilkinson, W. Xu, A.C.S. Readhead, T.J. Pearson, R.C. Vermeulen and G.B. Taylor (1995) The first Caltech-Jodrell Bank VLBI survey: II λ 18cm VLBI observations of 87 radio sources, *Ap.J.S.*, **98**, 1.
5. A.C.S. Readhead, G.B. Taylor, W. Xu, T.J. Pearson, P.N. Wilkinson and A.G. Polatidis (1996) The statistics and ages of Compact Symmetric Objects, *Ap.J.*, **460**, 612.
6. A.C.S. Readhead, G.B. Taylor, T.J. Pearson and P.N. Wilkinson (1996) Compact Symmetric Objects and the evolution of powerful extragalactic radio sources, *Ap.J.*, **460**, 634.
7. D.D. Thakkar, W. Xu, A.C.S. Readhead, T.J. Pearson, R.C. Vermeulen, G.B. Taylor, A.G. Polatidis and P.N. Wilkinson (1995) The first Caltech-Jodrell Bank VLBI survey: II λ 18cm VLBI observations of 25 radio sources, *Ap.J.S.*, **98**, 33.
8. P.N. Wilkinson and A.G. Polatidis and A.C.S. Readhead and W. Xu and T.J. Pearson (1994) Evidence for two-sided ejection :The Compact Symmetric Objects, *Ap.J.*, **432** ,L87
9. W. Xu, C.R. Lawrence, A.C.S. Readhead and T.J. Pearson (1994) Optical ID's and redshifts from the CJ survey, *A.J.*, **108**, 395.
10. W. Xu, A.C.S. Readhead, T.J. Pearson, P.N. Wilkinson and A.G. Polatidis (1995) The first Caltech-Jodrell Bank VLBI Survey of powerful radio sources: III. VLBI

ADAPTIVE OPTICS IMAGING OF THE PRE-MAIN SEQUENCE STAR YY ORIONIS

FRANÇOIS MÉNARD AND FABIEN MALBET
Laboratoire d'Astrophysique de l'Observatoire de Grenoble
414, rue de la Piscine
B.P. 53
38041 Grenoble Cedex 9
France

Abstract. We present high angular resolution near-infrared adaptive optics observations of the low mass pre-main sequence star YY Orionis. The results show that the circumstellar material surrounding the star is resolved and that it appears roughly circular in projection. Implications for the star formation scenarios, especially the physics/geometry of accretion, are discussed.

1. Introduction

Most of our current knowledge of the early phases of stellar evolution derives from observations of circumstellar (CS) material around young stellar objects (YSOs). Based on IR data, the current model put forward for low-luminosity embedded YSOs includes a central stellar core, surrounded by a massive accretion disk and a remnant infalling envelope (see for example the review by Shu et al. [1]). This embedded stage is frequently associated with energetic bipolar molecular outflows (e.g. [2]). As the system evolves, the CS envelope progressively dissipates and the core + disk system becomes optically visible, entering the so-called "T Tauri" phase. The disappearance of CS material must be caused by one (or both) of the two processes central to star formation, i.e., accretion and outflow. Therefore, investigating the spatial distribution of the CS material and the time scales for its evolution (clearing) via its dynamics offer means to understand these two processes which have recently appeared to be intimately related [3,4].

A-M. Lagrange et al. (eds.), High Angular Resolution in Astrophysics, 365–369.
© 1997 *Kluwer Academic Publishers.*

YY Orionis is the prototype of a subclass of T Tauri stars (TTS) that display, amongst other features, a strong UV excess compared to normal stars of similar spectral types, and inverse P Cygni profiles, variable with time, at the Balmer emission lines. The red-shifted absorption component of the P Cygni profile indicates that YY Ori stars are surrounded by infalling material, at least along some directions. The material seems to be infalling near free-fall velocities.

For the sake of conciseness, we will neglect rewiewing the current models available in the literature and describing the various accretion mechanisms. We will limit ourselves to say that, for now, we don't know exactly how accretion proceeds in pre-main sequence stars: from a spherical envelope? from an accretion disk? from both? Because various scenarios exist, we need observational tests to discriminate. YY Orionis are very good candidates because they show conspicuous signs of accretion. By studying their CS surroundings, i.e., shell/accretion disk, AND by combining with other results, we might finally understand better how accretion proceeds during star formation, a key problem nowadays...

2. Observations and Data Reduction

YY Orionis was observed in the near-infrared JHK bands with the ESO 3.6m telescope, La Silla, Chile, equipped with the adaptive optics system COME-ON-PLUS and the Sharp II camera on 1994 January 2. During observations, the average air mass was 1.14 and the local uncorrected seeing was excellent, better than $0\farcs6$, with very long coherence time. However, because YY Ori is rather faint in the optical, V=14.5 (variable), the sampling frequency of the WFS had to be reduced to 25Hz, far from COME-ON-PLUS' optimum capabilities. Nevertheless, as we will see in the next section, K-band images are close to being diffraction-limited, a spectacular achievement for COME-ON-PLUS.

PSF-reference observations were intertwined with observations of YY Ori in all filters every 5 minutes or so, insuring a decent monitoring of the seeing and AO correction evolution during the night. Images have been obtained for a total of about 8 minutes on source in each filter, after image selection.

Data reduction included standard procedures like dead pixels identification and removal, sky subtraction and flat fielding. Images were then selected. Good criteria in this case are hard to define. Selection on the sole basis of angular resolution (i.e., minimizing the FWHM) is not good as we suspect YY Ori to be extended for example. We chose to keep all the individual images of the data cube that had the following criteria within three (3) standard deviations of the average computed in the whole data cube: FWHM, Strehl ratio, radius for 50% flux, total flux, x- and y-centroid position (after recentering). The final single image was then obtained by summing all the selected images together. Obviously, this procedure does not minimize the FWHM, but it ensures us that we are looking at a final image that is representative of the data cube. The same selection procedure was applied to the PSF-reference and to YY Orionis for consistency.

3. Results

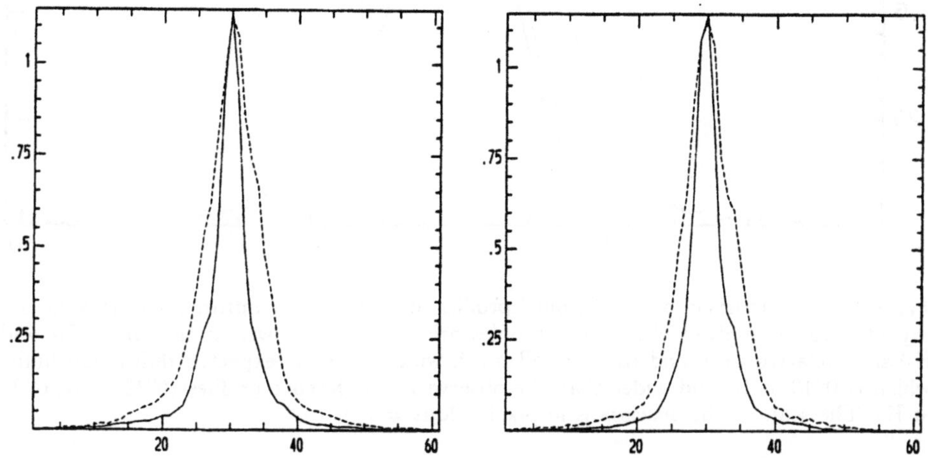

Figure 1. K-band intensity profiles of YY Orionis. *Left:* North-South cut; *Right:* East-West cut. In both figures the dashed line is YY Ori's profile while the full line is the PSF-referrence profile, both normalized to unity. The scale of the abscissa is in pixels ($20px = 1''$). The FWHM of the reference profiles are 0.″18 and 0.″19, and those of YY Ori are 0.″42 and 0.″40 in the N-S and E-W directions respectively.

As seen from Fig.1 and Table 1, YY Orionis is resolved. Come-On-Plus is expected to provided a better correction (actually diffraction limited given there is enough flux on the WFS) for long wavelengths, that is 2.2μm (K) and longward. Shortward of 2.2μm, the images are not expected to

TABLE 1. FWHM values (in arcsec)

wavelength (μm)	YY Ori N-S	YY Ori E-W	PSF-reference N-S	PSF-reference E-W
J (1.25)	0.46	0.54	0.40	0.44
H (1.65)	0.47	0.46	0.26	0.22
K (2.2)	0.42	0.40	0.18	0.19

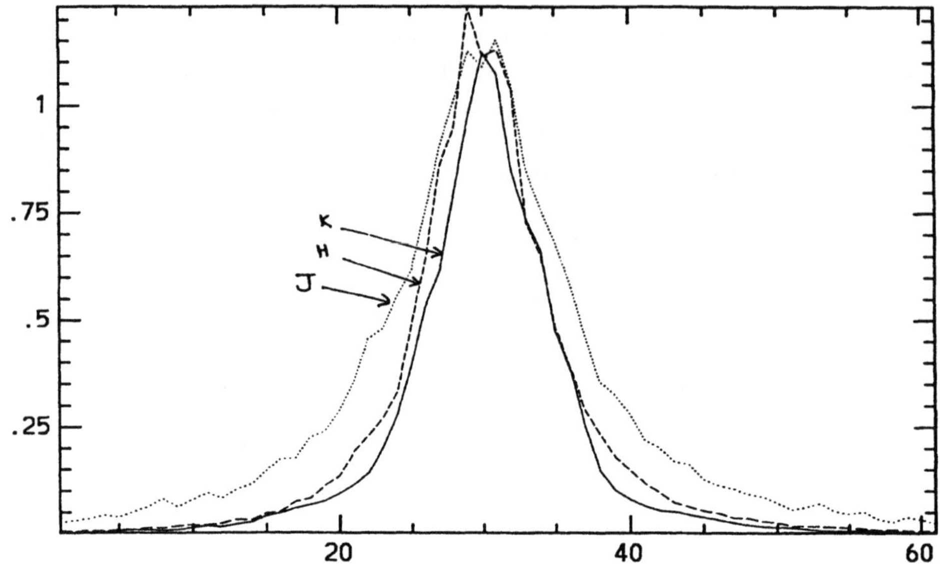

Figure 2. Comparison of YY Orionis' profiles at various near-infrared wavelengths. As expected for a resolved object, the profiles are sligthly narrower at the longer H- and K-bands relative to J, and they are all much wider than the expected diffraction limit, which is $0\rlap{.}''13$ at K, and wider than the observe PSF-reference profiles ($0\rlap{.}''19$ at K; $0\rlap{.}''24$ at H). The scale of the abscissa is in pixels ($20px = 1''$).

be diffraction-limited. At K, the PSF-reference shows a clear diffraction-limited spike on top of residual, uncorrected, natural seeing broadened wings. With an exactly identical instrumental set-up, YY Ori's profile is much broader, hence resolved. The undeconvolved FWHM of YY Ori is $\sim 0\rlap{.}''40$. At shorter wavelengths, the FWHM of YY Ori's remains roughly constant, i.e., $\sim 0\rlap{.}''45$ (see Tab.1 for exact values), while the PSF FWHM increases steadily, reaching almost that of YY Ori at J.

The broadening of the PSF profile at H and J reflects the relatively poorer correction provided by Come-On-Plus. The steadiness of the YY Ori profiles, *a fortiori* always larger than the PSF ones, argues strongly in favor of a resolved CS medium surrounding the star. Since we are most probably looking at scattered light from dust particles, the slight decrease of the apparent CS medium size from J to K (see Fig. 2) might reflect the scattering properties of the grains, i.e., their scattering efficiency decreases, producing a CS medium apparently smaller.

We estimate the error on the sizes quoted above to one pixel, i.e., $\pm 0\rlap{.}''05$. At that level, we have to claim that YY Ori's CS medium appears circular in projection.

4. Conclusions

The main conclusions of this study are:

• The circumstellar environment of YY orionis is resolved in the near infrared;

• It appears roughly circular, with FWHM of $\sim 0\overset{''}{.}45$, suggesting either a spherical envelope or an equatorial disk viewed pole-on. Both ideas are well supported by the very low integrated optical polarization observed.

• However, we know from the spectral energy distribution that YY Ori is surrounded by an accretion disk but, from monitoring, that it also exhibits large and sometimes quasi-periodical photometric modulations. These modulations [5], wherever they come from, either from the disk and/or the photosphere, are most likely not compatible with a pure pole-on view;

As a consequence we suggest the circumstellar environment we detect around YY Ori is not the disk itself, but rather a massive envelope that scatters a lot a optical/near-infrared light and that hides the disk from direct detection.

The results of this study will be presented in much more details in forecoming paper [6].

References

1. Shu, F., Adams, F., & Lizano, S. 1987, ARA&A, 25, 23
2. Lada, C.J. 1985, ARA&A, 23, 267
3. Cabrit, S., Edwards, S., Strom, S.E., & Strom, K.M. 1990, Ap.J., 354, 687
4. Hartigan, P., Edwards, S., & Ghandour, L. 1995, Ap.J., 452, 736
5. Bertout, C., Harder, S., Malbet, F., Mennessier, C. 1996, A.J., submitted
6. Ménard, F., Malbet, F. et al. 1996, A.& A., in preparation.

DETECTING DISKS AROUND YOUNG STELLAR OBJECTS WITH INFRARED INTERFEROMETERS

F. MALBET
Laboratoire d'Astrophysique, Observatoire de Grenoble
BP 53, F-38041 Grenoble cedex 9, France.
E-mail: malbet@gag.observ-gr.fr

Abstract. We have simulated accretion disks around young stellar objects with the aim of computing interferometric visibility curves for present and soon-to-be infared interferometers. We demonstrate that such disks are detectable with an interferometric layout like the VISA configuration of the VLTI or with present infrared interferometers with improved sensitivity.

1. Introduction

Many young stellar objects (YSOs) are believed to be surrounded by an accretion disk located in a flat axisymmetric envelope. Many evidence for the presence of disks are found in infrared/ultraviolet excesses, in visual extinction, in forbidden line profiles, etc... (cf. Bertout, 1989) but are still indirect ones. Recently disks have been detected in the millimetric range with interferometers (Dutrey *et al.*, 1994) at a spatial resolution of a few arcseconds corresponding to several hundred astronomical units. In order to observe closer to the star, we need to investigate the circumstellar environment at shorter wavelengths. However at visible wavelengths, spectral energy distributions of young stars are dominated by stellar radiation; disk radiation begins to dominate the spectra in the near-infrared. Observing YSO disks with infrared interferometers is therefore of importance for studying physical processes occuring in the close circumstellar environment. We have simulated 3 different models of circumstellar disks around young stars in order to compute their expected interferometric visibilities and conclude that observing disks around YSOs is achievable with current or soon-to-be infrared interferometers.

A-M. Lagrange et al. (eds.), High Angular Resolution in Astrophysics, 371–378.
© 1997 *Kluwer Academic Publishers.*

2. Radiation processes in accretion disk

The two major processes of light emission in disks around YSOs are:

- the scattered light from the central stellar source in the upper optically thin disk atmosphere;
- the thermal emission of the heated disk particles in the optically thick part of the disk.

Scattered light is emitted in the visible range and usually luminosity is faint compared to the star one, except when the disk is seen almost edge-on and the stellar light is absorbed by the circumstellar disk. Therefore the scattered light will be almost absent of systems seen with an inclination angle different of 90°. This is the case studied in this paper.

Thermal emission is dominated by the optically thick part of the disk. The heating comes mainly from two mechanisms:

- direct heating from the star by absorption of stellar photons. The part of energy is however always limited to the stellar luminosity;
- energy released in the accretion process by viscous friction of orbiting particles. The part of energy is independent of the stellar luminosity and can be much larger than the star luminosity. In the case of FU Orionis disks, the accretion luminosity is several order of magnitude larger than the typical luminosity of T Tauri stars.

Temperature in the disk ranges from 10K in the outer part to about 1500K in the inner part of the disk. We then compute the flux araising from the disk at a given radius by assuming black-body emission at the effective temperature calculated for this radius.

3. Models

We consider different models for 3 types of young stellar objects. More details is given in Malbet & Bertout (1995).

3.1. MODEL DESCRIPTIONS

1. Model 1 is a classical T Tauri star (CTTS) with a standard accretion disk. The temperature law in the disk is $T_D \propto r^{-3/4}$. Accretion rate is 10^{-7} $M_\odot$/yr. Outer disk radius is 50 AU. The object is located at the closest distance for young stars, i.e. 150 pc.
2. Model 2 is the same model as the previous one, except for the temperature law which is $T_D \propto r^{-1/2}$. This temperature law can be derived from flared passive disks (cf. Adams *et al.*, 1990; Beckwith *et al.*, 1990). Accretion rate is 10^{-6} $M_\odot$/yr. Outer disk radius is also 50 AU and distance 150 pc. The spectral energy distribution can then fit a spectra

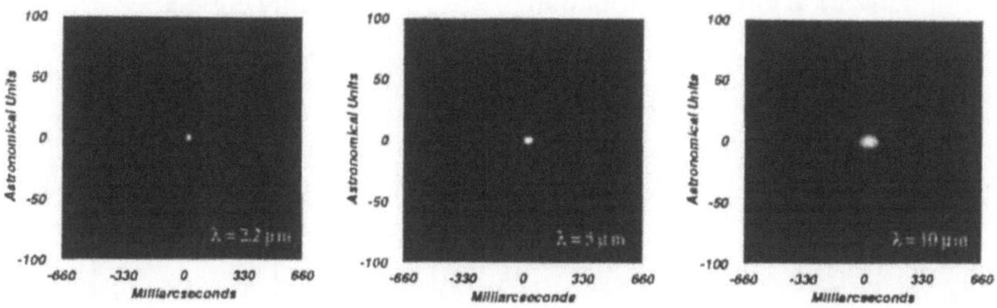

Figure 1. Synthetic images from the *classical T Tauri disk* model

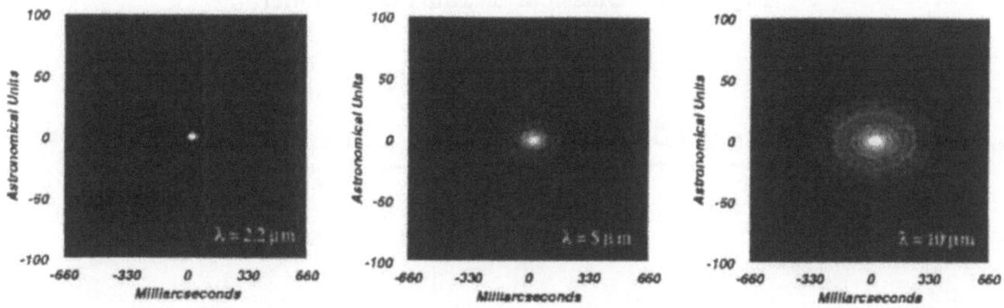

Figure 2. Synthetic images from the *flat spectrum T Tauri disk* model

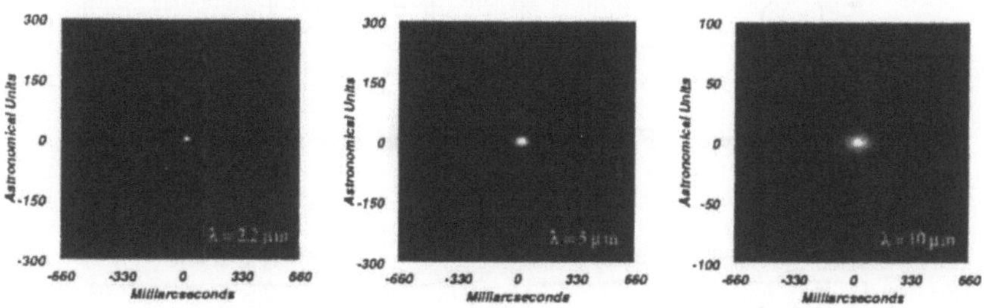

flat in the infrared as the ones observed in many objects like HL Tau, T Tau. We call this model the flat IR T Tauri star (FTTS).

3. Model 3 matches the case of FU Orionis objects (FUOr). It contains a massive standard accretion disk ($\dot{M} = 10^{-4}$ $M_\odot$/yr) which overwhelms the stellar emission. Outer disk radius is 150 AU and distance is 450 pc.

For all models we assume an 45° inclination of the disk plane to the line of sight.

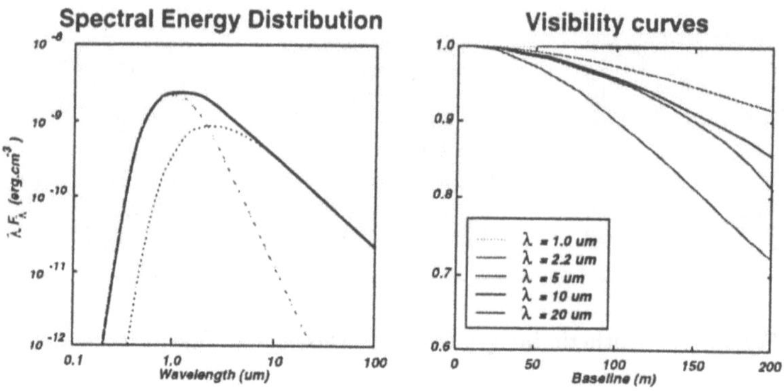

Figure 4. Classical T Tauri disk model

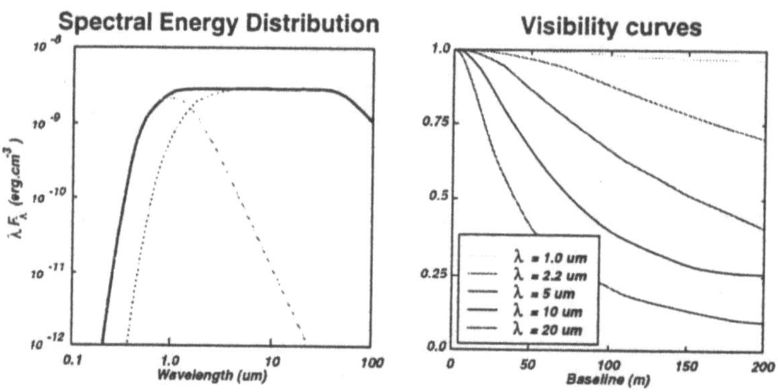

Figure 5. Flat spectrum T Tauri disk model

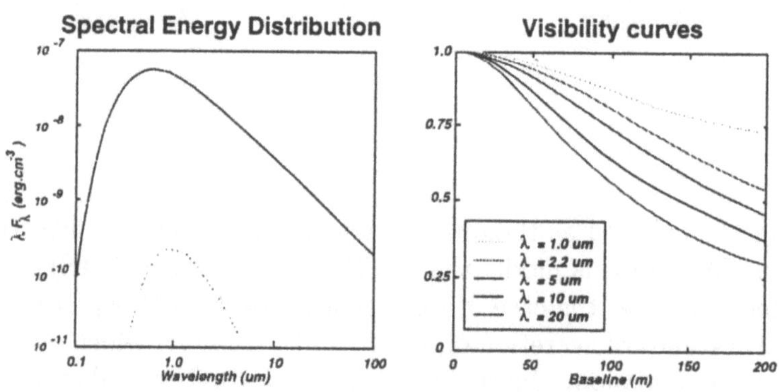

Figure 6. FU Orionis disk model

TABLE 1. **Disk spatial extension vs. spatial resolution of an 100m-baseline interferometer**

λ	CTTS (mas)	FTTS (mas)	FUOr (mas)	λ/B B=100m (mas)	$0.08\lambda/B$ B=100m (mas)
1 μm	-	-	0.1	1	0.08
2.2 μm	0.3	0.5	1	4.5	0.4
5 μm	1	3	3	10	0.8
10 μm	3	10	9	21	1.6
20 μm	7	40	20	41	3.3

3.2. RESULTS

We synthetize images from these three models at different wavelengths. Description of the different steps is given with more details in Malbet & Bertout (1995). Results are shown in Figs. 1, 2 and 3.

From these images, we derive visibility curves for baselines oriented along the major axis of the disks. They are shown on Figs 4, 5 and 6 together with the spectral energy distribution.

4. Discussion

4.1. OBSERVATION WAVELENGTHS

From spectral energy distributions and visibilities presented in Figs 4, 5 and 6, it is clear that disk emission dominates the overall flux when $\lambda > 2$ μm in CTTS, when $\lambda > 1.4$ μm in FTTS and at every wavelength in FUOr. However the reader must be aware that only thermal emission has been considered in these simulations. Ménard *et al.* (1996) show that the disk can also dominate the flux at shorter wavelength if the disk is seen almost edge-on. As a matter of fact, stellar scattered light in a disk seen edge-on is dominent compared to direct stellar light which is strongly extincted on the line of sight.

4.2. SPATIAL RESOLUTION

Spatial resolution seems not to be an issue for baselines ranging from 50 m to 100 m. One has to be careful about using the resolution criterion of λ/B. We prefer to use the $0.08\lambda/B$ criterion which gives the resolution at baseline B for the half width at half maximum of a gaussian function at the 0.9 level in the visibilities. In other words, extended objects with a gaussian shape of half width at half maximum R_d give a visibility of 0.9

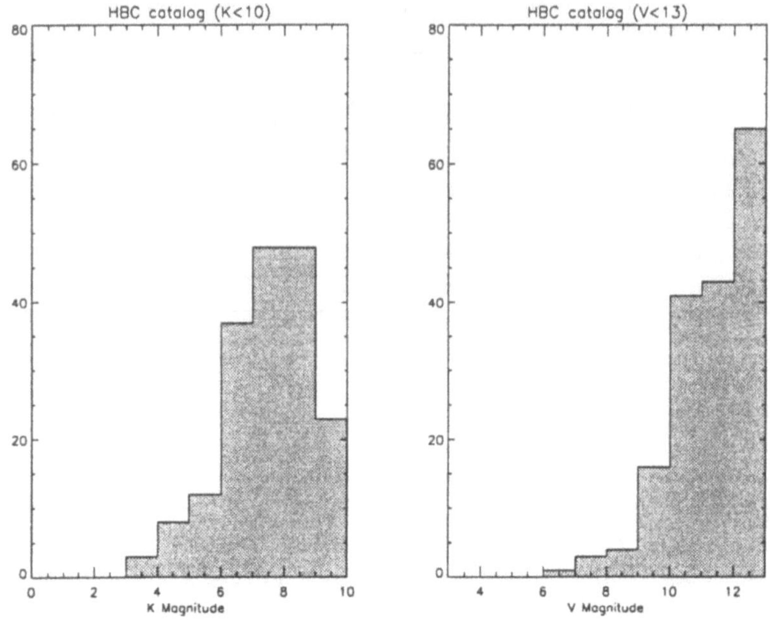

Figure 7. K- and V-magnitude histogram of young stars listed in the Herbig & Bell catalog.

at the baseline $0.08\lambda/R_d$. Table 1 shows that disks around young stellar objects are detectable at the 0.9 detection level.

4.3. DYNAMIC RANGE

If the contrast between the star and the disk is too high, the star will dominate the flux and the visibility will be constant equal to unity (the star is unresolved at $d > 100pc$). In the simulations we see that visibility curves decrease from unity by a few percent or even by a subtantial amount. In the observing mode where only data from one baseline are available, dynamic range is directly related to the decrease in the visibility. Therefore accuracies of the order of 1% in visibility measurements in our models will lead to disk detection.

4.4. SENSITIVITY

Figure 7 displays K-magnitude and V-magnitude histograms of young stellar objects extracted from the Herbig & Bell catalog (Herbig & Bell, 1988). We can conclude from this figure that infrared interferometers with limiting magnitudes of the order of $K > 6$ and $V > 10$ are necessary

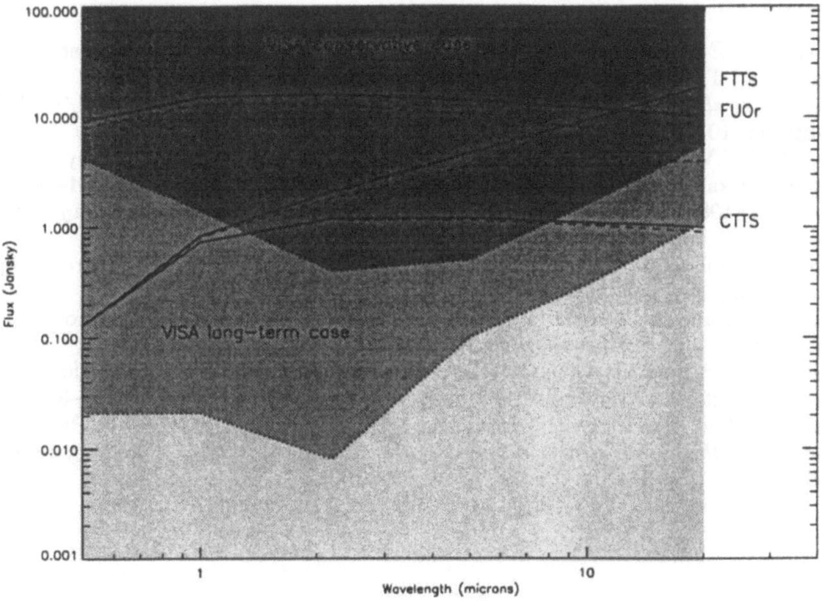

Figure 8. Disk fluxes compared to VISA expected sensitivity (conservative case and long-term case). Solid lines are total fluxes from the 3 different models (CTTS, FTTS, FUOr see text for details). Dashed lines are interferogram maximum fluxes for a 100m baseline (product of flux and visibility).

to start an observational study of these objects. Limiting magnitudes of current infrared interferometers ($K \approx 4-5$, $V \approx 8-10$) are below this threshold, but should be soon improved.

Figure 8 drawn both with specifications given in VLT Report 59b (Beckers *et al.*, 1989) and results from the simulations (cf. Malbet & Bertout, 1995) demonstrates that disks around young stellar objects are detectable with VISA sensitivity. The VISA configuration of the VLTI is therefore well suited for observing these objects even as a first light project.

5. Conclusion

The present study shows that studies of the very close environment of young stellar objects can be achieved with the ultimate sensitivity of current infrared interferometers or with conservative estimates of next generation infrared interferometers like the VLTI. The gain in sensitivity will permit to widen the field of application of astronomical interferometers to the domain of star and planet formation.

References

Adams, F.C., Emerson, J.P., and Fuller, G.A. (1990) Submillimeter photometry and disk masses of T Tauri disk systems, *Astrophysical Journal* **357**, 606–620.

Beckers, J. *et al.* (1989) The VLT Interferometer implementation plan, *ESO/VLT Report* **59b**, pp. 10–19.

Beckwith, S.V.W., Sargent, A.I., Chini, R.S., and Güsten R. (1990) A survey for circumstellar disks around young stellar objects, *Astronomical Journal* **99**, 924–945.

Bertout, C. (1989) T Tauri stars: wild as dust, *Annual Review of Astronomy and Astrophysics* **27**, 351–395.

Dutrey, A., Guilloteau, S., and Simon, M. (1994) Images of the GG Tauri rotating ring, *Astronomy & Astrophysics* **286**, 149–159.

Herbig, G.H., and Bell, K.R. (1988) Third catalog of emission-line stars of the Orion population, *Lick Observatory Bulletin* **1111**, 1–90.

Malbet, F., and Bertout, C. (1995) Detecting T Tauri disks with optical long baseline interferometry, *Astronomy & Astrophysics Supplement Series* **113**, 369–377.

Ménard, F., Malbet, F., Berger, J.-P., and Monin, J.-L. (in press) , in F. Paresce (ed.), *Science with the VLT Interferometer*, proc. of ESO workshop, Garching 18-21 June 1996.

DETECTION OF CIRCUMSTELLAR DUST SHELL AROUND SUPERGIANT TV GEM FROM MILLIARCSECOND RESOLUTION NEAR INFRARED OBSERVATIONS

SAM RAGLAND, CHANDRASEKHAR,T. AND ASHOK,N.M.
Physical Research Laboratory, Navrangpura,
Ahmedabad-380 009, India
e_mail:sam@prl.ernet.in

1. Introduction

High angular resolution (HAR) observations of evolved giants and supergiants in the infrared can answer some of the most interesting questions related to grain condensation, mass loss and hence stellar evolution. Lunar occultation is a powerful and productive technique at present for achieving HAR particularly in the near infrared region, because of the reduced background of scattered moon light compared to the optical region. The technique consists of recording the straight edge diffraction pattern of the star light produced by the sharp edge of the moon and obtaining HAR by mathematical modelling. The most attractive aspect of the method is its ability to achieve high one dimensional angular resolution down to 1−2 mas with an accuracy which can be as high as $\sim$ 0.1 mas. However careful consideration of various factors like optical bandwidth of the filter used, telescope aperture and time constant of recording system that influence the occultation light curve is necessary before such HAR can be achieved in practice.

A program of observing lunar occultations in the near infrared (1−5μm) is in progress at the Physical Research Laboratory, Ahmedabad, India to investigate the circumstellar dust envelope around late giants and supergiants. Several occultations have been successfully observed mainly in the K band. A rare opportunity to observe the lunar occultation of M1 supergiant TV Gem was successfully utilized. In this paper we report the detection of a circumstellar dust shell around the star consistent with the strong silicate emission feature and far infrared excess seen in IRAS observations.

A-M. Lagrange et al. (eds.), High Angular Resolution in Astrophysics, 379–382.
© 1997 *Kluwer Academic Publishers.*

2. Observations

The lunar occultation observations presented here were carried out at the 0.75 m telescope at Kavalur (78°, 49′ 45″E, 12° 34′ 35″N, 725 m) during early 1993. Event prediction was computed with a code developed by us which is accurate to few seconds. The occultation event reported here was a disappearance event at the dark limb of the moon.

The instrument used was a LN_2 cooled InSb based high speed infrared photometer. Occultation event reported here was observed in the standard K filter ($\lambda = 2.2\mu$m,$\Delta\lambda = 0.4\mu$m). A 2 mm circular diaphragm was used which corresponds to 42″ field on the sky. Data sampling was at a rate of 1 KHz for 30 seconds using a 16 bit high speed A/D converter (Keithley system 575). The absolute timing of the event was not recorded, since it is not relevant for the present work.

3. Data Analysis

Data analysis was performed using two different methods. The first method is the standard nonlinear least squares (NLS) technique and the second method is a model independent algorithm (MIA). In the case of NLS, a model is assumed for the one dimensional brightness distribution of the source (along the direction of occultation) with a set of physical parameters and the problem is to obtain best statistical estimations for these parameters along with other scaling parameters like source intensity, background intensity, velocity component of the moon along the direction of occultation and the time of geometric occultation.

The data reduction using NLS method can be used to model complicated geometry for the stellar photosphere and the circumstellar dust envelope. However, the NLS method demands the knowledge of the geometry of the source which is not always possible in practice. In such cases, MIA can be advantageously used. MIA is a composite algorithm which makes use of the NLS method and Lucy's deconvolution algorithm wherein a guess profile is assumed for the brightness profile of the source and is iteratively modified to obtain a better fit for the data.

4. Discussion

TV gem is a distant oxygen rich supergiant classified as M1 Iab. It is a short period semi regular pulsating variable of variability type SRc with a period of 182 days. The distance is estimated to be 1200pc. The circumstellar gaseous environment of TV Gem has been studied in the CO (1−0) and CO (2−1) lines in the millimeter region. From the CO velocity profile, an expansion velocity (V_e) of 12 km/s has been derived. The mass loss rate

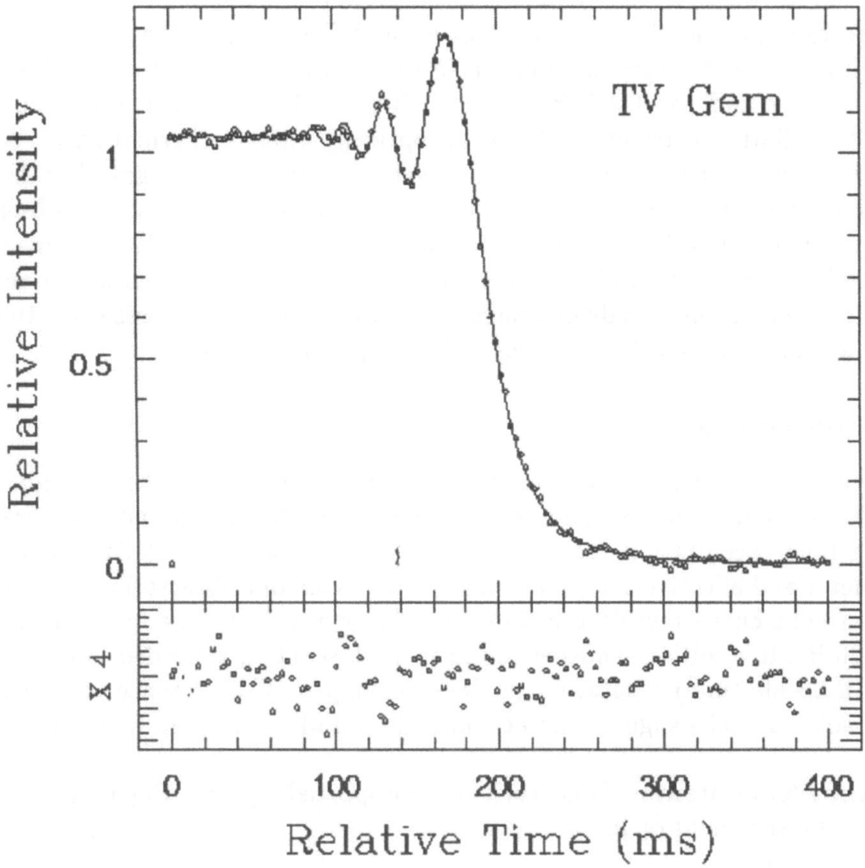

Figure 1. Occultation light curve of TV Gem (dots) fitted using MIA (solid line) is shown in the upper panel. Lower panel shows the residuals of the fit enlarged by a factor of 4.

($\dot{M} \propto V_e^2\ d^2$) for a distance of 1200pc to the source is $2\times10^{-6}\ M_\odot\ yr^{-1}$, assuming a $[CO/H_2]$ ratio of 5×10^{-4} typical of oxygen rich stars. The 9.7μm silicate emission feature seen in IRAS LRS and the infrared excess present in the IRAS photometric data for TV Gem unambiguously points to the presence of dusty circumstellar shell.

The observed light curve fitted using MIA is shown in Fig.1. The recovered brightness profile is fairly symmetric. For the best fit model we derive the FWHM of the shell extent to be 100 $\pm$ 20 mas and the star to shell flux ratio at 2.2μm to be 35. The stellar angular diameter can be accurately determined in this two component model. This value is 4.9$\pm$0.3 mas. From the above angular size and bolometric flux obtained from available photometric data, we derive an stellar effective temperature of 3670±125K.

To examine the dust distribution around TV Gem in the light of our occultation observations, an analysis of the 9.7μm silicate feature exhibited by this star and the far infrared shell flux has been carried out using a simple radiative transfer model of an optically thin isothermal shell. We find that the dust exists mainly in two well separated shells suggesting that the dust condensation in TV Gem is not a continuous process. Our observations, in conjunction with infrared spatial interferometric observations of multiple shells on a similar supergiant α Ori, suggest that mass loss and the consequent dust condensation is probably a sporadic process operating over a time scale of a few decades in late type supergiants.

5. Conclusions

We have investigated the circumstellar environment around the supergiant TV Gem from milliarcsecond resolution observations in the near infrared using lunar occultation technique. Accurate estimation of stellar angular diameter and effective temperature are made from our observations. Occultation light curve reveals the presence of a warm dust zone in the region 15–20 $R_\star$. In addition, another cooler shell at $\sim 500R_\star$ is needed to explain the IRAS photmetric data, particularly at 60 μm. The presence of dust in two isolated shells suggests an episodic or sporadic mass loss in TV Gem.

Acknowledgements: This work was supported by the Department of Space, Government of India.

COMBINED HST AND HAR IMAGING AND PHOTOMETRY OF ERUPTIVE BINARIES IN GLOBULAR CLUSTER CORES

Based upon Observations taken with the William Herschel Telescope on La Palma and the Hubble Space Telescope

RAYMOND BUTLER, ANDREW SHEARER AND
MICHAEL REDFERN
Applied Imaging Research Group,
Dept. of Physics,
University College Galway,
Galway,
Ireland.

Abstract. We have combined the complementary qualities of HST's high astrometric precision with TRIFFID's extended coverage and photometric precision to perform crowded-field photometry in the innermost region of the post core-collapse globular cluster M15. Our technique virtually eliminates the problem of extreme crowding which has hitherto hampered searches for the predicted binary population in globular cluster cores, and thereby provides an extension of the HST's capability. Our photometric study produced evidence of faint variable stars. We stress that it will require longer period photometry to uniquely identify the type of variable: should they be confirmed as eruptive binaries this would represent a vital step in our understanding of globular cluster evolution and dynamics. An interesting by-product of our program is that candidate bright cluster variables detected with the HST have been confirmed and monitored over longer periods.

1. Introduction - the Importance of Binaries in Globular Clusters

Cores of globular clusters (GCs) contain crucial information on the evolution of the cluster. They are also known to contain some of the more exotic species of stellar objects such as close interacting binaries, pulsars, x-ray sources and blue stragglers. The formation of close binaries by tidal interaction in the cores of post core-collapse clusters was expected theoretically (see Statler et al. 1987 & refs therein). Hut et al. (1992) is an excellent

A-M. Lagrange et al. (eds.), High Angular Resolution in Astrophysics, 383–389.
© 1997 *Kluwer Academic Publishers.*

review of the current status of binaries in GC cores. However, apart from the existence of the dozen or so X-ray binaries in GCs (see Charles 1989 & refs therein), which is a much higher density of sources than in the galactic plane, the anticipated large number of binaries in cluster cores have remained most elusive. These cores are too crowded for conventional ground-based studies of the CV variability signature, and only limited observations have so far been and will be possible with the HST; for this reason, only one certain CV has been discovered in a GC with the HST, by Paresce & De Marchi (1994); it was in the frequently observed HST/FOC calibration object, 47 Tuc. Spectroscopic techniques (*e.g.* v_{rad} determinations) are not currently feasible in compact cores, though Grindlay et. al. (1995) report 3 candidate CVs in the very close cluster NGC6397, on the basis of spectroscopy of Hα-bright stars with the HST.

Hodder et al. (1991) detected four erupting variables with magnitudes in the range 16 - 19 in M71. Scaling to M15 we would expect a range of 18-21. An unsuccessful search for CVs in M92, Shara et al. (1994), gave an upper limit of 7 erupting dwarf novae **outside** 6 core radii (they expected 11). However, about 60 such systems would be expected **within** 6 core radii, and for M15 the number would be higher still; the existence of a sizeable population of millisecond pulsars lends weight to this prediction.

2. Image Sharpening with the TRIFFID Camera

High angular resolution (HAR) imaging has been performed for a number of years using TRIFFID, the Galway/DIAS High Resolution Camera. It employs a pupil mask to reduce the effective aperture of a telescope so that it matches the prevailing seeing conditions at 3.5-4 R_0, enabling simple shift and add image sharpening, ie. tip/tilt correction. In order to maximise the signal this is carried out after the exposure (hence PEIS, Post-Exposure Image Sharpening) in the manner described by Redfern et al. (1989, 1992). Typically we obtain resolution improvements up to the theoretical maximum factor of two, or more where there is a significant telescope wobble component in the image motion.

TRIFFID has a simple optical design, which allows for either a single pupil with a large field of view or for 4 pupils with a reduced field of view. The latter is equivalent to observing with 4 reduced aperture telescopes simultaneously. Two 2-D imaging photon counting detectors are mounted at the red and blue focus of the camera. For these observations a MAMA detector from ESO (Cullum 1989; Timothy 1985) was used on the blue arm. On the red arm a CID based photon counting system from the Rutherford Appleton Laboratories (Carter et al. 1990) was used.

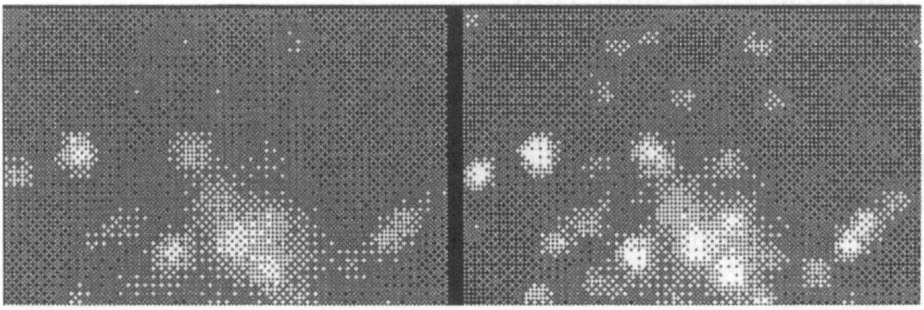

Figure 1. Raw coadded image of M15 core (left) and sharpened coadded image (right).

3. Observations and Data Reduction

3.1. OBSERVATIONS AND INITIAL REDUCTION

Observations of the core of M15 were made over 3 nights in June 1992 and 3 nights in July 1994 using the 4.2m WHT on La Palma. TRIFFID was mounted on the GHRIL optical bench at one of the Nasmyth focii, with the image derotator removed to maximise the light throughput - essential for the sharpening process as well for the photometry. As the position and arrival time of each photon was recorded, it was possible to derotate the images after the exposures were taken. In June 1992, TRIFFID was operated with 4 sub-pupils, a pixel scale of 0.075"/pixel (MAMA detector), and a field of view of 14", equivalent to at least 7 core radii. On all the nights the seeing was good and on one superb. In July 1994, 1 sub-pupil was used, with a field of view of 20"x75". Again, on all the nights the seeing was good.

Using software developed in Galway, (O'Byrne et al. 1991; O'Byrne 1993; Morris 1995) each sub-pupil of each exposure was flatfielded, derotated, sharpened, and exposure-corrected (at the edges) to produce a series of CCD-like images with resolution improved by factors of between 1.5 and 2, to values ranging from 0.38" to 0.60" (June 1992, 18 images) and 0.45" to about 0.7" (July 1994, 25 images). We split the longer exposures timewise to avoid phase-smearing in the photometry of candidate cluster variables. The 3 sharpest sub-pupil images in each June 1992 exposure were next coadded in IRAF. Figure 1 shows one of the images after coaddition of three of the sub-pupils: the left panel shows the raw image, whose resolution is 0.75", while the right panel shows the improvement to 0.45" after the shift and add image sharpening. HST resolution remains superior, but unlike TRIFFID it cannot be used for extended observations of variable objects, and its photometric calibration is poorer.

3.2. PHOTOMETRIC REDUCTION

We enhanced some features of the crowded-field photometry package Daophot2 (Stetson 1994), as implemented in the external IRAF package *daophotx*, to enable all the photometric reductions. (A full description of our photometric techniques will soon be published elsewhere.) Photometry was first performed on archival HST/WFPC2 (PC chip) images of M15 in two filters - F439W which approximates to B, and F555W which approximates to V. This provided a complete astrometric and photometric reference for 10722 stars over a 35"x35" field.

Each TRIFFID image was Maximum Entropy or Lucy restored using *mem* and *lucy* in STSDAS (Wu 1994, Lucy 1974). The restored images, although not photometrically reliable, provided accurate positional information because of their improved sharpness. This enabled accurate coordinate transformations to be determined using 50-230 common stars on each of our images and the HST/WFPC2 field. Transforming the HST starlist thus provided a positionally accurate starlist complete to magnitude B $\sim$ 21 for each of our images. This allowed the fitting of typically 90% *more* stars in the core region with an obvious improvement in photometric scatter.

4. Discussion

Figure 2 plots the photometry of one TRIFFID/MAMA image (using the transformed HST starlist) against the same stars measured on the HST/WFPC2 F439W image. Even at B=18 the scatter is $\pm$0.2mag, a result which is impressive considering that these stars lie within 7" of the cluster centre, where stellar densities rise from *12 to 20 stars per square arcsecond* in the HST starlist. Furthermore, some of the scatter is due to differences in the spectral responses of the two imaging systems - both approximate to B, but WFPC2/F439W is slightly bluer. A number of bright variables stand out immediately - including the X-ray binary AC211, (Auriere & Cordoni 1981), a component of the multiple AC214, and at least five Horizontal Branch variables.

Ferraro & Paresce (1993) detected 15 candidate RRLyrae variables in the core, V1 to V15, with HST/FOC. The HST coverage of the light curves was too short to determine their nature. Figure 3 shows the light-curve a typical one of these, V6, from a phase-dispersion minimisation analysis of the June 1992 TRIFFID data. This is a further indicator of the photometric quality.

For the binary search, all of the stars in the HST list were subtracted from the TRIFFID/MAMA images following their fitting. Then all of the subtracted images from a particular night were coadded to reinforce their common features. Figure 4 shows the results of this subtraction and coad-

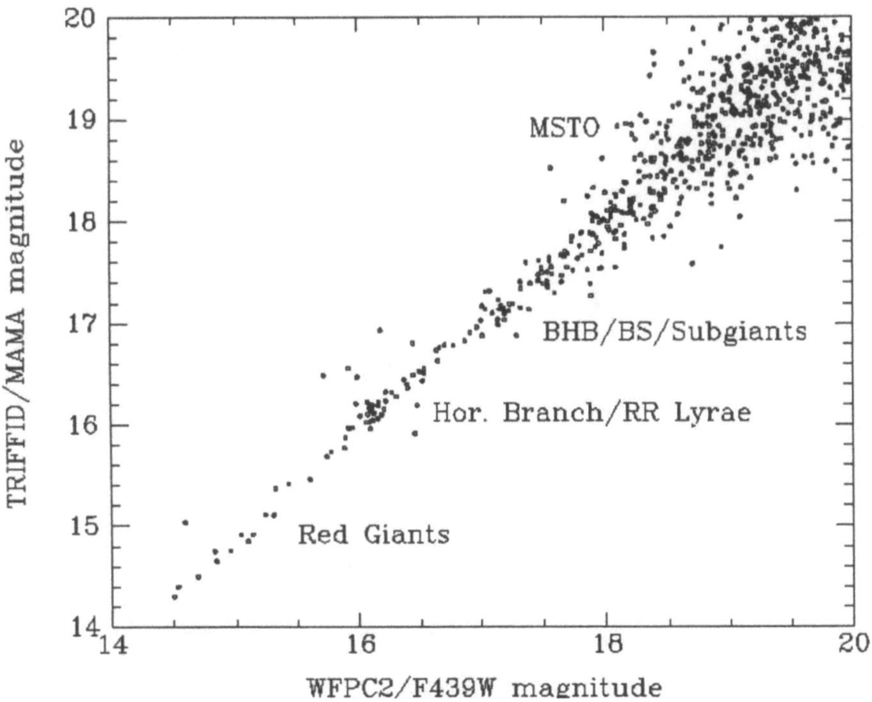

Figure 2. **Photometry of one TRIFFID/MAMA image vs. HST/WFPC2 F439W.**

dition, at high contrast. Large residuals are due to the reference stars used in the sharpening stage and to the residual light of the core-cusp. Away from these regions we identified a number of star-like objects which are consistently visible on each night. They are apparently stars whose brightness has increased from less than 21.5 (below the HST detection limit) to magnitude ~19-20. This is consistent with one type of eruptive binary system, dwarf novae, which at the distance of M15 would rise from magnitude 23-25 to 19-21. Furthermore the observed magnitude swings are not consistent with normal intrinisic variables. However, to uniquely identify the type of variable, longer period photometry would be required.

We have shown that the combination of HST astrometry and high resolution ground based photometry is capable of determining variability right into the cores of the most crowded GCs. We plan to start a survey of all compact GCs, using the TRIFFID camera on small telescopes. This technique should, in the near future, provide the first firm evidence of erupting binaries in cluster cores; success is most likely in other clusters which are several times closer than the remote M15.

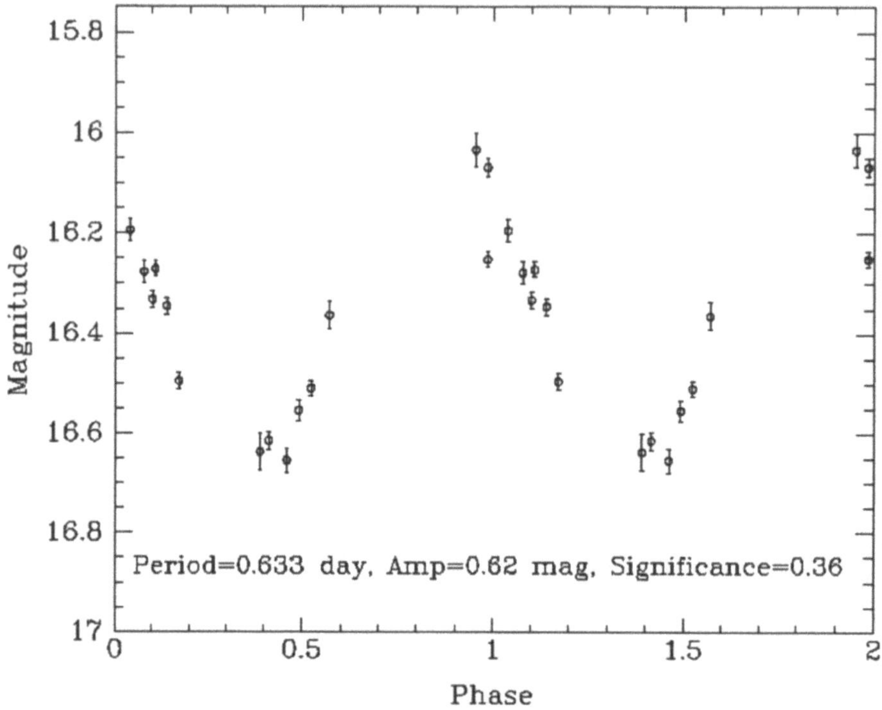

Figure 3. Phase Plot for RR Lyrae candidate V6 in M15.

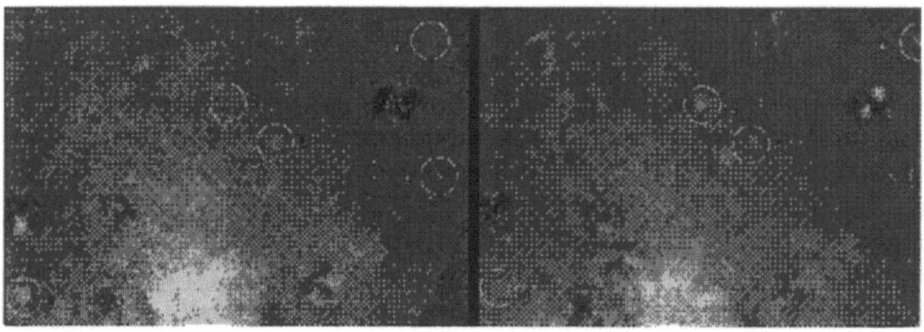

Figure 4. Starlike objects identified on coadded subtracted images: 26 June 1992 (left) and 27 June 1992 (right).

5. Acknowledgements

We would like to thank the Irish science agencies EOLAS and FORBAIRT for their support of this work. Pete Read and Martin Carter of the Rutherford Appleton Laboratory are thanked for the provision of their detector (the RAL-PCD). Martin Cullum and the European Southern Observatory are thanked for the provision of their MAMA detector. Lindsey Davis of NOAO is thanked for making the suggested changes to the *daophotx* code.

References

Auriere, M. and Cordoni, J.-P. (1981), *A & AS*, **46**, 347.

Carter, M. K., Cutler, R., Patchett, B. E., Read, P. D., Waltham, N. and van Breda, I. G. (1990), *SPIE*, **1235**, 644.

Charles, P. A. (1989), *ESA-SP*, **296**, X-ray Astronomy, ed. Hunt, J. and Battrick, B., vol 1, 103.

Ferraro, F. R. and Paresce, F. (1993), *AJ*, **106**, 154.

Grindlay, J. E., Cool, A. M., Callanan, P. J., Bailyn, C. D., Cohn, H. N. and Lugger, P. M. (1995), *ApJ*, **455**, L47.

Hodder, P. J. C., Nemec, J. M., Richer, H. B. and Fahlman, G. G. (1992), *AJ*, **103**, 460.

Hut, P., *et al.* (1992), *PASP*, **104**, 981.

Morris, P. W. (1995), *UCG Software Report 95S.*

O'Byrne, C., Jordan, B., O'Kane, P., Redfern, R. M., Shearer, A. and Wouts, R., 1991), *IAJ*, **20**, 282.

O'Byrne, C. (1993), *(private communication)*

Paresce, F. and De Marchi, F. (1994), *ApJ*, **247**, L33.

Redfern, R. M., Devaney, M. N., O'Kane, P., Ballestros Ramirez, E., Gomez Renasco, R. and Rosa, F. (1989), *MNRAS*, **238**, 791.

Redfern, R. M., *et al.* (1992), *Gemini*, **38**, 1.

Shara, M. M., Bergeron, L. E. and Moffat, A. F. J. (1994), *ApJ*, **429**, 767.

Statler, T. S., Ostriker, J. P. and Cohn, H. N. (1987), *ApJ*, **316**, 626.

Stetson, P. B. (1994), *PASP*, **106**, 250.

Timothy, J. B. and Bybee R. L. (1995), *Proc. SPIE*, **687**, 1090.

Wu, N. (1994), *Proceedings of STScI Workshop "The Restoration of HST Images and Spectra II"*, ed. Hanisch, R. J. and White, R. L., 58.

AN INDIRECT SEARCH FOR LOW-MASS COMPANIONS TO DISTANT STARS

M.V.SAZHIN AND I.A.TRIFALENKOV
Sternberg Astronomical Institute,
Universitetsky pr. 13, 119899 Moscow, Russia

Abstract. Since the observation of microlensing events toward the galactic bulge, the unique opportunity has opened up for astronomers: the use of microlensing as a space telescope with an extremely high angular resolution. Here we discuss the possibility of using this effect to search for an invisible companions to distant stars. These stellar systems move around a common barycenter, resulting in a non-uniform apparent motion of them in the sky. As a result, the light curve of this microlensing event becomes asymmetrical. Applying our analysis to available microlensing observations has led us to conclude that event OGLE2 can be a binary object with a massive dark companion.

1. Introduction

Recent observations of the effects of gravitational microlensing of stars in the Large Magellanic Clouds (Alcock et al., 1993; Aubourg et al., 1993) and in the Galactic Bulge (Udalski et al., 1994, Griest et al., 1995) has revealed a new population in our Galaxy – dark bodies with masses $\sim 0.1M_{\odot}$, presumably brown dwarfs which form the halo of our Galaxy.

Paczinsky (1986) has put forward the fundamental idea that our Galaxy has a spherical subsystem (halo) containing dark bodies, presumably neutron stars, brown dwarfs etc. down to low mass objects ("jupiters"). He has also proposed a method for observing such dark bodies on the basis of the effect of gravitational microlensing.

New opportunities which have been discussed in the literature (Blair and Sazhin, 1993) had appeared in connection with the discovery of microlensing events: the use of microlensing as an efficient extraterrestrial telescope

A-M. Lagrange et al. (eds.), High Angular Resolution in Astrophysics, 391–395.
© 1997 *Kluwer Academic Publishers.*

with extremely high angular resolution which is unattainable even to space telescopes. The principal amplification of star light occurs when the star is located inside the Einstein ring. The mean size of such a ring for the observed dark body population is $\sim 0.''001$. From the point of view of the observing method, microlensing is similar to lunar occultation because in both cases the stars are chosen by observational conditions but not by observer himself.

In the first section we consider the theory of microlensing for nonuniform motion of lensed star. Next, we discuss the main effect, that we need to take into account for mivcrolensing curve interpretation (parallax). Finally, we apply our analysis to really observed microlensisg events.

2. Microlensing with dark companion

Since the publication of the first papers (Refsdal, 1964; Paczinski, 1986), variations in stellar flux has been considered by many authors. If a massive body moves between background star and observer, so that the apparent distance between star and such body is small, visible flux from the star will be increased as a function of apparent distance. If I_0 is initial stellar flux, flux $I(t)$ for given time t can be calculated as:

$$I(t) = I_0 \cdot A(t)$$

where $A(t)$ is amplification coefficient:

$$A(t) = \frac{1}{2}(u + \frac{1}{u}) \tag{1}$$

and u - additional parameter:

$$u = \sqrt{1 + \frac{4 \cdot \theta_0^2}{\theta^2(t)}} \tag{2}$$

In last relation, $\theta(t)$ is angular separation between star and body and θ_0 is the size of Einstein ring. If D_1 and D_2 are the distances from the lensed star to gravitational lens and from the lens to observer respectively, than the size of Einstein ring can be found for mass of lens M as:

$$\theta_0^2 = \frac{4GM}{c^2} \cdot \frac{D_1}{(D_1 + D_2) \cdot D_2}$$

However, it will be more convinient to represent relation for $\theta(t)$ as function of $A(t)$. In this case we can write the next relation:

$$\theta^2(t) = \frac{2\theta_0^2}{\sqrt{A^2 - 1}(A + \sqrt{A^2 - 1})} \tag{3}$$

For single star, lensed by single lens, apparent motion of the star is uniform and the light curve of microlensing event is symmetrical. For apparent velocity Ω_0 and impact parameter θ_p, $\theta(t)$ can be represented as polynomial with a maximal order of two:

$$\theta^2(t) = \Omega^2 \cdot t^2 + \theta_p^2$$

Therefore, if the polynomial fit of the function $F(t)$

$$F(t) = \frac{2}{\sqrt{A^2-1}(A+\sqrt{A^2-1})}$$

have any significant coefficients with an order of more than two (corresponding to an asymmetric light curve of microlensing event), we can conclude, that the star moves non-uniformly relative to the lens.

3. Parallax effect in microlensing

Let us consider the main reasons for non-uniformity of the apparent motion. First factor, that we have to take into account is parallax. This effect has recently been observed by the MACHO group (Griest et al., 1995). Parallax effect is assotiated with the motion of the Earth around the Sun. As a result, relative equatorial coordinates of lens and lensed star ($\Delta\alpha$, $\Delta\delta$) can be represented as function of the solar coordinates (α_0,δ_0):

$$\Delta\alpha \cdot \cos(\delta) = (\pi_{lens} - \pi_{star})\cos(\delta_0) in(\alpha_0 - \alpha) \qquad (4)$$

$$\Delta\delta = (\pi_{lens} - \pi_{star}) \cdot (\sin(\delta_0)\cos(\delta) - \cos(\delta_0) sin(\delta)\cos(\alpha_0 - \alpha)) \qquad (5)$$

where π_{lens} and π_{star} are parallaxes of lens and star respectively, α, δ are equatorial coordinates of the lensed star.

Introducing ψ as angle between direction of stellar motion and $\vec{\alpha}$ in local grid of equatorial coordinates, we obtain next equations for $\theta^2(t)$:

$$\theta^2(t) = (\Omega \cdot t + \Delta x_{parallax})^2 + (\theta_p + \Delta y_{parallax})^2 \qquad (6)$$

where

$$\Delta x_{parallax} = \Delta\alpha\cos(\delta) \cdot \cos(\psi) + \Delta\delta \cdot \sin(\psi) \qquad (7)$$

$$\Delta y_{parallax} = -\Delta\alpha\cos(\delta) \cdot \sin(\psi) + \Delta\delta \cdot \cos(\psi) \qquad (8)$$

In this case for long-term (more than 90 days) events relation (3) can not be fitted by second order polynomial only.

4. Binary system with dark companion

The second reason of significant coefficient of order more than two in polynomial fitting for curve $\theta^2(t)$ (3) can be existence of a dark companion to lensed star. In this case, the stellar motion will also be non-uniform, and $\theta^2(t)$ will be described as:

$$\theta^2(t) = (\Omega \cdot t + x_{companion})^2 + (\theta_p + y_{companion})^2$$

where

$$x_{companion} = \varepsilon \ \cos\iota \cos(\omega t + \varphi) \quad (9)$$

$$y_{companion} = \varepsilon \ \sin(\omega t + \varphi)) \quad (10)$$

assuming $\varepsilon = \frac{m_p}{M}\frac{a}{l_*}$, ω- angular revolution velocity of companion, m_p and M are masses of companion and system "star+companion" respectively, a - orbital radius of companion, and l_* - distance to star.

We can treat this case as microlensing of double star (considered by Sazhin and Cherepashuk, 1994) with one of them being.

5. Analysis of the available microlensing curves

We apply this models to available data for observed microlensing events. Unfortunately only the first ten events, observed by OGLE group (Udalski et al., 1994) were accessible to us via Internet OGLE ftp site. Short-term events and events with significant differences in number of observations for ascending and decending branches of light curve were removed from this list. We analyse the remaining three events and found a significant asymmetry in curve (3) for OGLE2 event. We used Fisher criterium to estimate the probability that curve (3) obtained from OGLE2 observations is asymmetric. We conclude that with a probability of 75% the observed curve can not be fitted by two order polynomial. This give reasons to suggest the existence of a massive companion $\approx 0.05M_*$ to the star in OGLE2 system.

6. Conclusion

We reached the a following conclusions:

1. Polynomial analysis of equation (3) allows us to select candidates for companion searches in microlensing observations.

2. Effect of parallax must be take into account in analysis of microlensing light curve for long-term events.

3. Observations of OGLE2 event give reasons to suggest the existence a dark companion for this star.

7. Acknowledgements

This study was carried out within the framework of the program "Kosmomicrofizika" and was supported by the Center for Cosmoparticle Physics "Cosmion".

References

Alcock C., Akerlof C.W., Allsman R.A. et al. (1993) Nature, V. **365**, P.621.

Griest K., Alcock C., Allsman R.A., et al. (1995) Proc.of the Pascoc. Hopkins Symposium, Baltimore, MD, World Sci.

Blair D.G., Sazhin M.V.(1993) Astronomical and Astrophysical Transactions.**V.3**. P.191.

Aubourg E., Bareyre P.,Brehin S. et al.(1993) Nature. V.**365**. P.623.

Paczinsky B.(1986) Astrophys.J.1986.V.**304**. P.1.

Liebes R. (1964) Phys.Rev., **B133**, p.835.

Refsdal S.(1964) Mon.Not.R.astr.Soc., **128**,N4.

Sazhin M.V., Cherepashchuk A.M.(1994) Astronomy Letters. **V. 20**. P.613

Udalski A., Szymanski M., Kaluzny J., Kubiak M., Mateo M., Krzeminski W. (1994) Astrophys.J.Lett,V.**426**, P. L69.

SUBJECT INDEX

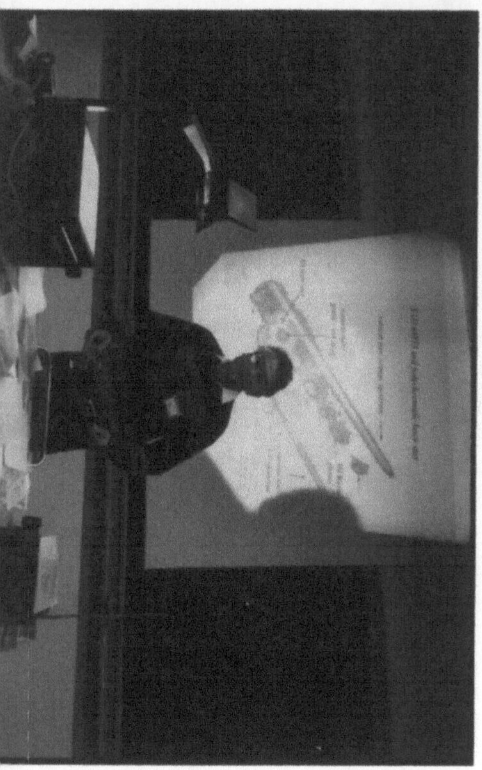

MIX
Papier aus verantwortungsvollen Quellen
Paper from responsible sources
FSC® C105338

If you have any concerns about our products,
you can contact us on
ProductSafety@springernature.com

In case Publisher is established outside the EU,
the EU authorized representative is:
Springer Nature Customer Service Center GmbH
Europaplatz 3, 69115 Heidelberg, Germany

Printed by Libri Plureos GmbH
in Hamburg, Germany